ELEMENTARY ALGEBRA

The Prindle, Weber and Schmidt Series in Mathematics

Althoen and Bumcrot, *Introduction to Discrete Mathematics*
Boye, Kavanaugh, and Williams, *Elementary Algebra*
Boye, Kavanaugh, and Williams, *Intermediate Algebra*
Burden and Faires, *Numerical Analysis,* Fourth Edition
Cass and O'Connor, *Fundamentals with Elements of Algebra*
Cullen, *Linear Algebra and Differential Equations,* Second Edition
Dick and Patton, *The Oregon State University Calculus Curriculum Project*
Eves, *In Mathematical Circles*
Eves, *Mathematical Circles Adieu*
Eves, *Mathematical Circles Revisited*
Eves, *Mathematical Circles Squared*
Eves, *Return to Mathematical Circles*
Fletcher, Hoyle, and Patty, *Foundations of Discrete Mathematics*
Fletcher and Patty, *Foundations of Higher Mathematics*
Gantner and Gantner, *Trigonometry*
Geltner and Peterson, *Geometry for College Students,* Second Edition
Gilbert and Gilbert, *Elements of Modern Algebra,* Second Edition
Gobran, *Beginning Algebra,* Fifth Edition
Gobran, *Intermediate Algebra,* Fourth Edition
Gordon, *Calculus and the Computer*
Hall, *Algebra for College Students*
Hall and Bennett, *College Algebra with Applications,* Second Edition
Hartfiel and Hobbs, *Elementary Linear Algebra*
Kaufmann, *Algebra for College Students,* Third Edition
Kaufmann, *Algebra with Trigonometry for College Students,* Second Edition
Kaufmann, *College Algebra,* Second Edition
Kaufmann, *College Algebra and Trigonometry,* Second Edition
Kaufmann, *Elementary Algebra for College Students,* Third Edition
Kaufmann, *Intermediate Algebra for College Students,* Third Edition
Kaufmann, *Precalculus,* Second Edition
Kaufmann, *Trigonometry*
Laufer, *Discrete Mathematics and Applied Modern Algebra*
Nicholson, *Elementary Linear Algebra with Applications,* Second Edition
Pence, *Calculus Activities for Graphic Calculators*
Pence, *Calculus Activities for the TI-81 Graphic Calculator*
Powers, *Elementary Differential Equations*
Powers, *Elementary Differential Equations with Boundary-Value Problems*
Powers, *Elementary Differential Equations with Linear Algebra*

Proga, *Arithmetic and Algebra*, Second Edition
Proga, *Basic Mathematics*, Third Edition
Rice and Strange, *Plane Trigonometry*, Fifth Edition
Schelin and Bange, *Mathematical Analysis for Business and Economics*, Second Edition
Strnad, *Introductary Algebra*
Swokowski, *Algebra and Trigonometry with Analytic Geometry*, Seventh Edition
Swokowski, *Calculus with Analytic Geometry*, Fifth Edition
Swokowski, *Calculus with Analytic Geometry*, Fourth Edition (Late Trigonometry)
Swokowski, *Calculus of a Single Variable*
Swokowski, *Fundamentals of College Algebra and Trigonometry*, Seventh Edition
Swokowski, *Fundamentals of College Algebra*, Seventh Edition
Swokowski, *Fundamentals of Trigonometry*, Seventh Edition
Swokowski, *Precalculus: Functions and Graphs*, Sixth Edition
Tan, *Applied Calculus*, Second Edition
Tan, *Applied Finite Mathematics*, Third Edition
Tan, *Calculus for the Managerial, Life, and Social Sciences*, Second Edition
Tan, *College Mathematics*, Second Edition
Trim, *Applied Partial Differential Equations*
Venit and Bishop, *Elementary Linear Algebra*, Third Edition
Venit and Bishop, *Elementary Linear Algebra*, Alternate Second Edition
Wiggins, *Problem Solver for Finite Mathematics and Calculus*
Willard, *Calculus and Its Applications*, Second Edition
Wood and Capell, *Arithmetic*
Wood, Capell, and Hall, *Developmental Mathematics*, Fourth Edition
Wood and Capell, *Intermediate Algebra*
Zill, *A First Course in Differential Equations with Applications*, Fourth Edition
Zill, *Calculus with Analytic Geometry*, Second Edition
Zill, *Differential Equations with Boundary-Value Problems*, Second Edition

The Prindle, Weber and Schmidt Series in Advanced Mathematics

Brabenec, *Introduction to Real Analysis*
Ehrlich, *Fundamental Concepts of Abstract Algebra*
Eves, *Foundations and Fundamental Concepts of Mathematics*, Third Edition
Keisler, *Elementary Calculus: An Infinitesimal Approach*, Second Edition
Kirkwood, *An Introduction to Real Analysis*
Ruckle, *Modern Analysis: Measure Theory and Functional Analysis with Applications*

ELEMENTARY ALGEBRA

Dale E. Boye
Ed Kavanaugh
Larry G. Williams

Schoolcraft College

PWS-KENT Publishing Company

BOSTON

PWS-KENT
Publishing Company

Copyright © 1991 by PWS-KENT Publishing Company.
All rights reserved. No part of this book may be reproduced, stored in a retrieval system, or transcribed, in any form or by any means—electronic, mechanical, photocopying, recording, or otherwise—without the prior written permission of PWS-KENT Publishing Company, 20 Park Plaza, Boston, Massachusetts 02116.

PWS-KENT Publishing Company is a division of Wadsworth, Inc.

Library of Congress Cataloging-in-Publication Data
Boye, Dale E.
    Elementary algebra / Dale E. Boye, Ed Kavanaugh, Larry G. Williams.
      p. cm.
    Includes index.
    ISBN 0-534-92319-4
    1. Algebra. I. Kavanaugh, Ed. II. Williams, Larry G. III. Title.
QA152.2.B68 1991
512.9—dc20   90-43563
                                                                                                                                                          CIP

Printed in the United States of America

1 2 3 4 5 6 7 8 9—95 94 93 92 91

Sponsoring editor: *Timothy L. Anderson*
Production editor: *Pamela Rockwell*
Interior design: *Julia Gecha*
Cover design: *Jean Hammond*
Production service: *Cece Munson, The Cooper Company*
Typesetting: Polyglot Pte Ltd
Cover printing: *Henry N. Sawyer Co., Inc.*
Printing and binding: *Arcata Graphics/Halliday Lithograph*

CONTENTS

PREFACE TO THE INSTRUCTOR   xi
PREFACE TO THE STUDENT   xv

Chapter One   BASIC CONCEPTS   1

1.1 Sets and Set Notation   1
1.2 Sets of Numbers: N, W, I, Q, I_1, R 5
1.3 Equality, Order, and the Number Line   10
1.4 Negative, Absolute Value, and Reciprocal of a Number   15
1.5 Properties of Real Numbers and Order of Operations   22
1.6 Addition and Subtraction of Signed Numbers   33
1.7 Multiplication and Division of Signed Numbers   41
1.8 Constants, Variables, Exponents, and Algebraic Expressions   50
1.9 Review and Chapter Test   57

Chapter Two   LINEAR EQUATIONS AND INEQUALITIES IN ONE VARIABLE   73

2.1 Applications of the Distributive Law   73
2.2 Preliminaries   80
2.3 Basic Definitions   88
2.4 Solving Linear Equations in One Variable   91
2.5 Literal Equations   104
2.6 Solving Inequalities   108
2.7 Applications   117
2.8 Review and Chapter Test   131

Chapter Three — LINEAR EQUATIONS AND INEQUALITIES IN TWO VARIABLES   142

- 3.1 The Cartesian Plane   142
- 3.2 Graphs of Linear Equations   147
- 3.3 Further Topics on Graphing   154
- 3.4 Slope and Slope-intercept Form   159
- 3.5 Linear Inequalities in Two Variables   167
- 3.6 Point-slope Form   174
- 3.7 Review and Chapter Test   179

Chapter Four — SYSTEMS OF EQUATIONS AND INEQUALITIES   188

- 4.1 Solving Systems of Equations by Graphing   189
- 4.2 Solving Systems of Equations by the Elimination Method   193
- 4.3 Solving Systems of Equations by the Substitution Method   200
- 4.4 Solving Applications with Two Equations in Two Variables   207
- 4.5 Systems of Inequalities   214
- 4.6 Review and Chapter Test   219

Chapter Five — POLYNOMIALS   229

- 5.1 Definitions   229
- 5.2 Addition and Subtraction of Polynomials   233
- 5.3 Multiplication of Polynomials   241
- 5.4 Special Products   249
- 5.5 Division of Polynomials   256
- 5.6 Factoring   266
- 5.7 Factoring Trinomials: $x^2 + bx + c$ 270
- 5.8 Factoring Trinomials: $ax^2 + bx + c, a \neq 1$ 275
- 5.9 Factoring Polynomials (Two Special Cases)   280
- 5.10 Factoring the Sum and Difference of Cubes and Grouping   285
- 5.11 Factoring Summary   289
- 5.12 Review and Chapter Test   292

Chapter Six — RATIONAL EXPRESSIONS   303

- 6.1 Fundamental Principle of Fractions   304
- 6.2 Multiplication and Division of Rational Expressions   311
- 6.3 Prime and Composite Numbers: Prime Factorization and Least Common Multiples   319
- 6.4 Addition and Subtraction of Rational Expressions   327
- 6.5 Complex Fractions   335
- 6.6 Fractional Equations   341
- 6.7 Ratio and Proportion   348
- 6.8 Applications   356
- 6.9 Review and Chapter Test   365

Contents

Chapter Seven EXPONENTS AND RADICALS 380

    7.1 Natural Number Exponents 380
    7.2 Integer Exponents 388
    7.3 Scientific Notation 399
    7.4 Roots and Rational Number Exponents 406
    7.5 Radical Form and Properties of Radicals 411
    7.6 Simplest Radical Form 416
    7.7 Addition and Subtraction of Radical Expressions 422
    7.8 Multiplication and Division of Radical Expressions 425
    7.9 Review and Chapter Test 431

Chapter Eight QUADRATIC EQUATIONS 440

    8.1 Solving Quadratic Equations by Factoring 440
    8.2 Solving Quadratic Equations by the Square Root Method 445
    8.3 Solving Quadratic Equations by Completing the Square 450
    8.4 The Quadratic Formula 455
    8.5 Further Topics 461
    8.6 Applications 465
    8.7 Solving Equations Containing Square Root Radicals 474
    8.8 The Graph of $y = ax^2 + bx + c$ (Optional) 478
    8.9 Review and Chapter Test 484

Chapter Nine RELATIONS AND FUNCTIONS 493

    9.1 Relations and Functions 493
    9.2 Function Notation 499
    9.3 Linear Functions 505
    9.4 Review and Chapter Test 510

SELECTED ANSWERS 517

INDEX 571

PREFACE TO THE INSTRUCTOR

This text, **Elementary Algebra**, is written for college students who have had no previous study of algebra. It is also written for those students who feel they need a review of basic algebraic concepts. **Elementary Algebra** is written with a traditional format.

Each chapter is partitioned into sections and presents from four to ten related topics. Each section includes definitions of new terms, a development of procedures and proofs of necessary theorems, many fully worked-out examples with explanations of steps, a summary where appropriate, and an exercise set. The first four chapters include a number of important concepts, including the properties of real numbers, evaluation of expressions, solving equations and inequalities in one variable, introduction to graphing in two dimensions, graphing of linear equations and inequalities in two variables, and solving systems of equations. In the remaining chapters, the student is introduced to polynomials, operations on polynomials, factoring polynomials, operations on rational expressions, exponents and radicals, solving quadratic equations, and, finally, relations and functions.

The text contains enough material for a class meeting five times per week for up to 17 weeks, allowing time for tests and review. Optional sections included at the end of some chapters may be skipped without loss of continuity, making the text suitable for a 3- or 4-hour course.

Some of the features of the text include:

1. Each chapter concludes with an extensive review section, complete with worked-out examples and an exercise set. These review sections should prove very helpful to the student preparing for individual tests or major exams.

2. In addition to the chapter review section, there is also a chapter test. This test is considered to be a typical test that the student might complete in approximately one class session (55 minutes).

3. A second color is used to highlight definitions, summaries of processes, and key steps in problem solving.

4. Definitions are clearly stated, and all concepts and procedures are presented using only definitions and methods introduced earlier in the text. The only assumed prerequisite is a satisfactory background in arithmetic.

5. Each exercise set, beginning with Exercise Set 2.1, includes several review exercises. These are problems from previous sections included for reinforcement and for review of topics that will be used in the subsequent section.

Supplements

A complete set of supplements designed to accompany this text is offered to adopters.

An Instructor's Manual with four sample review tests per chapter and answers to all even-numbered questions.

A Computerized Test Bank, EXPTest for IBM-PCs and compatibles, allows instructors to select test items for printed exams and to edit, add, and delete multiple-choice and open-ended questions to customize their tests further. An LXR computerized test bank is available for the Macintosh.

A Partial Solutions Manual for both students and instructors contains the solutions for every other odd exercise in the text.

A Developmental Mathematics Review Videotape Series created by Hope Florence, College of Charleston, covering the review of arithmetic (one tape), topics in elementary algebra (three tapes), and topics in intermediate algebra (one tape) is available to adopters of this text. The videotape series includes specialized worksheets that afford students the opportunity to work additional exercises.

Expert Algebra Tutor, tutorial software keyed to specific sections of the text, is available for IBM-PCs and compatible developmental-math-lab work stations.

Acknowledgments

We would like to express our deep gratitude to the following reviewers for their invaluable suggestions that contributed to the overall success of this textbook:

Jay Graening
University of Arkansas

Elizabeth Hodes
Santa Barbara City College

Billie James
University of South Dakota

Charlotte K. Lewis
University of New Orleans

Richard Semmler
Northern Virginia College

George T. Wales
Ferris State University

We also wish to thank our families, for their patience and support during the completion of this project, and the staff of Schoolcraft College, especially Candis Martin, Cathy Kiurski, Suzette Massard, and Joyce Boyce, for their ongoing aid and encouragement.

Many thanks to the staff at PWS-KENT, to our editor, Tim Anderson, for his continued guidance and advice, and to Pamela Rockwell, Barbara Lovenvirth, Cathie Griffin, and Diana Kelley, for their help with the development and production of this book.

PREFACE TO THE STUDENT

Studying mathematics has frequently been compared to studying music or learning how to play a sport, such as golf. They all require a good deal of concentration and much practice. You cannot do well simply by watching someone else. In a mathematics class, the classroom presentation may seem to make a topic fairly clear. However, reinforcing the lecture by doing the assigned homework regularly is essential for *you* to be able to do well.

We offer the following suggestions to make your experience in this course pleasant and successful:

1. Prior to class, finish the previous assignment and note those problems, definitions, or procedures you did not understand. Plan to get help with those problems in the next class meeting. Do not stockpile assignments, expecting to do them all at once or to cram them in before an exam.

2. Attend every class session. Arrive on time and have a pen or sharpened pencil and other necessary materials (straight edge, calculator, etc). In class, pay active attention. When your instructor suggests how he or she intends to solve a problem, try to anticipate the next step. We encourage you to ask questions. Take notes, especially on definitions, outlines of procedures, and example problems.

3. After class, make any necessary corrections in your notes. Reread all definitions and procedure summaries. Read the example problems, studying each step. Complete all assigned problems that you can. If you are unable to complete one or more problems, indicate these and be prepared to seek help from your instructor, teaching assistant, or tutor.

Remember, the more effort you put in early in the course, the easier you will find the later chapters. So be ready to begin the first day of class. Good luck, and we hope you enjoy your study of mathematics.

Dale E. Boye
Ed Kavanaugh
Larry G. Williams

CHAPTER ONE

BASIC CONCEPTS

This chapter presents a review of our number system, together with its arithmetic and algebraic properties. There are several definitions and some nomenclature to learn. You are encouraged to study the material of Chapter One carefully to increase your opportunities for success with the rest of the text.

1.1 Sets and Set Notation

Set notation is useful in representing the various types of numbers that we will study in the next section and in developing the concept of function in Chapter Nine. It may also be used to describe the solutions to equations and inequalities. We begin with a definition.

A **set** is a collection or group of objects. The objects contained in the set are called **elements** or **members** of the set. Examples of sets that we may encounter in our daily lives include:

the set of dishes from which we take our meals.

a shopping list of items to be purchased.

a team roster.

a class list of names of students taking a beginning algebra course.

a flock (set) of geese.

a herd (set) of cattle.

a set of encyclopedias.

Each of these sets represents a collection of objects.

We define two sets to be *equal* if they contain the same elements. This definition is not affected by duplication of elements or by the order in which the elements are listed. For example, for sets C and D listed next, we write $C = D$.

$$C = \{2, 1, 8, 1, 1, 2, 1\}$$
$$D = \{1, 2, 8\}$$

Sets may be described either by *listing* the elements in the set between a pair of braces, { }, or by giving a rule for membership in the set.

Example 1

Let A and B be the sets as described as follows:

By listing: $A = \{\text{January, June, July}\}$
By a rule: $B = \{\text{all months that begin with the letter J}\}$
Sets A and B contain the same elements even though they are defined in different ways. Therefore, we write

$$A = B \qquad \blacktriangleleft$$

Let $C = \{4, 6, 8, 10\}$. To indicate that 4 is an element of the set C, we write $4 \in C$. We write $5 \notin C$ to indicate the number 5 does not belong to the set C. To show that two or more objects belong to a set we write, for example, $4, 6 \in C$ or $4, 6, 10 \in C$.

The set containing no elements is called the **empty set** or the **null set**. The null set may be represented by the Greek letter phi, ϕ, or by a pair of empty braces, { }. That is,

$$\phi = \{\ \}$$

Example 2

Consider the following set, M:

M = the set of all beginning algebra students who are nine feet tall

Since there are no beginning algebra students who are nine feet tall, we conclude that the set M contains no elements. Therefore,

$$M = \phi \qquad \blacktriangleleft$$

A set may also be described using **set builder notation**, which combines the use of braces with a rule for membership in the set. When set builder notation is used, each symbol has a specific translation, as illustrated in Figure 1.1-1.

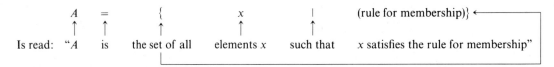

FIGURE 1.1-1 *Set Builder Notation*

1.1 Sets and Set Notation

Example 3

Describe each of the following sets using set builder notation:

A. $A = \{\text{April, June, September, November}\}$
We could express A as

$$A = \{M \mid M \text{ names a month containing exactly 30 days}\}$$

We read: A is the set of all elements M such that M names a month containing exactly 30 days.

B. $C = \{\text{red, orange, yellow, green, blue, indigo, violet}\}$
We could express C as

$$C = \{x \mid x \text{ is a color appearing in a rainbow}\}$$

C. $A = \{\text{Kavanaugh, Williams, Boye}\}$
We could express A as

$$A = \{x \mid x \text{ is the name of an author of this text}\}$$

D. $T = \{8, 9, 10, \ldots, 73, 74, 75\}$
We could express T as

$$T = \{W \mid W \text{ is a whole number between 8 and 75 inclusive}\}$$

E. $P = \{21, 22, 23, 24, \ldots\}$
We could express P as

$$P = \{y \mid y \text{ is a whole number greater than 20}\}$$

In Examples D and E, the "..." is called an ellipsis and is used to show unwritten but obvious members. ◀

Set builder notation is used in Chapter Nine to facilitate the development of the function concept.

Set A is called a **subset** of B if all the elements in A are also elements in B. We write $A \subseteq B$ to indicate that A is a subset of B, or A is contained in B. The null set is taken to be a subset of any set A. That is, $\phi \subseteq A$ for all sets A. Further, if A is a subset of B and B contains at least one element not in A, then A is a **proper subset** of B, and we write $A \subset B$.

Example 4

Let

$A = \{1, 2, 3, 4, 5, 6, 7\}$

$B = \{1, 3, 5, 7\}$

$C = \{2, 4, 6\}$

$D = \{0, 1, 2, 3\}$

Then $B \subseteq A$ since all the elements in B are also in A. Note also that we could correctly write $B \subset A$ (since A contains elements not in B).

Similarly, $C \subseteq A$ or $C \subset A$. By definition, every set is a subset of itself; so for any set A, we may always write $A \subseteq A$, but we must never write $A \subset A$.

Since all the elements in D are not also contained in A, D is not a subset of A. We write:

$$D \nsubseteq A$$

Note: it is *incorrect* to write $7 \subseteq A$ even though 7 belongs to A since 7 is not a set (as it is written) and, therefore, cannot be a subset. It is correct to write either

$$7 \in A$$

or $\quad \{7\} \subseteq A, \quad$ since $\{7\}$ is a set ◀

Exercise Set 1.1

Determine whether the statements in Problems 1–20 are true or false. The sets referred to in these problems are defined as follows:

$A = \{1, 2, 3, 4, 5, 6, 7, 8, 9, 10\}$
$B = \{1, 3, 7, 9\}$
$C = \{2, 4, 6, 8, 10\}$
$D = \{1, 3, 5, 7, 9\}$
$E = \{2, 10\}$
$F = \{1, 2, 7, 8\}$
$N = \{1, 2, 3, 4, 5, \ldots\}$
$W = \{0, 1, 2, 3, 4, 5, \ldots\}$

True or false?

1. $C \subseteq A$
2. $E \subseteq A$
3. $E \subseteq C$
4. $D \subseteq D$
5. $A \subset F$
6. $D \subset B$
7. $D \subset A$
8. $\phi \subset B$
9. $\phi \subset \phi$
10. $A \subset N$
11. $A \subseteq W$
12. $A \subseteq \phi$
13. $\{3, 7\} \subseteq B$
14. $5 \in E$
15. $8 \subset F$
16. $A \subset A$
17. $\{2, 10\} \subseteq E$
18. $2 \subseteq E$
19. $E \in C$
20. $1, 3 \subseteq B$

Use the listing method to specify each of the following sets. Refer to the sets listed at the beginning of this exercise set.

21. The set of numbers belonging to both B and D
22. The set of numbers belonging to both B and F
23. The set of numbers belonging to C or F

24. The set of numbers belonging to E or B
25. The set of numbers belonging to both B and E
26. The set of numbers belonging to both D and E
27. The set of numbers in A between 2 and 9
28. The set of numbers in A between 4 and 8
29. The set of numbers in A that are no more than 5
30. The set of numbers in A that are no less than 6
31. The set of numbers in A that are larger than 3 and less than 9
32. The set of numbers in A that are less than 8 and larger than 4

List the members of the following sets. Use the ellipsis (. . .) if the set is very large.

33. $A = \{m \mid m \text{ is a month whose spelling includes the letter } r\}$
34. $B = \{d \mid d \text{ is a day of the week whose spelling begins with the letter } T\}$
35. $C = \{x \mid x \text{ is a whole number between 7 and 70 inclusive}\}$
36. $D = \{y \mid y \text{ is a whole number between 10 and 1000 inclusive}\}$
37. $E = \{n \mid n \text{ is a whole number greater than 40}\}$
38. $F = \{t \mid t \text{ is a whole number greater than 65}\}$
39. What sets would be more conveniently described using the listing method? Using set builder notation?
40. Why must we not write $A \subset A$? Is a set not a proper subset of itself?

1.2 Sets of Numbers: N, W, I, Q, I_r, R

This section introduces the various kinds (sets) of numbers generally used. The first numbers we encounter are the **counting** or **natural numbers**. We designate the set of natural numbers with N.

$$N = \{1, 2, 3, 4, 5, \ldots\}$$

Since N contains an unlimited number of elements, N is called an **infinite set**.

If we extend N by including the number 0, we create the set of **whole numbers**, designated W.

$$W = \{0, 1, 2, 3, 4, 5, 6, \ldots\}$$

Note that $N \subset W$.

Recall from arithmetic that the result of addition is called a *sum*; the result of subtraction is called a *difference*; the result of multiplication is called a *product*; and the result of division is called a *quotient*. When letters are used to represent numbers, the product of a and b is written $a \cdot b$, $(a) \cdot (b)$, or ab.

The operations of addition and multiplication are always "safe" in W. That is, the sum or product of any two whole numbers is also a whole number. This is expressed symbolically as follows:

If $a \in W$, $b \in W$, then, $(a + b) \in W$

and $\quad ab \in W$

We say that W is **closed** with respect to the operations of addition and multiplication. Since the difference or quotient of two whole numbers will not always be a whole number (consider $2 - 5$ or $2 \div 3$), W is not closed with respect to subtraction or division.

To create a set in which we guarantee that the difference of two whole numbers will always be defined, we extend W by adding the negatives of the natural numbers, thereby forming the set of **integers**, designated I.

$$I = \{\ldots, -5, -4, -3, -2, -1, 0, 1, 2, 3, 4, 5, \ldots\}$$

Note that $W \subset I$ so we can write $N \subset W \subset I$.

The set of integers, I, is closed with respect to addition, multiplication, *and* subtraction. We will show how to find the sum, product, or difference of any two integers in subsequent sections. Symbolically,

If, $a, b \in I$, then

$\quad (a + b) \in I$,

$\quad \quad ab \in I$,

and $\quad (a - b) \in I$

The set I is not closed with respect to division (consider $2 \div 5$).

A set in which the quotient of two integers is defined is obtained by extending I to create the set of **rational numbers**, designated Q. We use set builder notation to define Q.

$$Q = \left\{ \frac{a}{b} \,\middle|\, a, b \in I; b \neq 0 \right\}$$

That is, the set of rational numbers is the set containing every number that can be written as a fraction whose numerator and denominator are integers. Some members of Q are:

$$\frac{3}{4}, \quad \frac{5}{2}, \quad 6 = \frac{6}{1}, \quad 0 = \frac{0}{1}, \quad -2 = \frac{-2}{1}, \quad 1\frac{3}{5} = \frac{8}{5}$$

1.2 Sets of Numbers: N, W, I, Q, I_r, R

Because any integer can be written as a quotient ($6 = \frac{6}{1}$), $I \subset Q$. We can write $N \subset W \subset I \subset Q$.

The set Q is closed with respect to the four arithmetic operations. That is, if a and b are rational numbers, then $a + b$, $a - b$, ab, and $a \div b$ (except for division by zero) are also rational numbers.

A rational number may also be defined as a number that can be written as a repeating decimal. For example,

$$\frac{1}{3} = 0.3333\ldots = 0.33\overline{3}$$

The line over 3, called an *overbar*, indicates that the digit 3 repeats forever.

$$\frac{2}{11} = 0.181818\ldots = 0.18\overline{18}$$

The digit pattern 18 repeats forever.

$$\frac{1}{5} = 0.2 = 0.20000\ldots = 0.20\overline{0}$$

A decimal whose repeating portion is zeros is called a *terminating decimal*, and the zeros are generally not written. We would generally write only:

$$\frac{1}{5} = 0.2$$

Any number that cannot be expressed as the quotient of two integers or as a repeating decimal does not belong to Q. Examples of such numbers include:

$$\sqrt{2}, \quad \sqrt[3]{5}, \quad \pi$$

Such numbers are called **irrational numbers**. We designate the set of irrational numbers I_r. That is,

$$I_r = \{x \mid x \notin Q\}$$

The sets Q and I_r are called **disjoint sets**, meaning they have no common elements. If we take all the elements of Q and I_r together, we obtain the set of **real numbers** designated R.

$$R = \{x \mid x \in Q \quad \text{or} \quad x \in I_r\}$$

We can write

$$N \subset W \subset I \subset Q \subset R$$

and $\quad I_r \subset R$

R is the number system that is traditionally studied in an elementary algebra course. Figure 1.2-1 pictures the subset relationships among the various sets of numbers.

FIGURE 1.2-1

The Set of Real Numbers and Its Subsets

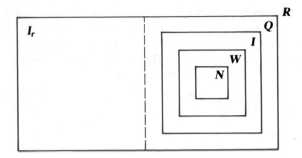

The real numbers can be represented graphically on the **number line**. (See Figure 1.2-2.)

FIGURE 1.2-2

Real Number Line

The number line extends forever (infinitely) in both directions. An arrowhead is drawn on the right end of the number line to show the direction of increasing numbers. An arbitrary point is chosen to represent the number 0 and is called the **origin**. Evenly spaced points are marked off on the line to the left and right of the origin. Those points to the right of 0 represent the positive integers, and those points to the left of 0 represent the negative integers. To represent a number on the number line, we appropriately place a dot on the line. The dot is called the **graph** or **plot** of that number, and the number is called the **coordinate** of the point.

Example 1

Draw a number line and graph the following real numbers:

$$\left\{2, -1.5, \pi \text{ (approx. 3.14)}, \frac{2}{3}, \sqrt{2} \text{ (approx. 1.4)}\right\}$$

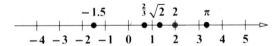

Generally we can only approximate the location of any real number, so labeling all graph points is important. ◀

Example 2

Graph each of the following sets on the real number line:

A. $\{x \mid x \in I \text{ and } x \text{ is between } -4 \text{ and } 5\}$

1.2 Sets of Numbers: N, W, I, Q, I_r, R

B. $\{x \mid x \in N \text{ and } x \text{ is between } -4 \text{ and } 5\}$

Note that all points representing negative integers and zero are excluded from the graph since N contains only positive integers.

C. $\{x \mid x \in I \text{ and } x \text{ is greater than } 4\}$

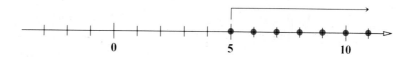

The extra arrow is drawn to indicate that all numbers greater than 4 are included (not just 5, 6, 7, 8, 9, 10, and 11 as the graph might tend to indicate).

Exercise Set 1.2

Graph the following sets of real numbers by drawing a number line and plotting points:

1. $A = \{x \mid x \in I \text{ and } x \text{ is between } -3 \text{ and } 6\}$
2. $D = \{x \mid x \text{ is an even integer and } x \text{ is between } -5 \text{ and } 8\}$
3. $B = \{x \mid x \in N \text{ and } x \text{ is between } -3 \text{ and } 6\}$
4. $E = \{x \mid x \in W, x \text{ is even, and } x \text{ is between } -5 \text{ and } 8\}$
5. $C = \{x \mid x \in W \text{ and } x \text{ is between } -3 \text{ and } 6\}$
6. $F = \{x \mid x \in N, x \text{ is even, and } x \text{ is between } -5 \text{ and } 8\}$
7. $G = \{x \mid x \in N \text{ and } x \text{ is no more than } 7\}$
8. $H = \{x \mid x \in W \text{ and } x \text{ is no more than } 7\}$
9. Let $A = \{3, -2, 0, \sqrt{3}, \frac{2}{3}, \pi, 4.67\}$. List all the elements from A that are:

    a. natural numbers   b. whole numbers
    c. rational numbers   d. integers
    e. irrational numbers

10. Let $B = \{-4, 2.5, -\frac{4}{3}, \sqrt{2}, 12.5\overline{5}\}$. List all the elements from B that are:

    a. natural numbers   b. whole numbers

c. rational numbers d. integers

e. irrational numbers

11. Let $C = \{\frac{1}{2}, -\frac{4}{3}, -\pi, \sqrt{5}, 1.3, -4.71\}$. List all the elements from C that are:

   a. natural numbers b. whole numbers
   c. rational numbers d. integers
   e. irrational numbers

12. Let $D = \{-1, 2.\overline{2}, -\sqrt{6}, \frac{5}{4}, 4, 3.8, -\frac{1}{9}\}$. List all the elements from D that are:

   a. natural numbers b. whole numbers
   c. rational numbers d. integers
   e. irrational numbers

Draw a number line from -10 to 10. Locate and label the points having the following coordinates:

13. $A: 6$
14. $B: 10$
15. $C: -4$
16. $D: -5$
17. $E: 2.5$
18. $F: 3.7$
19. $G: -7.3$
20. $H: -3.1$
21. $I: \frac{1}{2}$
22. $J: \frac{1}{4}$
23. $K: -\frac{3}{4}$
24. $L: -\frac{5}{6}$
25. $M: -\frac{9}{5}$
26. $N: -\frac{7}{3}$
27. $O: 3\frac{1}{4}$
28. $P: 2\frac{1}{3}$
29. $Q: -5\frac{1}{3}$
30. $R: -3\frac{1}{5}$
31. $S: \pi$ (approx. 3.14)
32. $T: -\sqrt{2}$ (approx. -1.41)
33. $U: 0.4\overline{4}$
34. $V: 1.66\overline{6}$
35. $W: -2.43\overline{43}$
36. $X: -0.7\overline{7}$
37. $Y: 0$
38. Given that $a \in Q$, justify the claim that $a \notin I_r$.

1.3 Equality, Order, and the Number Line

The real numbers, R, which were discussed in the previous section, form what is called an **ordered set** of numbers. That is, each real number is represented by its own point on the number line and is related to all other real numbers in a unique way. We will discuss the nature of this relationship.

1.3 Equality, Order, and the Number Line

We assume you are comfortable with the statements that follow:

3 is less than 8

8 is greater than 3

3 is equal to 3

3 is not equal to 8

It is customary to use symbols to express such relationships between numbers. The following chart lists the symbols along with their meanings (translations).

Symbol	Translation
$<$	is less than
$>$	is greater than
$=$	is equal to
$\neq$	is not equal to

We now write the preceding statements symbolically:

3 is less than 8: $3 < 8$

8 is greater than 3: $8 > 3$

3 is equal to 3: $3 = 3$

3 is not equal to 8: $3 \neq 8$

We also observe from the number line that the magnitude of a natural number is related to its position on the line. If we locate the numbers 3 and 8 on the number line, we observe that the smaller number 3 is located to the left of 8.

In general,

if $a < b$, then a is to the left of b.* Similarly, if $a > b$, then a is to the right of b.

These observations may be generalized to relate any two real numbers (not just natural numbers). That is, either one of the numbers is larger than the other or they are equal. The larger number is located to the right of the smaller number on the number line. The real numbers are well ordered.

---

*More formally, we should say "the graph of a is to the left of the graph of b on the number line." However, it is common practice to state this in the simpler form.

ONE Basic Concepts

Example 1

Consider the number line

```
       -3.5   -√3        0.75 √2         π      4.75
    ────●─────●──────────●──●───────────●────●──→
    -5 -4 -3 -2 -1    0    1    2    3    4    5
```

Insert the correct symbol (<, >, =) to make each of the following statements a true relationship:

	Statement		Solution	Position on number line
A.	1	5	$1 < 5$	1 is to the left of 5 on the line
B.	5	3	$5 > 3$	5 is to the right of 3 on the line
C.	0	4	$0 < 4$	0 is to the left of 4 on the line
D.	-4	0	$-4 < 0$	-4 is to the left of 0 on the line
E.	$\sqrt{2}$	-3.5	$\sqrt{2} > -3.5$	-3.5 is left of $\sqrt{2}$
F.	0	$-\sqrt{3}$	$0 > -\sqrt{3}$	
G.	-3	π	$-3 < \pi$	
H.	5	4.75	$5 > 4.75$	
I.	.75	.75	$.75 = .75$	Same point

Note: the symbols < or > may be thought of as "arrows" that point to the smaller number. ◀

Example 2

Consider the number line

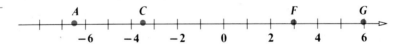

```
         A         C                       F         G
    ────●─────────●─────┼─────┼─────┼─────●─────┼───●──→
        -6       -4    -2     0     2     4         6
```

Let A, C, F, G represent numbers whose graphs are as shown. Determine the relationship between these numbers by inserting the symbol (<, >, or =) that makes the following statements true.

	Statement		Relationship	
A.	A	C	$A < C$	A is left of C
B.	F	C	$F > C$	F is right of C
C.	G	0	$G > 0$	
D.	F	3	$F = 3$	
E.	A	-7	$A > -7$	

◀

1.3 Equality, Order, and the Number Line

For the last topic in this section, we use the number line to develop the notion of **directed distance** from one point on the line to another. Let A and B represent numbers on the number line.

note: $A < B$

The directed distance from A to B is the number of units we must move to the right, the positive direction, to move from A to B (a positive number). Similarly, the directed distance from B to A is defined to be the number of units we must move to the left, the negative direction, to get from B to A (a negative number).

Example 3

Let A, B, and C label the points representing the numbers 0, 2, and 9, respectively, on a number line.

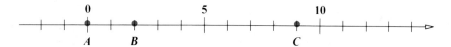

Find:

A. The directed distance from B to C.
The directed distance from B to C is 7 because we move 7 units to the right to get from B to C.

B. The directed distance from C to B.
The directed distance from C to B is -7 because we move 7 units to the left to get from C to B.

C. The directed distance from A to C.
The directed distance from A to C is 9 because we move 9 units to the right to get from A to C.

D. The directed distance from B to A.
The directed distance from B to A is -2 because we move 2 units to the left to get from B to A.

The notion of directed distance will be useful in the next section. ◀

Exercise Set 1.3

Consider the number line

ONE   Basic Concepts

Insert the symbol ($<$, $>$, $=$) that makes each of the following statements true:

1. 4   1
2. 0   6
3. 0 -3
4. 5.6   7
5. π 0
6. -4 0
7. -2 3
8. 4 π
9. π $-\pi$
10. -5 -1
11. -3 -7
12. -2.5 -1
13. 5.6   5
14. $-\pi$ -2.5
15. π $\sqrt{5}$
16. $-\pi$ -5

Consider the number line

Let A, B, C, D represent numbers whose graphs are as shown. Determine the relationship between these numbers by inserting the symbol ($<$, $>$, $=$) that makes the following statements true:

17. C A
18. B D
19. B C
20. D A
21. 0 B
22. D 0
23. A B
24. D C
25. -4 B
26. -1 A
27. C $-\dfrac{1}{2}$
28. A $\dfrac{4}{3}$
29. -5 B
30. A 1

Consider the number line

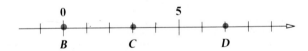

Find the directed distance:

31. from B to C
32. from B to D
33. from C to B
34. from D to B
35. from C to D
36. from D to C
37. from 2 to 5
38. from 4 to D
39. from 6 to 1
40. from 8 to B
41. from 4 to 4
42. from B to 8
43. from 0 to 8
44. from 7 to 0
45. from -1 to 4
46. from -1 to 6
47. from 4 to -1
48. from 6 to -1
49. from C to -1
50. from D to -1

Imagine a thermometer as a number line and a "directed distance" as a change in temperature. Find the change in temperature from:

51. −7 to 8 **52.** −4 to 9 **53.** 75 to 49 **54.** 82 to 69
55. 12 to −1 **56.** 12 to −12 **57.** −4 to 0 **58.** −8 to 0
59. 0 to −4 **60.** 0 to −8

61. Consider the number line:

Why is the directed distance from B to A negative, even though both A and B are positive?

1.4 Negative, Absolute Value, and Reciprocal of a Number

In this section we introduce three *unary* operations on numbers. A unary operation is an operation performed on a single number. The arithmetic operations of adding, subtracting, multiplying, or dividing require two numbers and so are called *binary* operations.

The Negative of a Number

The **negative of a number** is defined as the mirror image of that number with respect to the origin on the number line. Thus it is the number that is the same distance from the origin on the other side. By this definition, the negative of a positive number is a negative number; the negative of a negative number is a positive number; and the negative of 0 is 0.

Example 1

Consider the number line

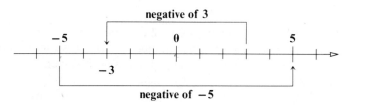

The negative of 3 is given by −3 since −3 is the same distance to the left of 0 as 3 is to the right.
The negative of −5 is given by 5 since 5 is the same distance to the right of 0 as −5 is to the left. ◀

The negative of a number is symbolized by placing a "$-$" sign in front of the number. If a number is represented by a letter such as n, then its negative is written "$-n$." For example, the negative of positive 3 is written $-(+3)$, and by definition, $-(+3) = -3$. Similarly, the negative of negative 5 is written $-(-5)$, and by definition, $-(-5) = +5$. Notice that taking the negative of a number simply changes the sign of the number. To avoid the use of double signs, the number itself is enclosed in parentheses.

Other than the traditional meaning of subtraction, there are two distinct uses of the "$-$" sign:

1. as part of the number itself, such as -5, and

2. as an operation on a number (taking its negative), such as $-(-5) = 5$.

Example 2

Shown on the number line are the negative of $+4$ and the negative of -6.

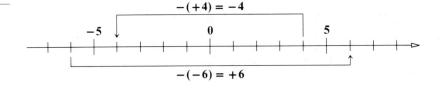

Example 3

Write the negative of each of the following numbers:

	Number	Solution
A.	$+4$	$-(+4) = -4$
B.	-6	$-(-6) = +6$
C.	0	$-(0) = 0$
D.	x	$-(x) = -x$
E.	$2\frac{1}{2}$	$-(2\frac{1}{2}) = -2\frac{1}{2}$
F.	-3.7	$-(-3.7) = +3.7$

Although, for emphasis, we have written positive numbers with a plus sign, in general practice, a positive number is written without a preceding "$+$" sign. That is, $+5 = 5$. To find the negative of a positive number, change the preceding sign from "$+$," written or unwritten, to "$-$." To find the negative of a negative number, change the preceding sign from "$-$" to an unwritten "$+$."

G.	-7	$-(-7) = 7$	
H.	5	$-(5) = -5$	*Note:* the unwritten "$+$" in front of the 5 was changed to "$-$"

The Absolute Value of a Number

It is often convenient to discuss the magnitude or size of a number without regard to its sign. To do so, we define the **absolute value** of the number.

> **DEFINITION**
>
> The *absolute value* of a number is the distance between the graph of the number and the origin.

The absolute value of a number is symbolized by enclosing the number in a pair of vertical lines. Thus, the absolute value of the number represented by the letter n is written $|n|$. By definition, the absolute value of a positive number is the number itself; the absolute value of a negative number is that number with its sign changed to positive. Symbolically:

$$|n| = \begin{cases} n & \text{if } n \text{ is a positive number or zero} \\ -n & \text{if } n \text{ is a negative number} \end{cases}$$

We observe: *the absolute value of a number is never negative.*

Example 4

Find the absolute value of each of the following numbers:

Number	Solution	
A. $+43$	$|+43| = |43| = 43$	
B. -2.8	$|-2.8| = 2.8$	
C. 96	$|96| = 96$	
D. -9	$|-9| = 9$	
E. 0	$|0| = 0$	
F. -2.17	$|-2.17| = 2.17$	
G. y	$|y| = |y|$	Cannot be simplified further since the sign of y is unknown.

The Reciprocal of a Number

The properties of real numbers (to be introduced in Section 1.5) generally are expressed only for addition and multiplication. In order to extend these properties to include division, we need a manner of expressing division as a form of multiplication. For this process we introduce the **reciprocal** of a number.

> **DEFINITION**
>
> For a, b, not equal to zero, the *reciprocal* of the number $\frac{a}{b}$ is given by $\frac{b}{a}$.
> (Recall from arithmetic that any number can be written as itself over 1. Symbolically, $a = \frac{a}{1}$.

Example 5

Find the reciprocal of each of the following numbers:

	Number	Reciprocal
A.	$\frac{7}{11}$	$\frac{11}{7}$
B.	$\frac{1}{5}$	$\frac{5}{1} = 5$
C.	$\frac{1}{9}$	$\frac{9}{1} = 9$
D.	4	$\frac{1}{4}$
E.	10	$\frac{1}{10}$

◀

The number 0 does not have a reciprocal since if we attempt to name the reciprocal of 0 by $\frac{1}{0}$ we obtain an undefined expression (division by 0 is never defined). The reciprocal of a negative number is also negative. For example, the reciprocal of $-\frac{4}{3}$ is $-\frac{3}{4}$. The reciprocal of -2 is $-\frac{1}{2}$. This concept will be discussed in more detail in Section 1.7, but it is mentioned here to support the following statement: Every real number, except 0, has a unique reciprocal.

Example 6

In the following chart, the negative, absolute value, and reciprocal of each given number are shown.

Number (x)	Negative ($-x$)	Absolute Value $\|x\|$	Reciprocal $\frac{1}{x}$
5	-5	5	$\frac{1}{5}$
$\frac{2}{3}$	$-\frac{2}{3}$	$\frac{2}{3}$	$\frac{3}{2}$
$+4$	-4	4	$\frac{1}{4}$

1.4 Negative, Absolute Value, and Reciprocal of a Number

Number (x)	Negative ($-x$)	Absolute Value $\|x\|$	Reciprocal $\dfrac{1}{x}$
0	0	0	undefined
$-\dfrac{3}{4}$	$\dfrac{3}{4}$	$\dfrac{3}{4}$	$-\dfrac{4}{3}$
-7	7	7	$-\dfrac{1}{7}$

Each of these operations is a unary operation (performed on only one number). It is not uncommon to perform more than one operation on a given number. In such a case we evaluate the expression by proceeding from the innermost to the outermost operation, as illustrated in the next example. ◀

Example 7

Find the value of each expression:

A. $-(+4)$

$\qquad -(+4) = -(4) = -4 \qquad$ Negative of 4

B. $-(-7)$

$\qquad -(-7) = 7 \qquad$ Negative of -7

C. $-|4|$

$\qquad -|4| = -(4) = -4 \qquad$ Absolute value is performed first

D. $-|-8|$

$\qquad -|-8| = -(8) = -8 \qquad$ Absolute value first

E. $-(-0)$

$\qquad -(-0) = -(0) = 0$

F. $-|-(-6)|$

$\qquad -|-(-6)| = -|6| \qquad$ Negative of -6

$\qquad \qquad \qquad \; = -(6) \qquad$ Absolute value of 6

$\qquad \qquad \qquad \; = -6 \qquad$ Negative of 6 ◀

Exercise Set 1.4

1. Find the negative of each number:

   a. -7 b. $+4$ c. 0 d. -5.8 e. 8

2. Find the negative of each number:

   a. -3.7 b. $+7$ c. $-4\tfrac{1}{2}$ d. 10

3. Find the absolute value of each number:
   a. +4   b. 6   c. −5   d. 0   e. −7.1

4. Find the absolute value of each number:
   a. −9   b. $+\frac{1}{2}$ c. −0.8   d. 3.7   e. −11

5. Find the value of each expression:
   a. −(−4)   b. −(+5)   c. −(30)   d. −(−7)

6. Find the value of each expression:
   a. −(−6)   b. −(+4)   c. −(−8)   d. −(8)

7. Find the value of each expression:
   a. −(+6)   b. −|−6| c. −|23| d. |−23| e. −|0|

8. Find the value of each expression:
   a. −(+5)   b. |−5| c. −|−46| d. |46| e. −(−0)

9. Find the reciprocal of each number:
   a. +4   b. −6   c. $\frac{7}{8}$ d. $-\frac{1}{4}$ e. $\frac{1}{3}$

10. Find the reciprocal of each number:
    a. −3   b. +3   c. 0   d. $-\frac{2}{3}$ e. $\frac{1}{4}$

11. Complete the chart for each given number:

Number (x)	Negative (−x)	Absolute Value |x|	Reciprocal $\frac{1}{x}$
+3			
5			
$\frac{2}{3}$			
$\frac{7}{9}$			
0			
−4			
$-\frac{5}{6}$			
$-\frac{1}{3}$			
1			

1.4 Negative, Absolute Value, and Reciprocal of a Number

12. Complete the chart for each given number:

Number (x)	Negative (−x)	Absolute Value |x|	Reciprocal $\frac{1}{x}$
8			
+2			
$\frac{7}{5}$			
$\frac{6}{11}$			
0			
−5			
$\frac{7}{8}$			
$-\frac{1}{9}$			
1			

13. Complete the chart for each given number:

Number (x)	Negative (−x)	Absolute Value |x|	Reciprocal $\frac{1}{x}$
+6			
4			
2.15			
$\frac{3}{4}$			
−0			
−5			
$-\frac{1}{6}$			
−3.25			
−0.4			
$-\frac{7}{8}$			

14. Complete the chart for each given number:

Number (x)	Negative ($-x$)	Absolute Value $|x|$	Reciprocal $\frac{1}{x}$
$+12$			
2			
$+\frac{7}{3}$			
1.6			
-1			
-2.1			
$-\frac{1}{2}$			
$-\frac{6}{7}$			
$\frac{1}{10}$			

15. Is the negative of a number always negative? Discuss.

16. Is the absolute value of a number always positive? Discuss.

17. Can any number be its own negative? Its own absolute value? Its own reciprocal? Explain.

1.5 Properties of Real Numbers and Order of Operations

One of the objectives of algebra is to learn to manipulate numbers and the symbols that represent numbers. Therefore, we must understand the properties of the numbers being used, which in our case is the set of real numbers. These are properties that you have used since your introduction to arithmetic and, because they are intuitive in nature, you assume them to be true without proof. However, some standardization of the language describing these properties is required. This section provides a list of some of the basic properties of the real numbers, along with a description of each.

1. *Closure Property* The real numbers are **closed** with respect to both multiplication and addition. That is, both the product and sum of any two real numbers are also real numbers. Symbolically:

   If $a, b \in R$, then $(a + b) \in R$ and $a \cdot b \in R$

   In other words, if a and b are any two real numbers, then:
   1. $a + b$ is a real number
   2. $a \cdot b$ is a real number

1.5 Properties of Real Numbers and Order of Operations

2. *Commutative Property* The product or sum of any two real numbers is **commutative**. That is, the product or sum can be taken in either order. Symbolically:

$$\text{If } a, b \in R, \text{ then } a + b = b + a \text{ and } a \cdot b = b \cdot a$$

Example 1

Illustrate the commutative properties with the numbers 8 and 3.

$$8 + 3 = 3 + 8 \qquad 8 \cdot 3 = 3 \cdot 8$$
$$11 = 11 \qquad 24 = 24$$

◀

3. *Associative Property* The product or sum of any three real numbers is **associative**. That is, the product or sum of three numbers is taken two at a time in any order. Symbolically:

If $a, b, c \in R$, then

$$a + b + c = (a + b) + c = a + (b + c) \quad \text{and}$$
$$a \cdot b \cdot c = (a \cdot b) \cdot c = a \cdot (b \cdot c)$$

The operation inside the parentheses is performed first.

Example 2

Illustrate the associative properties with the numbers 4, 7, and 12.

$$4 + 7 + 12 = (4 + 7) + 12 = 4 + (7 + 12)$$
$$11 + 12 = 4 + 19$$
$$23 = 23$$
$$4 \cdot 7 \cdot 12 = (4 \cdot 7) \cdot 12 = 4 \cdot (7 \cdot 12)$$
$$28 \cdot 12 = 4 \cdot 84$$
$$336 = 336$$

◀

4. *Additive Identity* When the number 0 is added to any real number, the sum is that real number. Zero is called the **additive identity** for the real numbers. Symbolically:

If $a \in R$, then

$$a + 0 = 0 + a = a$$

Example 3

Consider

$$3 + 0 = 3$$
$$0 + (-2) = -2$$
$$0 + 0 = 0$$

◀

5. *Multiplicative Identity* When the number 1 is multiplied by any real number, the product is that real number. The number 1 is called the

multiplicative identity for the real numbers. Symbolically:

If $a \in R$, then
$$a \cdot 1 = 1 \cdot a = a$$

Example 4

Consider
$$4 \cdot 1 = 4$$
$$(-3) \cdot 1 = -3$$
$$1 \cdot 1 = 1$$ ◀

6. *Additive Inverse* Every real number has a unique **additive inverse**. The additive inverse of the real number a is a number that when added to a yields 0. It is denoted by $-a$. Symbolically:

If $a \in R$, then
$$-a \in R \text{ and } a + (-a) = -a + a = 0$$

Note that the additive inverse of a number is the same as the *negative* of the number (see Section 1.4). Addition of signed numbers will be introduced in the next section.

Example 5

Find the additive inverse of each of the following real numbers:

A. 5

The additive inverse of 5 is -5, and $5 + (-5) = (-5) + 5 = 0$.

B. $-\frac{1}{2}$

The additive inverse of $-\frac{1}{2}$ is $-(-\frac{1}{2})$ or $\frac{1}{2}$, and
$$-\frac{1}{2} + \frac{1}{2} = \frac{1}{2} + \left(-\frac{1}{2}\right) = 0$$

C. 0

The additive inverse of 0 is -0 or 0, and $0 + (-0) = -0 + 0 = 0$. ◀

7. *Multiplicative Inverse* Every real number, except 0, has a unique **multiplicative inverse**. The multiplicative inverse of the non-zero real number a is a number that when multiplied by a yields 1. The multiplicative inverse is denoted $\frac{1}{a}$. The multiplicative inverse of a number is the same as the *reciprocal* of the number (see Section 1.4). Multiplication of signed numbers will be introduced in Section 1.7. Symbolically:

If $a \in R$, $a \neq 0$, then
$$\frac{1}{a} \in R \text{ and } a \cdot \frac{1}{a} = \frac{1}{a} \cdot a = 1$$

1.5 Properties of Real Numbers and Order of Operations

Example 6 Find the multiplicative inverse of each of the following real numbers:

A. -4

The multiplicative inverse of -4 is $\frac{1}{-4}$ or $-\frac{1}{4}$, and

$$(-4)\left(-\frac{1}{4}\right) = \left(-\frac{1}{4}\right)(-4) = 1$$

B. $\frac{2}{3}$

The multiplicative inverse or reciprocal of $\frac{2}{3}$ is $\frac{3}{2}$ and

$$\frac{2}{3} \cdot \frac{3}{2} = \frac{3}{2} \cdot \frac{2}{3} = 1$$

◀

8. *Distributive Property* Multiplication is **distributive** over addition in the real numbers. Symbolically:

If $a, b, c \in R$, then

$a(b + c) = ab + ac$, or

$(b + c)a = ba + ca$

In geometry, we show that the area (A) of a rectangle is found by multiplying the length (L) times the width (W). See Figure 1.5-1.

FIGURE 1.5-1
Area of a Rectangle

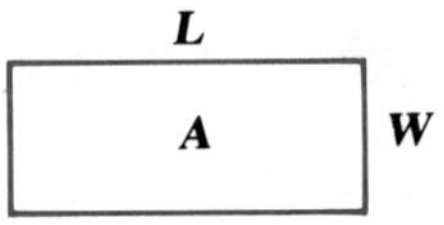

[rectangle diagram with L on top, W on right, A inside] $A = L \cdot W$

We can interpret the distributive law geometrically by noting in Figure 1.5-2 that the total area of the large rectangle is the sum of the areas of the two smaller rectangles.

FIGURE 1.5-2
The Distributive Property

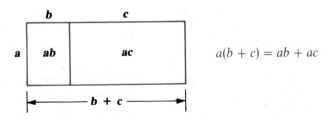

Total area: $A = L \cdot W = (b + c)a = a(b + c)$
Area of "left" rectangle: $A = L \cdot W = ab$
Area of "right" rectangle: $A = L \cdot W = ac$

Example 7

Verify the distributive property in each of the following statements by evaluating both sides and also by drawing the appropriate rectangles:

A. $3(5 + 2) = 3 \cdot 5 + 3 \cdot 2$
$3(5 + 2) = 3 \cdot 5 + 3 \cdot 2$
$3(7) = 15 + 6$
$21 = 21$ True

Total area: $3(5 + 2)$
$= 3(7)$
$= 21$

Sum of small areas: $3 \cdot 5 + 3 \cdot 2$
$= 15 + 6$
$= 21$

B. $(7 + 3)2 = 7 \cdot 2 + 3 \cdot 2$
$(7 + 3) \cdot 2 = 7 \cdot 2 + 3 \cdot 2$
$10 \cdot 2 = 14 + 6$
$20 = 20$

Total area: $(7 + 3)2$
$= 10 \cdot 2$
$= 20$

Sum of small areas: $7 \cdot 2 + 3 \cdot 2$
$= 14 + 6$
$= 20$

◀

9. **Zero in Multiplication**   The product of any real number with 0 is 0. Symbolically:

If $a \in R$, then

$$a \cdot 0 = 0 \cdot a = 0$$

Example 8

Consider

$$(-5) \cdot 0 = 0$$

$$0 \cdot \frac{2}{3} = 0$$

$$0 \cdot 0 = 0$$

◀

Order of Operations

When we attempt to evaluate an expression such as $2 \cdot 5 + 4$, we see that the order in which operations are performed affects the answer we obtain. For example, if we multiply first, the preceding expression becomes

$$2 \cdot 5 + 4 = 10 + 4 = 14$$

However, if we add first, the expression becomes

$$2 \cdot 5 + 4 = 2 \cdot 9 = 18$$

1.5 Properties of Real Numbers and Order of Operations

Therefore, there must be some standardization as to what order of operations will be followed.

> **ORDER OF OPERATIONS AGREEMENT**
>
> When evaluating an arithmetic expression, first perform all multiplications and divisions, working from left to right in the order in which these operations appear. Second, perform all additions and subtractions, also working from left to right. The order in which these operations are performed may be altered with the use of "grouping symbols," which are discussed later.

With this agreement, the expression $2 \cdot 5 + 4$ is evaluated by multiplying first. The correct value is

$$2 \cdot 5 + 4 = 10 + 4 = 14$$

Example 9

Evaluate each of the following expressions:

A. $3 + 6 \cdot 2$

$\quad 3 + 6 \cdot 2 = 3 + 12 \qquad$ Multiply first

$\qquad \qquad \quad = 15 \qquad$ Then add

B. $3 \cdot 2 - 4 \div 2 + 15$

$\quad 3 \cdot 2 - 4 \div 2 + 15 = 6 - 2 + 15 \qquad$ Multiply/divide first left to right

$\qquad \qquad \qquad \qquad = 4 + 15 \qquad$ Subtract first

$\qquad \qquad \qquad \qquad = 19 \qquad$ Add ◀

Example 10

Using the formula for the area of a rectangle $A = L \cdot W$, write an expression for the shaded area of Figure 1.5-3, and then evaluate that expression.

FIGURE 1.5-3

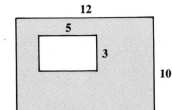

Outside rectangle ⟶   ⟵ Inside rectangle

$A = 12 \cdot 10 - 5 \cdot 3$

$A = 120 - 15 \qquad$ Multiply first

$A = 105$

◀

Grouping Symbols

We use grouping symbols to indicate when an arithmetic expression is to be evaluated in an order different from that of the order of operations agreement. Three frequently used grouping symbols are *parentheses*, indicated by ( ), *brackets*, indicated by [ ], and *braces*, indicated by { }.

Traditionally, parentheses are used as the innermost grouping symbols, brackets next, and braces the outermost. Expressions are evaluated from the innermost grouping symbols to the outermost, using the order of operations agreement. Any operations remaining to be performed after the grouping symbols have been removed are then performed according to the order of operations agreement.

The fraction bar is also considered a symbol of grouping. To simplify a fraction, first evaluate the numerator and denominator separately, then complete the division of the numerator by the denominator.

Example 11

Evaluate each of the following expressions:

A. $4(5 + 3) - 3(5 - 3)$

$$4(5 + 3) - 3(5 - 3) = 4(8) - 3(2) \quad \text{Parentheses first}$$
$$= 32 - 6 \quad \text{Multiply}$$
$$= 26 \quad \text{Subtract}$$

B. $2 + 3[12 - (3 + 4)]$

$$2 + 3[12 - (3 + 4)] = 2 + 3[12 - 7] \quad \text{Parentheses first}$$
$$= 2 + 3[5] \quad \text{Brackets second}$$
$$= 2 + 15 \quad \text{Multiply}$$
$$= 17 \quad \text{Add}$$

C. $15 - 2[4 + 1]$

$$15 - 2[4 + 1] = 15 - 2[5] \quad \text{Symbols of grouping first}$$
$$= 15 - 10 \quad \text{Multiply}$$
$$= 5 \quad \text{Subtract}$$

D. $3 + \{5 \cdot [15 \div (8 - 3)] + 2\}$

$$3 + \{5 \cdot [15 \div (8 - 3)] + 2\}$$
$$= 3 + \{5 \cdot [15 \div 5] + 2\} \quad \text{Parentheses first}$$
$$= 3 + \{5[3] + 2\} \quad \text{Brackets second}$$
$$= 3 + \{15 + 2\} \quad \text{Multiply inside braces first}$$
$$= 3 + 17 \quad \text{Add inside braces}$$
$$= 20 \quad \text{Add}$$

1.5 Properties of Real Numbers and Order of Operations

E. $\dfrac{6 + 4 \cdot 9}{20 - 7 \cdot 2}$

$\dfrac{6 + 4 \cdot 9}{20 - 7 \cdot 2} = \dfrac{6 + 36}{20 - 14}$ Evaluate numerator and denominator

$= \dfrac{42}{6}$

$= 7$ Divide

Exercise Set 1.5

State the property of real numbers illustrated by each of the following equations:

1. $4 \cdot 3 = 3 \cdot 4$
2. $2 + 5 = 5 + 2$
3. $2 \cdot 5 + 10 = 10 + 2 \cdot 5$
4. $5 \cdot 4 = 4 \cdot 5$
5. $4 + 5 = 9$
6. $3 + 7 = 10$ closure
7. $3(4 + 3) = 12 + 9$
8. $(2 + 5) \cdot 3 = 6 + 15$
9. $5[3 + (-3)] = 5 \cdot 0$
10. $\dfrac{3}{4} \cdot \dfrac{4}{3} = 1$
11. $9 \cdot 1 = 9$
12. $0 + 7 = 7$
13. $4 + (-4) = 0$
14. $\dfrac{7}{8} \cdot 1 = \dfrac{7}{8}$
15. $\dfrac{2}{3} \cdot \dfrac{3}{2} = 1$
16. $(-7) + 7 = 0$
17. $\dfrac{2}{3} + 0 = \dfrac{2}{3}$
18. $(-8) \cdot 1 = -8$
19. $\dfrac{2}{3} \cdot 1 = \dfrac{2}{3}$
20. $5[(-5) + 5] = 5 \cdot (0)$
21. $6 + (8 + 3) = (6 + 8) + 3$
22. $(2 + 5) + 9 = 2 + (5 + 9)$
23. $3 + (4 + 7) = (3 + 7) + 4$ EXPLAIN
24. $(9 + 2) + 4 = 9 + (4 + 2)$

Name the property of real numbers illustrated (assume all variables represent real numbers):

25. a. $7 + 3 = 3 + 7$
    b. $a \cdot 4 = 4 \cdot a$

c. $2 + 5 + 7 = 2 + 12$
d. If $a + (b + c) = a$, then $(b + c) = 0$.
e. $2xy$ is a real number.
f. $2 \cdot (8 - 3) \div 7$ is a real number.

26. a. $2 \cdot 6 \cdot 5 = (2 \cdot 6) \cdot 5$
b. $3(x + 7) = 3x + 21$
c. If $x(5 - y) = 5 - y$, then $x = 1$.
d. $3 + (5 + x) = (3 + 5) + x = 8 + x$

27. a. $2 \cdot 4 \cdot 5 = 2 \cdot 5 \cdot 4$
b. $(x + t) + 5 = 5 + (x + t)$
c. $5 - 6 + v$ is a real number.
d. $2 \cdot x \cdot 8 = 2 \cdot 8 \cdot x = 16x$

28. a. $5(3 + y) = 15 + 5y$
b. $(2x + y) \cdot 1 = 2x + y$
c. $5y + 0 = 5y$
d. $7 \cdot \frac{1}{4} \cdot 8 = 7 \cdot 2 = 14$

29. a. $(x + 2) \cdot 6 = 6(x + 2)$
b. $ab + ac = a(b + c)$
c. $ba + ca = ab + ac$
d. $2x + 5 = 5 + 2x$

30. a. $a(xy) = (xy)a$
b. $a(x + y) = ax + ay$
c. $(2x + 3y) = 1 \cdot (2x + 3y)$
d. $2x + 3x = (2 + 3)x = 5x$

31. a. If $x + m = 0$, then $m = -x$.
b. $a(2 + b) = a(b + 2)$
c. $2a + 3a = (2 + 3)a$
d. If $xm = 1$, then $m = \frac{1}{x}$.

32. a. If $a + b = 0$, then $b = -a$.
b. $3(2 + a) = 3 \cdot 2 + 3a$
c. $2x + 5x = (2 + 5)x$

1.5 Properties of Real Numbers and Order of Operations

     d. If $ab = 1$, then $b = \frac{1}{a}$.

33. a. $3(x + y) = (x + y)3$
    b. $3(x + y) = 3(y + x)$
    c. $a + (2 + x) = (a + 2) + x$
    d. $a + (2 + x) = a + (x + 2)$

34. a. $5(x + z) = 5x + 5z$
    b. $5(x + z) = (x + z)5$
    c. $5(x + z) = 5(z + x)$
    d. $5(x + z)$ is a real number.

35. a. If $2 + m = 2$, then $m = 0$.
    b. If $2 + m = 0$, then $m = -2$.
    c. If $2x = 2$, then $x = 1$.
    d. If $2x = 1$, then $x = \frac{1}{2}$.

36. a. If $5 + y = 0$, then $y = -5$.
    b. If $5 + y = 5$, then $y = 0$.
    c. If $5y = 1$, then $y = \frac{1}{5}$.
    d. If $5y = 5$, then $y = 1$.

Find the additive inverse and the multiplicative inverse of each of the following real numbers:

	Number (x)	Additive Inverse	Multiplicative Inverse
37.	7		
38.	$\frac{2}{3}$		
39.	0		
40.	$-\frac{3}{4}$		
41.	-5		
42.	π		
43.	$\sqrt{2}$		
44.	1		

Evaluate each of the following expressions using the order of operations agreement and rules for grouping symbols:

45. $3 \cdot 6 - 2$
46. $4 \cdot 7 - 5$
47. $4 + 2 \cdot 6$
48. $2 + 3 \cdot 5$
49. $4 \cdot 2 + 6 \cdot 3$
50. $5 \cdot 3 + 2 \cdot 7$
51. $(5 + 3) \cdot 4$
52. $(7 + 4) \cdot 3$
53. $3(4 + 2 + 5)$
54. $4(3 + 6 + 2)$
55. $5(7 - 3)$
56. $8(8 - 4)$
57. $(8 - 3)(8 + 3)$
58. $(9 - 4)(9 + 4)$
59. $3 + 2(5 + 3)$
60. $7 + 4(3 + 2)$
61. $24 - 6 \div 2 + 7$
62. $36 - 12 \div 4 + 2$
63. $(24 - 6) \div 2 + 7$
64. $(36 - 12) \div 4 + 2$
65. $(24 - 6) \div (2 + 7)$
66. $(36 - 12) \div (4 + 2)$
67. $2[3 + 4(5 + 2)]$
68. $4[5 + 3(4 + 3)]$
69. $13 + 2\{8 - [(4 - 2) + (5 - 1)]\}$
70. $18 + 3\{7 + [(8 - 3) + (7 - 4)]\}$
71. $4[5 + 2(5 + 2 \cdot 6)]$
72. $5[3 + 4(7 - 2 \cdot 3)]$
73. $\dfrac{6 \cdot 4 + 3 \cdot 5}{4 \cdot 9 - 3 \cdot 11}$
74. $\dfrac{5 \cdot 8 + 2 \cdot 15}{4 \cdot 6 - 7 \cdot 2}$
75. $\dfrac{60 - [12 - (8 - 2)]}{20 \div 5 + 5}$
76. $\dfrac{55 - [18 - (11 - 4)]}{3 + 16 \div 2}$

Write an expression for the shaded area in each figure, then evaluate that expression:

77.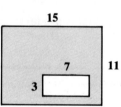

78.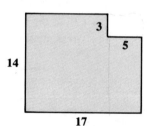

79.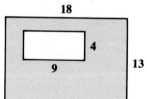

80.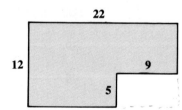

Write an expression for the shaded area:

83.

84.

1.6 Addition and Subtraction of Signed Numbers

Although we assume you can perform addition with the positive integers, such as $3 + 4 = 7$, we show such addition by using the number line. Our goal is to create a rule for addition that will include addition of both positive and negative integers. That is, we study the addition of signed numbers. By *signed numbers* we mean any set of numbers that includes both positive and negative numbers. Numbers that are added (and subtracted) together are called the *terms* of the expression.

To add the integers 2 and 3, draw an arrow starting at the origin and extending 2 units in the positive direction (to the right) to represent the number 2. Then draw a second arrow starting at the tip of the first arrow and extending 3 units to the right to represent the addition of 3 to 2.

$$2 + 3 = 5$$

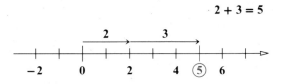

The position of the tip of the second arrow indicates that the sum of 2 and 3 is 5, which reinforces what we already know from arithmetic. We also use this procedure to determine the sum of two negative integers.

Example 1

Find $(-3) + (-4)$ on the number line.

We draw an arrow starting at the origin and extending 3 units in the negative direction (to the left) to represent the number -3. From the tip of the first arrow, we draw a second arrow extending 4 units to the left to represent the addition of -4 to -3.

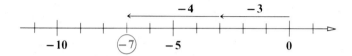

That is, $(-3) + (-4) = -7$ ◀

Notice that the sum of two negative integers is obtained essentially the same way as the sum of two positive integers. The only distinction is the sign of the sum. To add numbers having the same sign, add the absolute values of the numbers, then give the sum the same sign as the common sign of the numbers being added. This observation is stated more formally as follows.

RULE FOR ADDITION OF NUMBERS WITH LIKE SIGNS

To add two or more numbers having the same sign, add their absolute values and attach their common sign to the sum.

Example 2

Use the rule for addition of numbers with like signs to find the following sums:

A. $2 + 5$

$$2 + 5 = (+2) + (+5) \quad \text{Both numbers are positive}$$
$$= +(|+2| + |+5|) \quad \text{Add the absolute values and attach the common sign}$$
$$= +(2 + 5)$$
$$= +7$$
$$= 7$$

B. $(-3) + (-5)$

$$(-3) + (-5) = -(|-3| + |-5|) \quad \text{Add the absolute values and attach the common sign}$$
$$= -(3 + 5)$$
$$= -(8)$$
$$= -8 \quad ◀$$

1.6 Addition and Subtraction of Signed Numbers

Example 3 illustrates that the rule for addition can be applied less formally.

Example 3

Find each of the following sums:

A. $5 + 7$

$$5 + 7 = +12 = 12 \qquad \text{Add, attach common sign (positive)}$$

B. $(-4) + (-5)$

$$(-4) + (-5) = -(4 + 5) \qquad \text{Add absolute values}$$
$$= -9 \qquad \text{Attach common sign (negative)}$$

C. $(-2) + (-5) + (-8)$

$$(-2) + (-5) + (-8) = -(2 + 5 + 8)$$
$$= -15 \qquad \blacktriangleleft$$

The number line is also used to illustrate the addition of integers having opposite signs.

Example 4

Use the number line to find the following sums:

A. $7 + (-3)$

Draw an arrow starting at the origin and extending 7 units to the *right* to represent the number 7. From the tip of this arrow, draw a second arrow extending 3 units to the *left* to represent the addition of -3 to 7.

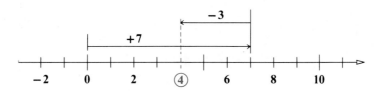

Therefore, $7 + (-3) = 4$.

B. $(-5) + 2$

Draw the first arrow starting at the origin and extending 5 units to the left. The second arrow starts at the tip of the first and extends 2 units to the right.

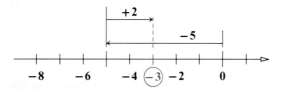

Therefore, $(-5) + 2 = -3$. $\blacktriangleleft$

From Example 4, we observe that the *sum* of two numbers having opposite signs is actually related to the *difference* of the absolute values of the numbers.

RULE FOR ADDITION OF NUMBERS WITH OPPOSITE SIGNS

To add two numbers having opposite signs, subtract their absolute values, smaller from larger, and attach the sign of the number with the greater absolute value.

Rule paraphrased: Subtract the smaller absolute value from the larger. Then give the result the sign of the number having the larger absolute value.

Use the rule for addition of numbers with opposite signs to find the following sums:

Example 5

A. $-3 + 7$

$$-3 + 7 = (-3) + (+7) \quad \text{Numbers have opposite signs}$$
$$= +(|7| - |-3|) \quad \text{Subtract smaller absolute value from larger}$$
$$\quad\quad\quad\quad\quad\quad \text{Attach "+" since } |7| \text{ is larger than } |-3|$$
$$= +(7 - 3)$$
$$= +(4)$$
$$= 4$$

B. $5 + (-8)$

$$5 + (-8) = (+5) + (-8)$$
$$= -(|-8| - |5|) \quad \text{Subtract smaller absolute value from larger}$$
$$\quad\quad\quad\quad\quad\quad \text{Attach "}-\text{" since } |-8| \text{ is larger than } |5|$$
$$= -(8 - 5)$$
$$= -(3) \quad \blacktriangleleft$$

Example 6 illustrates a less formal application of this rule for addition.

Example 6

Find each of the following sums:

A. $4 + (-9)$

$$4 + (-9) = -(9 - 4) \quad \text{Subtract 4 (the smaller) from 9 (the larger)}$$
$$= -5 \quad \text{Attach sign of the larger number } (-9)$$

1.6 Addition and Subtraction of Signed Numbers

B. $-4 + 9$

$\quad -4 + 9 = +(9 - 4) \qquad$ Subtract 4 from 9

$\qquad\qquad\;\; = 5 \qquad\qquad$ Attach sign of the larger signed number (9) ◀

Subtraction

The rule for subtraction of signed numbers is developed by studying a specific example. If we consider the subtraction $8 - 3$, the result is 5. The same answer is also obtained if we consider the addition $8 + (-3)$. That is, subtracting 3 is equivalent to adding the opposite (additive inverse) of 3.

RULE FOR SUBTRACTION OF NUMBERS

For any real numbers x and y,

$$x - y = x + (-y)$$

Rule paraphrased: To subtract a number, add its opposite (additive inverse).

Example 7 Find each of the following differences:

A. $7 - 3$

$\quad 7 - 3 = 7 + (-3) \qquad$ Change to addition

$\qquad\quad\; = 4 \qquad\qquad$ Addition of numbers with opposite signs

B. $-5 - 2$

$\quad -5 - 2 = -5 + (-2) \qquad$ Change to addition

$\qquad\quad\;\; = -(5 + 2) \qquad$ Addition of numbers with like signs

$\qquad\quad\;\; = -7$

C. $4 - (-3)$

$\quad 4 - (-3) = 4 + 3 \qquad$ Change to addition

$\qquad\qquad = 7 \qquad\qquad$ Like signs

D. $-5 - (-8)$

$\quad -5 - (-8) = -5 + (+8) \qquad$ Change to addition

$\qquad\qquad\;\; = -5 + 8$

$\qquad\qquad\;\; = +(8 - 5) \qquad$ Opposite signs

$\qquad\qquad\;\; = +3$

$\qquad\qquad\;\; = 3$

E. $0 - 7$

$$0 - 7 = 0 + (-7) \qquad \text{Change to addition}$$
$$= -7 \qquad \text{Additive identity} \qquad \blacktriangleleft$$

It is often necessary to evaluate expressions containing several additions and subtractions.

Example 8

Evaluate each of the following expressions:

A. $7 - 2 - (-4)$

$$7 - 2 - (-4) = 7 + (-2) + (4) \qquad \text{Change each subtraction to addition}$$
$$= 5 + 4 \qquad \text{Add, left to right}$$
$$= 9$$

B. $-2 + 5 - 6 - (-3) + (-4)$

$$-2 + 5 - 6 - (-3) + (-4)$$
$$= -2 + 5 + (-6) + (3) + (-4) \qquad \text{Change each subtraction to addition}$$
$$= 3 + (-6) + 3 + (-4) \qquad \text{Add, left to right}$$
$$= -3 + 3 + (-4)$$
$$= 0 + (-4)$$
$$= -4$$

C. $-3 - \{5 - [7 - (-3)]\}$

We must evaluate expressions within grouping symbols innermost to outermost to evaluate this expression.

$$-3 - \{5 - [7 - (-3)]\}$$
$$= -3 - \{5 - [7 + 3]\} \qquad \text{Simplify brackets}$$
$$= -3 - \{5 + [-10]\} \qquad \text{Change to addition}$$
$$= -3 - \{-5\} \qquad \text{Add, to remove brackets}$$
$$= -3 + \{5\} \qquad \text{Change to addition}$$
$$= 2 \qquad \text{Add} \qquad \blacktriangleleft$$

In Section 1.3 we found the directed distance from A to B by counting. It can be shown that the directed distance, D, from A to B is found using subtraction:

$$D = B - A$$

1.6 Addition and Subtraction of Signed Numbers

Example 9

Let A, B, and C represent numbers -8, -3, and 4, respectively, on a number line. Find the following directed distances:

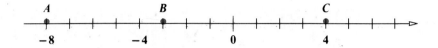

A. From B to C.
Directed distance from B to C:

$$D = C - B$$
$$= 4 - (-3)$$
$$= 4 + 3$$
$$= 7$$

B. From C to A.
Directed distance from C to A:

$$D = A - C$$
$$= -8 - 4$$
$$= -8 + (-4)$$
$$= -12$$

C. From A to B.
Directed distance from A to B:

$$D = B - A$$
$$= -3 - (-8)$$
$$= -3 + 8$$
$$= 5$$

D. From B to A.
Directed distance from B to A:

$$D = A - B$$
$$= -8 - (-3)$$
$$= -8 + 3$$
$$= -5$$

Verify each of these results by counting.

◀

Exercise Set 1.6

Add or subtract as indicated using the number line:

1. $(+7)+(+2)$
2. $(+5)+(+4)$
3. $8+6$
4. $4+7$
5. $(-5)+9$
6. $(-3)+7$
7. $2+(-6)$
8. $3+(-8)$
9. $(-9)+6$
10. $(-7)+4$

Add or subtract as indicated:

11. $-8+3$
12. $-9+2$
13. $(-8)+(-6)$
14. $(-3)+(-9)$
15. $-7+(-5)$
16. $-6+(-9)$
17. $11-3$
18. $13-5$
19. $(-9)-(+5)$
20. $(+8)-(+2)$
21. $-7-4$
22. $-8-5$
23. $-8-(-3)$
24. $-9-(-5)$
25. $7-9$
26. $8-11$
27. $-8-(-8)$
28. $-9-(-3)$
29. $7-(-3)$
30. $10-(-4)$
31. $-(|-4|-|-5|)$
32. $-(|-7|-|-5|)$
33. $-(|-7|-|-2|)$
34. $-(|-5|-|-9|)$
35. $7-4+3-5$
36. $8-2+5-6$
37. $-4-2+3-7$
38. $-5+4-3-5$
39. $6-[4-3]$
40. $7-[7-4]$
41. $-5-[-3-(-4)]$
42. $-6-[-2-(-5)]$
43. $[3-(-2)]-[-5+7]$
44. $[-4+8]-[3-(-5)]$
45. $(-4)+(+4)$
46. $(+7)+(-7)$
47. $(+3)+(-6)+(-4)$
48. $(-9)-(-4)+(+7)-(2)$
49. $(-6)+(+8)-(-4)-(8)$
50. $(+7)-(5)-(-5)-(6)$
51. $(-23)+(11)-(-7)+(-21)$
52. $(15)-(-7)+(+18)-(4)$
53. $(22)-(-6)-(+9)+(+9)+(+4)$
54. $(-31)+(+20)-(-15)-(+13)$
55. $(+17)+(0)+(-5)+(6)+(-16)$
56. $(+7)-(-7)+(+34)-(-12)$
57. $(6)-(9)+(8)-(-5)+(0)-(6)$
58. $(11)-(9)-(5)-12+10$
59. $8+13-12-(-7)-(+20)$

1.7 Multiplication and Division of Signed Numbers

60. $(-20) - 14 + 11 - (-9) - (-7) + 15$
61. $(-30) - (250) + (+90) - (-50) + 65$
62. $(22) - 17 - (+22) + (-27) - (-2) + 0 - 12$

Consider the number line

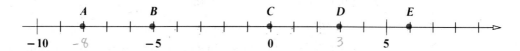

Find the directed distance:

63. from B to D
64. from A to E
65. from B to A
66. from E to D
67. from C to B
68. from C to A
69. from D to -4
70. from E to -6
71. from E to A
72. from D to B

Evaluate the following expressions:

73. $(4 - 3) - 5$
74. $(6 - 2) - 7$
75. $4 - (3 - 5)$
76. $6 - (2 - 7)$
77. $4 - 3 - 5$
78. $6 - 2 - 7$
79. $(-3 + 2) - (-3 - 4)$
80. $(-5 + 3) - (-4 - 2)$
81. $2 + [3 - (-4 - 2)]$
82. $3 + [2 - (-3 - 4)]$

1.7 Multiplication and Division of Signed Numbers

Multiplication

Multiplication is defined as repeated addition. Numbers that are multiplied together are called *factors*, and the result is called their *product*. For example, in the multiplication $3 \cdot 5 = 15$, we say that 15 is the product of 3 and 5 and that 3 and 5 are factors of 15. Because of the definition, we can use our rules for the addition of signed numbers (from Section 1.6) to develop the rules for multiplication. We interpret the product of 3 times 4 as the sum of three 4s:

$$3 \cdot 4 = 4 + 4 + 4 = 12$$

In general, the product of two positive numbers is positive since the repeated addition of a positive term with itself will yield a positive sum. If we also show the product $3 \cdot 0$ using repeated addition, we support the rule for multiplication by 0. (See Section 1.5.)

$$3 \cdot 0 = 0 + 0 + 0 = 0$$

In general, if a is any real number, then

$$a \cdot 0 = 0 \cdot a = 0$$

Now let's consider the product of two numbers having opposite signs. We examine the product $2 \cdot (-5)$ using repeated addition.

$$2 \cdot (-5) = (-5) + (-5) = -10$$

Similarly, $4 \cdot (-3)$ may be evaluated as

$$4 \cdot (-3) = (-3) + (-3) + (-3) + (-3) = -12$$

When the negative factor is written first, we use the commutative property of multiplication together with repeated addition to evaluate the product.

Example 1

Evaluate: $(-4) \cdot 5$

$(-4) \cdot 5 = 5 \cdot (-4)$ Commutative property

$ = (-4) + (-4) + (-4) + (-4) + (-4)$ Repeated addition

$ = -20$ ◀

These examples suggest that

The product of any two numbers having opposite signs is negative.

Example 2

Evaluate each of the following expressions:

A. $(-5) \cdot 4$

$ (-5) \cdot 4 = -20$

B. $7 \cdot (-3)$

$ 7 \cdot (-3) = -21$

C. $(-6) \cdot (3) + (8)(-6)$

$ (-6)(3) + (8)(-6) = -18 + (-48)$ Multiply first, then add

$ = -66$

1.7 Multiplication and Division of Signed Numbers

D. $(-4)(5) - (-4)(6)$

$(-4)(5) - (-4)(6) = -20 - (-24)$ Multiply first, then subtract
$= -20 + 24$
$= 4$

E. $(-1)(8)$

$(-1)(8) = -8$ ◀

We now look at the product of two negative numbers. We consider the expression

$(-4)[3 + (-3)]$

to help us develop the rule. To do this, we follow the order of operations agreement and evaluate the bracket first.

$(-4)[3 + (-3)] = (-4)[0]$ Additive inverses
$= 0$ Multiplication by zero

On the other hand, if we evaluate this expression using the distributive property, we encounter the product of two negative numbers.

$(-4)[3 + (-3)] = (-4) \cdot 3 + (-4) \cdot (-3)$
$= -12 + (-4) \cdot (-3)$

Since this expression is the same as zero, from above, then

$-12 + (-4)(-3) = 0$

This is possible only if $(-4)(-3) = 12$. This example suggests that

---

The product of two negative numbers is positive.

Example 3

Evaluate each of the following expressions:

A. $(-5) \cdot (-6)$

$(-5) \cdot (-6) = 30$

B. $(-7) \cdot (-3)$

$(-7) \cdot (-3) = 21$

C. $(-3) \cdot (-2) + (-4) \cdot (-5)$

$(-3) \cdot (-2) + (-4) \cdot (-5) = 6 + 20$ Multiply first
$= 26$ Add

D. $(-5)(-7) - (-3)(-8)$

$$(-5)(-7) - (-3)(-8)$$
$$= 35 - (24) \qquad \text{Multiply first}$$
$$= 35 + (-24) \qquad \text{Subtract (change to addition)}$$
$$= 11 \qquad \blacktriangleleft$$

Recall from the properties of real numbers that $a \cdot 1 = 1 \cdot a = a$. For example:

$$1 \cdot 5 = 5 \cdot 1 = 5$$
$$1 \cdot 0 = 0 \cdot 1 = 0$$

Similarly,

$$1 \cdot (-6) = (-6) \cdot 1 = -6$$
$$(-3) \cdot 1 = 1 \cdot (-3) = -3$$

Now, let's consider multiplication by -1.

$$-1 \cdot 6 = -6 \qquad \text{Product of numbers with opposite signs}$$
$$-1 \cdot (-3) = 3 \qquad \text{Product of numbers with like signs}$$
$$-1 \cdot 0 = 0$$

Observe that multiplying a number by -1 reverses the sign of that number. Also, when a number is preceded by a negative sign, it is equivalent to multiplying the number by -1. For example:

$$-10 = (-1) \cdot 10$$
$$-25 = (-1) \cdot 25$$

Similarly, $-(-5)$ is the same as $(-1) \cdot (-5)$. That is, $-(-5) = (-1)(-5) = 5$. This reinforces that the negative of a negative number is positive.

RULES FOR MULTIPLICATION OF SIGNED NUMBERS

1. The product of two numbers having the same sign is positive.

    Example: (a) $3 \cdot 4 = 12$ (b) $(-5) \cdot (-4) = 20$

2. The product of two numbers having opposite signs is negative.

    Example: (a) $(-5)(6) = -30$ (b) $(6)(-4) = -24$

1.7 Multiplication and Division of Signed Numbers

3. The product of any number with 0 is 0.

    Example: (a) $0 \cdot 7 = 0$ (b) $0 \cdot (-8) = 0$

    (c) $0 \cdot 0 = 0$

4. The product of any number with 1 is that number

    Example: (a) $1 \cdot 5 = 5$ (b) $1 \cdot (-6) = -6$

5. Forming the product of any real number with -1 is equivalent to reversing the sign of that number (taking its negative).

    Example: (a) $(-1) \cdot 8 = -8$ (b) $(-1) \cdot (-7) = 7$

Example 4

Evaluate each of the following products:

A. $5 \cdot 3$

$5 \cdot 3 = 15$ Like signs

B. $(-6)(4)$

$(-6)(4) = -24$ Opposite signs

C. $(-5)(-6)$

$(-5)(-6) = 30$ Like signs

D. $(0)(-7)$

$(0)(-7) = 0$ Multiplication by 0 ◀

Finally, observe that *multiplication is distributive over subtraction* as well as addition.

Example 5

Show: $5(8 - 3) = 5 \cdot 8 - 5 \cdot 3$

$5(8 - 3) = 5 \cdot 8 - 5 \cdot 3$

$5(5) = 40 - 15$

$25 = 25$ ◀

Example 6

Evaluate: $3(5 - 4 + 3 - 2 - 1)$

$3(5 - 4 + 3 - 2 - 1) = 3 \cdot 5 - 3 \cdot 4 + 3 \cdot 3 - 3 \cdot 2 - 3 \cdot 1$

$= 15 - 12 + 9 - 6 - 3$

$= 24 - 21$

   Sum of all positives   Sum of all negatives

$= 3$ ◀

Note in Example 6 that when several numbers having mixed signs are added it is convenient to form the sum of all the positive numbers and the sum of all the negative numbers separately. Then, add these two numbers (having opposite signs) as the last step.

Example 7

Evaluate: $6 - 7 + 5 + 8 - 3 - 4 + 3 - 2$

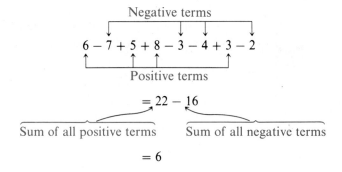

$= 22 - 16$

$= 6$ ◀

Division

In arithmetic, the division a divided by b is written $a \div b$. In algebra, we prefer to show division as a fraction, writing a/b or $\frac{a}{b}$. Division of real numbers may be defined using multiplication. Let a and b be any real numbers where $b \neq 0$. Then the unique **quotient** of a divided by b may be defined by:

$$\frac{a}{b} = c \quad \text{if} \quad c \cdot b = a$$

where a is the dividend, b is the divisor, and c is the quotient of a and b. The definition could be presented as follows:

$$\frac{\text{dividend}}{\text{divisor}} = \text{quotient if (quotient)} \cdot \text{(divisor)} = \text{dividend}$$

You probably recall checking your division problems in arithmetic using this definition.

1.7 Multiplication and Division of Signed Numbers

Example 8

Evaluate each of the following quotients:

A. $24 \div 6$

$$\frac{24}{6} = 4, \text{ since } 4 \cdot 6 = 24$$

B. $(-18) \div (-3)$

$$\frac{-18}{-3} = 6, \text{ since } 6 \cdot (-3) = -18$$

C. $12 \div (-4)$

$$\frac{12}{-4} = -3, \text{ since } (-3) \cdot (-4) = 12$$

D. $(-16) \div 8$

$$\frac{-16}{8} = -2, \text{ since } (-2) \cdot 8 = -16$$ ◀

The quotients in Example 8 suggest that the rules for signed numbers in division are very similar to those for multiplication. The answer is *positive* if the two numbers have the *same sign*; the answer is *negative* if the two numbers have *opposite signs*.

Example 9

Evaluate each of the following quotients:

A. $\dfrac{28}{-7}$

$$\frac{28}{-7} = -4 \qquad \text{Opposite signs}$$

B. $\dfrac{-35}{-7}$

$$\frac{-35}{-7} = 5 \qquad \text{Same signs}$$

C. $\dfrac{-21}{3}$

$$\frac{-21}{3} = -7 \qquad \text{Opposite signs}$$ ◀

Zero and Division

Consider the quotient $0 \div 8$. If we assume the quotient of 0 and 8 is some number c, then by definition:

$$\frac{0}{8} = c \quad \text{if} \quad c \cdot 8 = 0$$

However, $c \cdot 8 = 0$ only if $c = 0$. Therefore, $\frac{0}{8} = 0$. That is, the quotient of 0 divided by a non-zero number is zero.

Let's now examine the quotient $8 \div 0$. If we assume the quotient of 8 and 0 is some number d, then by definition:

$$\frac{8}{0} = d \quad \text{if} \quad d \cdot 0 = 8$$

However, $d \cdot 0 = 0$ for all possible values of d. Therefore, no number d can satisfy the division statement that requires $d \cdot 0 = 8$. This division is therefore undefined.

Finally, we consider the quotient $0 \div 0$. If we assume the quotient of 0 and 0 is some number e, then by definition:

$$\frac{0}{0} = e \quad \text{if} \quad e \cdot 0 = 0$$

However, this last statement is true for *any* real number e. Consequently, there is no *unique* quotient for $0 \div 0$. This division is also undefined.

The previous discussion shows that, in general, division by zero is undefined.

RULES FOR DIVISION OF SIGNED NUMBERS

1. The quotient of two numbers having the same sign is positive.
2. The quotient of two numbers having opposite signs is negative.
3. Zero divided by a non-zero number is zero.
4. Division by zero is undefined.

Recall that when a fraction bar is used to indicate division, it is a symbol of grouping. Therefore, the entire dividend (top number or expression) and divisor (bottom number or expression) are to be evaluated first, and then the division is to be completed.

Example 10

Evaluate each of the following quotients if possible:

A. $\dfrac{0}{-5} = 0$

1.7 Multiplication and Division of Signed Numbers

B. $\dfrac{7}{0}$ is undefined

C. $\dfrac{-9}{3+(-3)} = \dfrac{-9}{0}$ is undefined

D. $\dfrac{-5+5}{6} = \dfrac{0}{6} = 0$

E. $\dfrac{4+(-4)}{-6+6} = \dfrac{0}{0}$ is undefined

We can now evaluate expressions containing signed numbers and involving any combination of the four basic operations of arithmetic. The order of operations agreement still applies.

Exercise Set 1.7

Evaluate each of the following expressions:

1. $(7)(-4)$
2. $(-5)(8)$
3. $(-5)(-9)$
4. $(-7)(-8)$
5. $(12) \div (-4)$
6. $(-18) \div (6)$
7. $(-36) \div (-9)$
8. $(-48) \div (-8)$
9. $\dfrac{18}{-3}$
10. $\dfrac{-24}{6}$
11. $\dfrac{0}{5}$
12. $\dfrac{-7}{0}$
13. $\dfrac{4}{0}$
14. $\dfrac{0}{12}$
15. $\dfrac{-39}{-13}$
16. $\dfrac{-42}{-14}$
17. $\dfrac{-51}{17}$
18. $\dfrac{-42}{21}$
19. $\dfrac{36}{-9}$
20. $\dfrac{24}{-8}$
21. $(-4)(-3)(2)$
22. $(-5)(3)(-2)$
23. $6 - \left(\dfrac{12}{-3}\right)$
24. $7 - \left(\dfrac{-18}{6}\right)$
25. $(-3)(7) - (4)(-3)$
26. $(4)(-5) - (-6)(4)$
27. $\dfrac{28}{-7} - (-3)(-2)$
28. $(-5)(-3) - \dfrac{14}{-2}$
29. $\left(\dfrac{-4}{2}\right)\left(\dfrac{-9}{-3}\right)$
30. $\left(\dfrac{6}{-3}\right)\left(\dfrac{-14}{7}\right)$
31. $\left(\dfrac{12}{-4}\right)\left(\dfrac{18}{-6}\right)$
32. $\left(\dfrac{-18}{3}\right)\left(\dfrac{-27}{-3}\right)$
33. $\dfrac{(6)(-4) - (-3)(2)}{-8 - (-2)}$

34. $\dfrac{(-5)(4) - (-2)(4)}{2 - 8}$

35. $\dfrac{(-3)(-5) - (-4)(-3)}{-2 - (-1)}$

36. $\dfrac{(-5)(-6) - (-3)(-4)}{-6 - (-3)}$

37. $\dfrac{-3 - (6)(-2) + 3}{(2)(-2) + 4}$

38. $\dfrac{5 - (-4)(-6) - 5}{(-4)(3) - (6)(-2)}$

39. $-5 - 2\{-3 - [(4-6) - 3]\}$

40. $-8 - 3\{-4 - [-5 - (8-3)]\}$

41. $-8 - 3\{-4 - [-4 - (-7 - (-6))]\}$

42. $-5 - 2\{-3 - [-6 - (8 - (-3))]\}$

43. $\dfrac{4 - 2[3 - (-2)]}{6 - (7 - 4)}$

44. $\dfrac{6 + 3[6 - (1 - 5)]}{4 - 2(-4)}$

45. $\dfrac{7 - \{6 - [3 - (4 - 5)]\}}{19 - 2[3 - (-4)]}$ *answer is corrected in back*

46. $\dfrac{2(-5) - 3(-8)}{3(5) + 2(-4)}$

Find the value of each of the following numbers (review absolute value in Section 1.4 if necessary):

47. $-(+5)$ 48. $-(+2)$ 49. $-(-6)$ 50. $-(-4)$

51. $-[-(+3)]$ 52. $-[-(+9)]$ 53. $-[-(-7)]$ 54. $-[-(-2)]$

55. $|-12|$ 56. $|-15|$ 57. $-|-6|$ 58. $-|-7|$

59. $|0|$ 60. $-(0)$

61. We've seen in this section that multiplication can be defined as repeated addition. In a similar way division can be defined as repeated subtraction. Explain this concept.

1.8 Constants, Variables, Exponents, and Algebraic Expressions

Two different types of symbols are used to represent numbers when we write mathematical expressions: constants and variables. A **constant** is a symbol (or group of symbols) that names a single specific number. Examples of constants include:

$$\text{\fiveroman}, V, 5, 7-2, \text{FIVE}, \text{\raisebox{0pt}{✋}}, \sqrt{3}, |-2|, \pi, \text{ or } -6$$

A **variable** is a symbol (or group of symbols) that represents various constants. For example, the formula for the perimeter of (distance around) a rectangle is $P = 2L + 2W$.

where P, L, and W are all variables representing the perimeter, length, and width, respectively, for a rectangle. The symbol "2" in the formula is a constant. The variables P, L, and W assume various values when the formula is used to find the perimeter for different rectangles.

The formula for the circumference of (distance around) a circle is $C = 2\pi r$.

In this formula, both "2" and "π" are constants, while C and r are variables representing the circumference and radius of the circle, respectively.

When an amount of money, called the principal (P), is invested at a rate of interest (R) for a time period (T), the amount of interest (I) earned by the investment is given by the simple interest formula

$$I = PRT$$

In this formula, all four symbols are variables. The interest, I, may be calculated for different accounts by allowing P, R, and T to assume different constant values. We must always be aware of which symbols represent constants and which represent variables in any formula or expression.

Exponential Notation

An expression written in the form b^p where b is any real number and p is a natural number is said to be in **exponential form**. The number b is called the **base**, p is called the **exponent**, and b^p is read "base b raised to the power p" or more simply "b to the p." The form b^p is a symbolic way of expressing the product formed by using the number b as a factor p number of times. For example, 2^4 means $2 \cdot 2 \cdot 2 \cdot 2 = 16$; 4^3 means $4 \cdot 4 \cdot 4 = 64$; and $(a+b)^2$ means $(a+b)(a+b)$.

A number written without an exponent has an assumed exponent of 1. For example, $3^1 = 3$; a is the same as a^1; and $(x + y)^1 = (x + y)$.

Raising a number to the power 2 is called *squaring the number*, and we read a^2 as "*a* square," or "*a* squared." Raising a number to the power 3 is called *cubing the number*, and we read a^3 as "*a* cube" or "*a* cubed."

Powers are commonly used in expressions with more than one factor. In such expressions, each exponent applies only to the base by which it is written. For example, ab^3 means $a \cdot b \cdot b \cdot b$, and *not* $(ab)(ab)(ab)$. Similarly, $3x^3y^2 = 3xxxyy$. We call $3x^3y^2$ the *exponential form* and $3xxxyy$ the *expanded form*.

Example 1

Write each of the following exponential forms in expanded form and evaluate:

A. $2^3 \cdot 3 \cdot 5$

$$2^3 \cdot 3 \cdot 5 = 2 \cdot 2 \cdot 2 \cdot 3 \cdot 5$$
$$= 120$$

B. $2 \cdot 3 \cdot 5^3$

$$2 \cdot 3 \cdot 5^3 = 2 \cdot 3 \cdot 5 \cdot 5 \cdot 5$$
$$= 750$$

C. $2 \cdot 3^3 \cdot 5$

$$2 \cdot 3^3 \cdot 5 = 2 \cdot 3 \cdot 3 \cdot 3 \cdot 5$$
$$= 270 \qquad \blacktriangleleft$$

Example 2

A. Write a^2b^3c in expanded form.

$$a^2b^3c = a \cdot a \cdot b \cdot b \cdot b \cdot c$$
$$= aabbbc$$

B. Write $2xxyyyyzz$ in exponential form.

$$2xxyyyyzz = 2 \cdot x \cdot x \cdot y \cdot y \cdot y \cdot y \cdot z \cdot z$$
$$= 2x^2y^4z^2 \qquad \blacktriangleleft$$

In these examples note again that multiplication may be represented with or without the center dot.

Algebraic Expressions

An algebraic expression is a symbolic form that includes any combination of constants, variables, operations, and/or symbols of grouping. Operations include the four arithmetic operations as well as absolute value, negative, raising

1.8 Constants, Variables, Exponents, and Algebraic Expressions

to powers, taking roots, and others to be introduced in later sections. Some frequently used symbols of grouping include parentheses ( ), brackets [ ], braces { }, absolute value symbols | |, and fraction bars —. The following are several examples of algebraic expressions:

$$2 + a \qquad 3x - |x - y - 5| \qquad 2\{x + [5 - 2(x - 1)]\}$$

$$\frac{2x + y}{5} \qquad 2(L + W) \qquad \frac{4}{3}\pi r^3 \qquad \sqrt{x + y}$$

We frequently must determine the numeric value of an algebraic expression for given values of the variable(s). We use the following steps to *evaluate an expression*.

To evaluate an algebraic expression:

1. Replace each variable in the expression with its given value.
2. Evaluate each expression within symbols of grouping, proceeding from the innermost to the outermost symbols.

Note: It is sometimes necessary to use Steps 3, 4, and 5 in evaluating expressions within symbols of grouping (Step 2).

3. Evaluate all powers and roots (to be introduced later) as they are encountered from left to right.
4. Perform all multiplications and divisions as they are encountered from left to right.
5. Perform all additions and subtractions as they are encountered from left to right.

Observe that the rule for evaluating an expression is an expansion of the order of operations agreement (Section 1.5).

Example 3

Evaluate each algebraic expression for the given values of the variables:

A. $2x + 3y$ for $x = 6$ and $y = 3$

$\qquad 2x + 3y = 2(6) + 3(3) \qquad$ Substitute

$\qquad \qquad \quad = 12 + 9 \qquad \qquad$ Multiply first

$\qquad \qquad \quad = 21$

B. LWH for $L = 8$, $W = 2$, $H = -3$

$\qquad LWH = (8)(2)(-3) \qquad$ Substitute

$\qquad \qquad \ \ = (16)(-3) \qquad \ \ $ Multiply left to right

$\qquad \qquad \ \ = -48$

C. $P = 2L + 2W$ Find P if $L = 7$, $W = 5$.

$P = 2L + 2W$

$\quad = 2(7) + 2(5)$ \qquad Substitute

$\quad = 14 + 10$ \qquad Multiply first

$\quad = 24$ \qquad Add

D. $C = 2\pi r$ Find C if $r = 3$.

$C = 2\pi r$

$\quad = 2\pi(3)$ \qquad Substitute

$\quad = 2 \cdot 3\pi$ \qquad Commutative and associative properties

$\quad = 6\pi$

E. $I = PRT$ Find I if $P = 5000$, $R = 0.08$, $T = 2$.

$I = PRT$

$\quad = (5000)(0.08)(2)$ \qquad Substitute

$\quad = (400)(2)$

$\quad = 800$

F. $2(a - b) - 3(b - 2a)$ for $a = -2$, $b = 4$

$2(a - b) - 3(b - 2a)$

$\quad = 2(-2 - 4) - 3[4 - 2(-2)]$ \qquad Substitute

$\quad = 2(-6) - 3(4 + 4)$ \qquad Evaluate parentheses first

$\quad = 2(-6) - 3(8)$

$\quad = -12 - 24$ \qquad Multiply

$\quad = -36$ \qquad Subtract

G. $5\{a + [6 - 2(8 - 3b)]\}$ for $a = 4$ and $b = 2$

$5\{a + [6 - 2(8 - 3b)]\}$

$\quad = 5\{4 + [6 - 2(8 - 3 \cdot 2)]\}$ \qquad Substitute

$\quad = 5\{4 + [6 - 2(2)]\}$ \qquad Evaluate parentheses first

$\quad = 5\{4 + [6 - 4]\}$ \qquad Multiply in bracket

$\quad = 5\{4 + 2\}$ \qquad Evaluate brackets

$\quad = 5\{6\}$ \qquad Evaluate braces

$\quad = 30$

1.8 Constants, Variables, Exponents, and Algebraic Expressions

H. $\dfrac{4 + |-x|}{2} + x$ for $x = 8$

$\dfrac{4 + |-x|}{2} + x = \dfrac{4 + |-8|}{2} + 8$ Substitute

$= \dfrac{4 + 8}{2} + 8$ Absolute value

$= \dfrac{12}{2} + 8$ Fraction bar

$= 6 + 8$ Divide

$= 14$ Add

I. $\dfrac{2x^2 + 4y^2}{x + y}$ for $x = 6$, $y = 2$

$\dfrac{2x^2 + 4y^2}{x + y} = \dfrac{2(6)^2 + 4(2)^2}{6 + 2}$ Substitute

Note: The fraction bar is a symbol of grouping, so we separately evaluate the numerator and denominator.

$= \dfrac{2(36) + 4(4)}{8}$ Exponents first

$= \dfrac{72 + 16}{8}$

$= \dfrac{88}{8}$

$= 11$ ◀

Exercise Set 1.8

In each expression or formula, identify the constants and the variables:

1. $3x$
2. $2L + 2W$
3. πr^2
4. $P = 4s$
5. $A = \dfrac{1}{2}bh$
6. $s = 16t^2$
7. $V = \pi r^3 h$
8. $x^2 + 3x - 5$
9. $v = 48 - 32t^2$
10. $2x + 3y$
11. $y = mx + b$
12. $2x^y$

Evaluate each expression:

13. $3 \cdot 4 + 2 \cdot 1$
14. $\dfrac{6 + 2 \cdot 4}{7}$
15. $2(3)^2 - 3(-2)^2$
16. $4(-2)^2 - (-4)^2$
17. $\dfrac{10 - 2 \cdot 3}{2}$
18. $8 \cdot 2 - 2 \cdot 3$
19. $\dfrac{|-6| + 2}{2}$
20. $\dfrac{5 + |-3|}{4}$
21. $\dfrac{(3)^2 - 5}{2}$
22. $\dfrac{(-3)^2 - 3(-5)}{2 - 8}$
23. $24 \div 6 \div 2 + 4$
24. $36 \div 9 + 3 \cdot 2$
25. $3\{2 + [5 - (3 - 2)]\}$
26. $4\{8 - [5 - (4 - 2)]\}$

Evaluate each of the following expressions for $a = 12$, $b = -2$, $c = 5$, and $d = -4$:

27. $|d|$
28. $-d$
29. $a + b$
30. $a - b$
31. ab
32. $\dfrac{a}{b}$
33. $c + 2b$
34. $5 + ac$
35. $\dfrac{(a - b)}{7}$
36. $2 + \dfrac{a}{2}$
37. $3c - 2b$
38. ab^2c
39. $b^3(c + 4)$
40. $|-c|$
41. $-(-d)$
42. $-|d|$
43. $|-d|$
44. $16 - 4 - b$
45. $a - b$
46. $3(a - c)$
47. $3a^2 - 3c$
48. $-(c - b)$
49. $|c - b|$
50. $\dfrac{a - 2b^3}{2}$
51. $\dfrac{a}{2}$
52. $ab - 5$
53. $a + b + c$
54. $(a + 2b) + (a - 2b)$
55. $3[a - (b^2 + c)]$
56. $3[(a^2 - b) + c]$
57. $2\{a + [c - (1 + b)]\}$
58. $2\{a - [c - (b + d)]\}$

59. The area (A) of a ring is given by $A = \pi(R^2 - r^2)$ where R and r are the radii of the outside and inside circles of the ring, respectively. Find the area of the ring drawn below ($\pi = 3.14$).

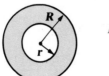

$R = 8$
$r = 4$

1.9 Review and Chapter Test

have him do 60

60. The volume (V) of a sphere is given by $V = \frac{4}{3}\pi r^3$ where r is the radius of the sphere. Find the volume of a sphere having radius 3 inches. Use $\pi = 3.14$.

61. The volume (V) of a rectangular box is given by $V = LWH$ where L is the length, W is the width, and H is the height of the box. Find the volume of a rectangular room that measures 30 feet long by 8 feet high by 20 feet wide.

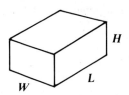

1.9 Review and Chapter Test

(1.1) A *set* is a collection or group of objects. The objects contained in the set are called the *elements* of the set. A set may be described by either listing the elements of the set between a pair of braces, { }, or by giving a *rule* for membership in the set. The *empty set* (*null set*) is the set containing no elements and is represented by ϕ, or { }.

Example 1

Consider the set $A = \{1, 3, 5, 7, 9, 11, 13\}$.

A. We write $3 \in A$ to indicate 3 is an element of the set A, and $2 \notin A$ to indicate that 2 is not an element of the set A.

B. The set A may be described by giving a rule for membership:

$A = \{\text{the set of all odd counting numbers less than } 15\}$ ◀

When *set builder notation* is used, each symbol has a specific translation.

$A = \{\quad\quad x \quad\quad | \quad\quad \text{rule for membership} \quad\quad\}$

Is read as A is the set of all elements x such that x satisfies the rule for membership

Set A is a *subset* of a set B if all the elements in A are also in B. We write $A \subseteq B$. The set A is a *proper subset* of set B if A is a subset of B and B contains at least one member not in A. We write $A \subset B$.

Example 2

Let $A = \{1, 3, 5, 7, 9, 11, 13\}$
$B = \{3, 7, 11\}$
$C = \{1, 2, 3, 4\}$

Then:

A. $B \subset A$ B is a proper subset of A.
B. $C \nsubseteq A$ C is *not* contained in A.
C. $\phi \subseteq A$ The empty set is a subset of every set.
D. $C \subseteq C$ Every set is a subset of itself.
E. $C \not\subset C$ C is not a proper subset of itself. ◀

(1.2) The set of *counting* or *natural numbers* is defined by

$$N = \{1, 2, 3, 4, 5, \ldots\}$$

The *whole numbers* are defined by

$$W = \{0, 1, 2, 3, 4, 5, \ldots\}$$

W is *closed* with respect to addition and multiplication since the sum or product of any two whole numbers is also a whole number.
The *integers* are defined by

$$I = \{\ldots, -3, -2, -1, 0, 1, 2, 3, \ldots\}$$

I is closed with respect to addition, subtraction, and multiplication.
The *rational numbers* are defined by

$$Q = \left\{\frac{a}{b} \,\middle|\, a, b \in I; b \neq 0\right\}$$

Q is closed with respect to all four arithmetic operations (division by zero excluded). Observe that:

$$N \subset W \subset I \subset Q$$

The set of *irrational numbers* is defined by

$$I_r = \{x \mid x \notin Q\}$$

An irrational number is any number which does *not* belong to Q.
The *real numbers* are defined by

$$R = \{x \mid x \in Q \text{ or } x \in I_r\}$$

We can write: $N \subset W \subset I \subset Q \subset R$ and
$I_r \subset R$

1.9 Review and Chapter Test

Example 3

Let $A = \{-3, -\frac{1}{2}, -1.1, 0, \frac{4}{3}, \sqrt{2}, 1.25, \sqrt[3]{5}\}$. List all the elements from A that are:

A. Natural numbers   None

B. Whole numbers   0

C. Integers $-3, 0$

D. Rational numbers $-3, -\frac{1}{2}, -1.1, 0, \frac{4}{3}, 1.25$

E. Irrational numbers $\sqrt{2}, \sqrt[3]{5}$

F. Real numbers   All elements of A

Real numbers can be represented graphically on the *number line*.

Example 4

Draw a number line and graph the following numbers:

$-3, -\frac{1}{2}, -1.1, 0, \frac{4}{3}, \sqrt{2}$ (approx. 1.41), 1.25, $\sqrt[3]{5}$ (approx. 1.71)

(1.3) The real numbers form an *ordered set* of numbers. Each real number is represented by its own point on the number line and relates to all other real numbers in a unique way. Given any two real numbers, then one of the numbers is larger than the other or they are equal. If $a < b$ (a is less than b), then a is to the left of b on the number line. Similarly, if $a > b$ (a is greater than b), then a is to the right of b.

Example 5

Consider the number line

A. Let A, B, C, D, E represent numbers whose graphs are as shown. Insert the symbol ($<$, $>$, or $=$) that makes the following statements true.

	Statement		Relationship	
1.	A	D	$A < D$	A is left of D
2.	E	C	$E > C$	E is right of C
3.	B	-1	$B = -1$	Same point

B. The *directed distance from B to D* is 7 because we move 7 units to the right to get from B to D.

C. The *directed distance from D to B* is −7 because we move 7 units to the left (negative direction) to get from D to B. ◀

(1.4) The *negative of a number* is its mirror image with respect to the origin on the number line.

Example 6

Consider the number line

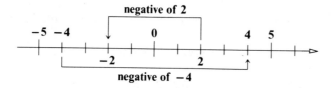

The negative of 2 is −2. The negative of −4 is 4. The negative of 0 is 0. In general, the negative of the number, x, is represented by $-x$.

The *absolute value* of a number is the distance between the graph of the number and the origin on the number line. Symbolically:

$$|n| = \begin{cases} n \text{ if } n \text{ is positive or zero} \\ -n \text{ if } n \text{ is negative} \end{cases}$$

◀

The absolute value of a number is never negative.

Example 7

Find the absolute value of each of the following numbers:

Number	Absolute value
A. $+8$	$|+8| = |8| = 8$
B. -9.3	$|-9.3| = 9.3$
C. 0	$|0| = 0$
D. $\dfrac{7}{8}$	$\left|\dfrac{7}{8}\right| = \dfrac{7}{8}$
E. x	$|x|$ Cannot be simplified further since the sign of x is unknown.

◀

For a, b not equal to zero, the *reciprocal* of the number $\frac{a}{b}$ is given by $\frac{b}{a}$. The number $0 = \frac{0}{1}$ does not have a reciprocal since $\frac{1}{0}$ is undefined.

1.9 Review and Chapter Test

Example 8

Find the reciprocal of each of the following numbers where possible:

Number	Reciprocal
A. 3	$\frac{1}{3}$
B. -8	$-\frac{1}{8}$
C. $\frac{2}{5}$	$\frac{5}{2}$
D. $-\frac{3}{4}$	$-\frac{4}{3}$
E. 0	undefined
F. $\frac{1}{7}$	7
G. $-\frac{1}{3}$	-3

◀

(1.5) The following are several basic properties of the real numbers: Let $a, b, c \in R$. Then,

1. $(a + b) \in R$ and $a \cdot b \in R$
   R is *closed* with respect to addition and multiplication.

2. $a + b = b + a$
   $a \cdot b = b \cdot a$
   Multiplication and addition are *commutative* in R.

3. $a + b + c = (a + b) + c = a + (b + c)$
   $a \cdot b \cdot c = a \cdot (b \cdot c) = (a \cdot b) \cdot c$
   Multiplication and addition are *associative* in R.

4. $a + 0 = 0 + a = a$
   0 is the *additive identity*.

5. $a \cdot 1 = 1 \cdot a = a$
   1 is the *multiplicative identity*.

6. $a + (-a) = (-a) + a = 0$
   $-a$ is the *additive inverse* of a.
   Every real number has an additive inverse, which is also called the negative of that number.

7. $a \cdot \frac{1}{a} = \frac{1}{a} \cdot a = 1, a \neq 0$
   $\frac{1}{a}$ is the *multiplicative inverse* of a.

Every non-zero real number has a multiplicative inverse, which is also called the reciprocal of the number.

8. $a(b + c) = ab + ac$
   $(b + c)a = ba + ca$
   Multiplication is *distributive* over addition.

9. $0 \cdot a = a \cdot 0 = 0$
   Multiplication by zero.

Order of Operations Agreement

When evaluating an algebraic expression, first perform all multiplications and divisions working from left to right in the order in which these operations appear. Second, perform all additions and subtractions, also working from left to right. Note the order in which these operations are performed may be altered with the use of "grouping symbols."

Example 9

Evaluate each of the following expressions:

A. $4 \cdot 3 + 2 \cdot 5$

$\quad 4 \cdot 3 + 2 \cdot 5 = 12 + 10 \quad$ Multiply first left to right

$\qquad\qquad\qquad = 22 \qquad\qquad$ Then add

B. $2 \cdot 4 + 3 \cdot 7 - 12$

$\quad 2 \cdot 4 + 3 \cdot 7 - 12 = 8 + 21 - 12 \quad$ Multiply first left to right

$\qquad\qquad\qquad\qquad = 29 - 12 \qquad$ Add next

$\qquad\qquad\qquad\qquad = 17 \qquad\qquad$ Subtract ◀

We use grouping symbols to indicate when an arithmetic expression is to be evaluated in an order different from that of the order of operations agreement. Three commonly used grouping symbols are *parentheses*, indicated by ( ), *brackets*, indicated by [ ], and *braces*, indicated by { }. The fraction bar is also considered a symbol of grouping. Expressions are evaluated from the innermost grouping symbols to the outermost. Traditionally parentheses are used as the innermost grouping symbol, brackets next, and braces the outermost. Any operations remaining to be performed after the grouping symbols have been removed are performed according to the order of operations agreement.

Example 10

Evaluate: $3 + 4[8 - (7 - 5)]$

$\quad 3 + 4[8 - (7 - 5)] = 3 + 4[8 - 2] \qquad$ Parentheses first

$\qquad\qquad\qquad\qquad = 3 + 4[6] \qquad\quad$ Brackets second

$\qquad\qquad\qquad\qquad = 3 + 24 \qquad\quad\;$ Multiply

$\qquad\qquad\qquad\qquad = 27 \qquad\qquad\;\,$ Add ◀

1.9 Review and Chapter Test

(1.6) Rules for Addition of Signed Numbers

To add two or more numbers having the same sign, add their absolute values and attach their common sign to the sum.

To add two numbers having opposite signs, subtract their absolute values, smaller from larger, and attach the sign of the number with the greater absolute value.

Example 11

Find each of the following sums:

A. $(-3) + (-5) + (-4)$

$(-3) + (-5) + (-4)$
$= -(|-3| + |-5| + |-4|)$ Add the absolute values and attach the common sign
$= -(3 + 5 + 4)$
$= -(12) = -12$

or more informally,

$(-3) + (-5) + (-4)$
$= -(3 + 5 + 4)$ Add absolute values
$= -12$ Attach common sign (negative)

B. $(-7) + 3$

$(-7) + 3 = (-7) + (+3)$ Numbers have opposite signs
$= -(|-7| - |+3|)$ Subtract smaller absolute value from larger
   Attach "$-$" since $|-7|$ is larger than $|3|$
$= -(7 - 3)$
$= -(4) = -4$

or more informally,

$(-7) + 3 = -(7 - 3)$ Subtract absolute values,
$= -(4)$ Attach sign of larger signed number (-7)
$= -4$

◀

Rule for Subtraction of Numbers

For any real numbers x and y.

$x - y = x + (-y)$

That is, to subtract a number, add its opposite.

Example 12

Find each of the following differences:

A. $-6 - 3$

$$-6 - 3 = -6 + (-3) \quad \text{Change to addition}$$
$$= -(6 + 3) \quad \text{Addition of like signs}$$
$$= -9$$

B. $-2 - (-5)$

$$-2 - (-5) = -2 + (+5) \quad \text{Change to addition}$$
$$= +(5 - 2) \quad \text{Addition of opposite signs}$$
$$= +(3)$$
$$= 3 \quad \blacktriangleleft$$

Example 13

Evaluate: $-5 - (-3) + (-2)$

$$-5 - (-3) + (-2) = -5 + (+3) + (-2) \quad \text{Change the subtraction to addition}$$
$$= -5 + 3 + (-2)$$
$$= -2 + (-2) \quad \text{Add, left to right}$$
$$= -4 \quad \blacktriangleleft$$

(1.7) Rules for multiplication of signed numbers

1. The product of two numbers having the same sign is positive.
2. The product of two numbers having opposite signs is negative.
3. The product of any number with zero is zero.
4. The product of any number with 1 is that number.
5. Forming the product of any number with -1 is equivalent to reversing the sign of that number.

Example 14

Evaluate each of the following products:

A. $4 \cdot 7$

$$4 \cdot 7 = 28 \quad \text{Like signs}$$

B. $(-3) \cdot (5)$

$$(-3) \cdot (5) = -15 \quad \text{Opposite signs}$$

C. $(-4) \cdot (-8)$

$$(-4)(-8) = 32 \quad \text{Like signs}$$

D. $(-3) \cdot (0)$

$$(-3) \cdot (0) = 0 \quad \text{Multiplication by 0}$$

1.9 Review and Chapter Test

E. $(1) \cdot (-8)$

$(1) \cdot (-8) = -8$ Multiplication by 1

F. $(-1) \cdot (6)$

$(-1) \cdot (6) = -6$ Multiplication by -1 ◀

Multiplication is distributive over both addition and subtraction.

Example 15

Evaluate: $4(5 - 3 + 2 + 1 - 8)$

$$4(5 - 3 + 2 + 1 - 8) = 4 \cdot 5 - 4 \cdot 3 + 4 \cdot 2 + 4 \cdot 1 - 4 \cdot 8$$
$$= 20 - 12 + 8 + 4 - 32$$
$$= 32 - 44$$

    ↗   ↑— Sum of all negatives

Sum of all positives

$$= -12$$ ◀

Rules for division of signed numbers

1. The quotient of two numbers having the same sign is positive.
2. The quotient of two numbers having opposite signs is negative.
3. Zero divided by any non-zero number is zero.
※ 4. Division by zero is undefined.

Example 16

Evaluate the following quotients when possible:

A. $\dfrac{24}{-4}$

$\dfrac{24}{-4} = -6$ Opposite signs

B. $\dfrac{-18}{-6}$

$\dfrac{-18}{-6} = 3$ Like signs

C. $\dfrac{-3 + (3)}{8}$

$\dfrac{-3 + (3)}{8} = \dfrac{0}{8}$

$= 0$ Zero divided by a non-zero number

D. $\dfrac{-7+(-3)}{2+(-2)}$

$\dfrac{-7+(-3)}{2+(-2)} = \dfrac{-10}{0}$ Undefined

E. $\dfrac{0}{-6+6}$

$\dfrac{0}{-6+6} = \dfrac{0}{0}$ Undefined ◀

(1.8) A *constant* is a symbol (or a group of symbols) that names a single specific number. A *variable* is a symbol (or group of symbols) that represents various constants.

Example 17

In the formula for the area of a triangle, "$\tfrac{1}{2}$" is a constant while A, b, and h are variables representing the area, length of base, and height of the triangle, respectively.

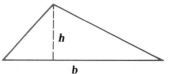

$A = \dfrac{1}{2}bh$ ◀

The expression b^p is a symbolic way of expressing the product formed by using the number b as a factor p number of times. b^p is called *exponential form*, and $b \cdot b \cdots b$ is called *expanded form*. The number b is called the *base*, p is called the *exponent*, and b^p is read "base b raised to the power p," or more simply, "b to the p."

Example 18

Write each expression in expanded form and evaluate if possible:

A. $4^3 = 4 \cdot 4 \cdot 4 = 64$

B. $3^4 = 3 \cdot 3 \cdot 3 \cdot 3 = 81$

C. $3^1 = 3$

D. $(x + y)^2 = (x + y)(x + y)$ ◀

Example 19

A. Write $3x^2y^3c$ in expanded form

$3x^2y^3c = 3 \cdot x \cdot x \cdot y \cdot y \cdot y \cdot c$

$= 3xxyyyc$

B. Write $3aabcccdd$ in exponential form

$3aabcccdd = 3 \cdot a \cdot a \cdot b \cdot c \cdot c \cdot c \cdot d \cdot d$

$= 3a^2bc^3d^2$ ◀

1.9 Review and Chapter Test

An algebraic expression is a symbolic form that includes any combination of constants, variables, operations, and/or symbols of grouping.
To evaluate an algebraic expression:

1. Replace each variable in the expression with its given value.
2. Evaluate each expression within symbols of grouping, proceeding from the innermost to the outermost.

Note: It is sometimes necessary to use Steps 3, 4, and 5 in evaluating expressions within symbols of grouping (Step 2).

3. Evaluate all powers and roots as they are encountered from left to right.
4. Perform all multiplications and divisions as they are encountered from left to right.
5. Perform all additions and subtractions as they are encountered from left to right.

Example 20

Evaluate each algebraic expression for the given values of the variables:

A. $2x - 3y$ for $x = 3$ and $y = -3$

$2x - 3y = 2(3) - 3(-3)$ Substitute
$\quad\quad\quad = 6 + 9$ Multiply first
$\quad\quad\quad = 15$ Add

B. $P = 2L + 2W$ Find P if $L = 8$ and $W = 5$.

$P = 2L + 2W$
$\quad = 2(8) + 2(5)$ Substitute
$\quad = 16 + 10$ Multiply first
$\quad = 26$ Add

C. $2(x - 2y) + 3(2x + y)$ for $x = -3$ and $y = 2$

$2(x - 2y) + 3(2x + y)$
$\quad = 2[(-3) - 2(2)] + 3[2(-3) + 2]$ Substitute
$\quad = 2(-3 - 4) + 3(-6 + 2)$ Evaluate
$\quad = 2(-7) + 3(-4)$
$\quad = -14 - 12$ Multiply
$\quad = -26$ Subtract

Exercise Set 1.9 Review

Let $A = \{1, 2, 3, 4, 5, 6, 7, 8\}$
$B = \{1, 3, 4, 8\}$
$C = \{2, 3, 5, 8\}$
$D = \{2, 7\}$

Use the listing method to specify each of the following sets:

1. The set of numbers belonging to both B and C.
2. The set of numbers belonging to B or C.
3. The set of numbers belonging to both B and D.

In Problems 4–9, determine whether the statement is true or false. Refer to the sets listed for Problems 1–3:

4. $B \subset A$
5. $D \in B$
6. $D \subset B$
7. $\phi \subset D$
8. $C \subset A$
9. $B \subset B$

Graph the following sets of real numbers by drawing a number line and plotting points:

10. $A = \{x \mid x \in N \text{ and } x \text{ is between } -4 \text{ and } 5\}$
11. $B = \{x \mid x \in W \text{ and } x \text{ is between } -4 \text{ and } 5\}$
12. $C = \{x \mid x \in I \text{ and } x \text{ is between } -4 \text{ and } 5\}$
13. $D = \{x \mid x \in N \text{ and } x \text{ is no more than } 5\}$
14. $E = \{x \mid x \in W \text{ and } x \text{ is no more than } 5\}$
15. Let $M = \{-\pi, 2, 0, \frac{1}{2}, 2.3, \sqrt{2}\}$. List all the elements from M that are:

    a. natural numbers
    b. whole numbers
    c. integers
    d. rational numbers
    e. irrational numbers

Consider the number line

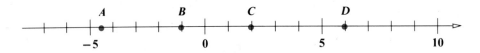

Determine the relationship between the numbers by inserting the symbol ($<$, $>$, or $=$) that makes the following statements true:

16. $A \quad C$
17. $B \quad A$
18. $0 \quad B$
19. $0 \quad D$
20. $A \quad 0$
21. $C \quad B$

1.9 Review and Chapter Test

Consider the number line

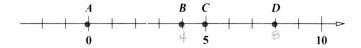

Find the directed distance:

22. from B to D D−B
23. from D to B B−D
24. from B to A
25. from C to B
26. from B to B
27. from C to 2
28. from 5 to C

29. Complete the following chart for each number.

Number (x)	Negative (−x)	Absolute Value ($|x|$)	Reciprocal $\frac{1}{x}$
+7			
2			
0			
−4			
$\frac{3}{8}$			
$-\frac{5}{6}$			
$\frac{1}{7}$			
$\frac{1}{-9}$			

Evaluate each of the following expressions using the order of operations agreement and rules for grouping symbols:

30. $14 - 3 \cdot 4$
31. $7 \cdot 3 - 2$
32. $28 - 8 \div 2 + 4$
33. $6 + 2(5 - 3)$
34. $3[4 + 3(5 - 1)]$
35. $(24 - 3) \div (2 + 5)$

State the property of real numbers illustrated by each of the following statements:

36. $5 \cdot 6 = 6 \cdot 5$
37. $(4 + 3) + 8 = 7 + 8$
38. $3(a + b) = 3a + 3b$
39. If $3 + m = 0$, then $m = -3$.

In Problems 40–45, add or subtract as indicated:

40. $7 + (-11)$
41. $7 - (-11)$
42. $-(1) - 5$
43. $[4 - (-3)] - [-2 + 5]$
44. $13 - (-5) + (-3) - (+3)$
45. $8 - [5 - 2]$

Consider the number line

Find the directed distance:

46. from B to A
47. from B to C
48. from A to C
49. from B to -6

Evaluate each of the expressions in Problems 50–57:

50. $(-12) \div \left(\dfrac{-8}{2}\right)$
51. $\left(\dfrac{-18}{-3}\right) \cdot \left(\dfrac{12}{-6}\right)$
52. $\dfrac{7}{0}$
53. $\dfrac{(-6)(2) - (-3)(-4)}{-3 - (-6)}$
54. $\dfrac{(12)(-2) + (-4)(-6)}{8 - 3}$
55. $-3 - \{-2 - [-3 - (8 + 2)]\}$
56. $-[-(-3)]$
57. $-|-3|$

Evaluate each of the following expressions for $x = 8$, $y = -4$, $z = -6$:

58. $-(-z)$
59. $x + z$
60. $-3 + \dfrac{x}{y}$
61. $3(z - y)$
62. $|x - z|$
63. $xy - z$
64. $2\{x - [3 + (z + x)]\}$
65. $-(x - y)$
66. $-|y|$
67. $-|-y|$
68. $x^2 + 2y^2$
69. $\dfrac{x^2}{y - z}$

Chapter 1 Review Test

1. Let $A = \{1, 2, 3, 4, 5, 6, 7\}$, $B = \{2, 4, 6\}$. Indicate each statement as true or false:

   a. $5 \in A$ T
   b. $5 \notin B$ T
   c. $A \subseteq B$ F
   d. $B \subseteq A$ T
   e. $\phi \subseteq B$ T
   f. $B \subseteq \phi$ F

2. From the set $C = \{-6, -\frac{3}{2}, 0, 2, \frac{9}{2}, \sqrt{6}\}$ list the elements that are:

   a. natural numbers 2
   b. whole numbers 2
   c. integers $-6, 0, 2$
   d. rational numbers $-\frac{3}{2}, \frac{9}{2}, -6, 0, 2$
   e. irrational numbers $\sqrt{6}$
   f. real numbers all

3. Draw a number line and graph the elements of set C in Problem 2. Use $\sqrt{6} = 2.45$.

4. Find the directed distance:

   a. from -3 to 6 $6 - (-3) = 9$
   b. from 0 to -8 $-8 - 0 = -8$

5. For each number given, show its negative, its absolute value, and its reciprocal.

   a. $\dfrac{-3}{4}$
   b. $\dfrac{5}{2}$
   c. 0
   d. $\sqrt{5}$
   e. -7

6. Insert the correct symbol $<$ or $>$ between each pair of numbers.

   a. $-\dfrac{5}{8}$ 2
   b. $\dfrac{5}{8}$ -2
   c. $-\dfrac{5}{8}$ 0
   d. $\dfrac{5}{8}$ 0

7. Assume $x, y \in R$
   Match the statement in the first column with the most appropriate property named in the second column.

   a. $3 + (4 + x) = (3 + 4) + x$ 1. closure
   b. $x \cdot 6 = 6x$ 2. commutative property
   c. $x\left(\dfrac{1}{x}\right) = 1$ 3. associative property
   d. $x + 2y \in R$ 4. identity
   e. $y + 0 = 0 + y$ 5. inverse
   f. $5(x + 2) = 5x + 10$ 6. distributive

g. $1 \cdot (x + 3) = x + 3$
h. $6(3y) = (6 \cdot 3)y$
i. $y + (-y) = 0$
j. $2 + (8 + y) = 2 + (y + 8)$

8. Evaluate each expression:
   a. $4 + 2 \cdot 5$
   b. $9 + 2[5 - (4 - 1)]$

9. Evaluate each of the following expressions for $k = 4, m = -8$:
   a. $k + m$
   b. $k - m$
   c. km
   d. $\dfrac{m}{k}$
   e. $3k - 2m$
   f. $\dfrac{2k}{m + 8}$
   g. $\dfrac{k - 4}{3m}$
   h. $\dfrac{3m}{k - 2}$
   i. $m^2 - 3k$
   j. $(m + 3)^2 - (k + 2)^2$

10. a. Write $3 \cdot 3 \cdot x \cdot y \cdot y \cdot y \cdot z \cdot z$ in exponential form.
    b. Write $2^2 \cdot 5m^4n^2$ in expanded form.

CHAPTER TWO

LINEAR EQUATIONS AND INEQUALITIES IN ONE VARIABLE

One of the important goals in a study of algebra is learning to solve various types of problems. To do this, we must develop the skills needed to analyze a problem, express it as an equation, and then solve the equation.

In this chapter, after a few preliminaries, we discuss solving linear equations in one variable. Later, this skill is extended as we discuss solving linear inequalities in one variable, a process similar to that for solving equations. In Section 2.7 we use these skills to solve application problems. Other types of equations, inequalities, and related application problems are solved in later chapters.

2.1 Applications of the Distributive Law

In this section, we introduce some uses of the distributive law and some new terminology that will be used later in this chapter in the important process of solving equations and inequalities.

Removing Parentheses

One of the ways of writing the distributive law is:

$a(b + c) = ab + ac$

This can be interpreted to mean that if an expression in parentheses (or other symbols of grouping) is preceded by a factor, then we remove the parentheses by multiplying each term inside by this factor.

Example 1

Remove parentheses:

A. $4(x + 5) = 4x + 20$

B. $5(x - 2) = 5x - 10$

C. $4 + 3(x - y) = 4 + 3x - 3y$

D. $8x - 2(y - 4) = 8x - 2y + 8$ ◀

To have parentheses immediately preceded by a "+" is equivalent to their being preceded by +1. Such parentheses are removed by multiplying each term inside by +1. However, to multiply a quantity by +1 is to leave it unchanged. Recall $1 \cdot a = a$. Mechanically, parentheses preceded by "+" are removed by omitting parentheses and writing each term inside with its sign unchanged.

Example 2

Remove parentheses:

A. $5x + (3y - 6) = 5x + 3y - 6$

B. $3a + (x^2 - 2x + 3) = 3a + x^2 - 2x + 3$ ◀

Recall $-1 \cdot a = -a$. Parentheses that are immediately preceded by a "−" are removed by multiplying each term inside by −1. However, when you multiply a quantity by −1, it reverses the sign. Mechanically, parentheses preceded by a "−" are removed by omitting the parentheses and writing each term inside with its sign reversed.

Example 3

Remove parentheses:

A. $5x - (3y - 4) = 5x - 3y + 4$

B. $3a - (x^2 - 2x + 3) = 3a - x^2 + 2x - 3$ ◀

In summary:

TO REMOVE PARENTHESES

A. Preceded by a factor, multiply each term inside by the factor.

B. Preceded by +, write each term inside with its sign unchanged.

C. Preceded by −, write each term inside with its sign reversed.

Factoring

Recall that when numbers are multiplied, the result is called the *product*, and the numbers that were multiplied together are the *factors*.

2.1 Applications of the Distributive Law

$$3 \cdot 7 = 21$$

21 is the *product* of 3 and 7
3 and 7 are *factors* of 21

When an algebraic expression is written as a product of two or more expressions, the procedure is known as *factoring* the expression. One form of factoring is suggested by using the distributive law in reverse:

$$ba + ca = (b + c)a$$

4·2 + 3·2 = (4+3)2

Here we say the common factor a has been removed or has been factored out.

Example 4 Remove the indicated factor from each expression:

A. $4x^2 + 2y; 2$

$$4x^2 + 2y = 2 \cdot (2x^2) + 2(y)$$
$$= 2 \cdot (2x^2 + y)$$

B. $5x^2 + 3x; x$

$$5x^2 + 3x = (5x) \cdot x + (3) \cdot x$$
$$= (5x + 3)x$$

8x² ?

C. $P + Prt; P$

Here we emphasize that P is a *factor* of each term by writing the first term as $P \cdot 1$

$$P + Prt = P \cdot 1 + P(rt)$$
$$= P(1 + rt)$$

2Prt ? ◀

Terms/Like Terms

The parts of an expression that are added together are called the *terms* of that expression. From our definition of subtraction $a - b = a + (-b)$, we can write the expression

$$3x + 5y - 6z - 8$$

in the form:

$$3x + 5y + (-6z) + (-8)$$

This expression has four terms:

$$3x, 5y, -6z, \text{ and } -8$$

Each term of an expression generally is the product of a constant, called the *numerical coefficient* (or simply the coefficient) and a letter or product of letters called the *literal part*. A term that has no literal part is called a *constant term*. The terms along with the numerical coefficient and literal part of each term of the expression $3x + 5y - 6z - 8$ are shown in Figure 2.1-1.

FIGURE 2.1-1

	$3x + 5y - 6z - 8$		
Term Numbers	Term	Numerical Coefficient	Literal Part
1st	$3x$	3	x
2nd	$5y$	5	y
3rd	$-6z$	-6	z
4th	-8	-8	none

If a term is written without a numerical coefficient, the numerical coefficient is understood to be 1. Recall $a = 1 \cdot a$ and $-a = (-1) \cdot a$. In the expression $5x^4 + x^3 - 3x^2 - x + 8$, the second term has a numerical coefficient of 1, and the fourth term has a numerical coefficient of -1.

Like terms (also called *similar terms*) of an expression are terms that have *identical literal parts*. Constant terms are considered as like terms.

In the expression:

$2x + y - 6x + 7 + 5x - 4y + x - 3$

$2x, -6x, 5x,$ and x are like terms,

y and $-4y$ are like terms.

7 and -3 are like terms.

In the expression:

$6x^2 + 2x - 5 + 3x + 14$

$2x$ and $3x$ are like terms.

-5 and 14 are like terms.

We simplify expressions by grouping like terms together and removing the literal part as a common factor (distributive property in reverse). We call the process *combining like terms*.

Example 5

Simplify the following expression by combining like terms:

$2x + 5y + 4x - 3y$

$= (2x + 4x) + (5y - 3y)$ Commutative and associative properties

$= (2 + 4)x + (5 - 3)y$ Factor

$= 6x + 2y$ Simplify

◀

In practice, the middle steps of the preceding example are not shown. To simplify expressions, combine like terms by adding their numerical coefficients, leaving the literal part unchanged.

Example 6

Each of the following expressions has been simplified where possible by combining like terms:

A. $3x + 5y - 6z - 8 = 3x + 5y - 6z - 8$
   This expression has no like terms, so no further simplification is possible.

B. $2x + y - 6x + 7y = -4x + 8y$

C. $2x^2 + 3x + 5 - x^2 - 7x - 5 = x^2 - 4x$
   The sum of the constants $+5 + (-5) = 0$, which is not written in the simplified expression. Note x^2 and x are not identical so terms such as $2x^2$ and $3x$ are not like terms.

D. $6y^3 + 5y^2 - 2y + 4y^2 - 8y + 11 = 6y^3 + 9y^2 - 10y + 11$

E. $5a + 4b - 5a + 3b = 7b$ ◀

Removing Parentheses and Combining Like Terms

In the preceding paragraphs of this section, we have discussed the two processes of removing parentheses and combining like terms. In many applications, we commonly do both, first removing parentheses, and then combining like terms in the resulting expression, as illustrated in the following example.

Example 7

In each of the following examples, remove parentheses (or other symbols of grouping), and then combine like terms. Note that when symbols of grouping appear within other symbols of grouping, we remove the *innermost symbols first*, and then proceed to the outermost.

A. $2x + (x^2 - 3x + 2)$

$$2x + (x^2 - 3x + 2)$$
$$= 2x + x^2 - 3x + 2 \qquad \text{Remove parentheses}$$
$$= x^2 - x + 2 \qquad \text{Combine like terms}$$

B. $(x^2 + 5x - 8) - (3x^2 - 4x + 1)$

$$(x^2 + 5x - 8) - (3x^2 - 4x + 1)$$
$$= x^2 + 5x - 8 - 3x^2 + 4x - 1$$
$$= -2x^2 + 9x - 9$$

C. $x^2 - [2x - (3x - 4)]$

$x^2 - [2x - (3x - 4)]$
$= x^2 - [2x - 3x + 4]$ Remove innermost first
$= x^2 - 2x + 3x - 4$ Remove outermost
$= x^2 + x - 4$

D. $x^2 - 3(x^2 + 2x - 5) - (2x - x^2)$

$x^2 - 3(x^2 + 2x - 5) - (2x - x^2)$
$= x^2 - 3x^2 - 6x + 15 - 2x + x^2$
$= -x^2 - 8x + 15$

E. $5p - 2[p - 4(2p + 1)]$

$5p - 2[p - 4(2p + 1)]$
$= 5p - 2[p - 8p - 4]$
$= 5p - 2p + 16p + 8$
$= 19p + 8$

F. $a^2 + 3a - 2[4 - a(a + 3)]$

$a^2 + 3a - 2[4 - a(a + 3)]$
$= a^2 + 3a - 2[4 - a^2 - 3a]$
$= a^2 + 3a - 8 + 2a^2 + 6a$
$= 3a^2 + 9a - 8$

◀

Exercise Set 2.1

Remove parentheses in each expression:

1. $3(x + 4)$
2. $2(x + 5)$
3. $5(x - 3)$
4. $3(x - 7)$
5. $x(2x + 6)$
6. $a(3a + 4)$
7. $3y(2y + 6)$
8. $2y(2y + 9)$
9. $-6(3m - 1)$
10. $-4(2u - 3)$
11. $5a + (3m - 1)$
12. $6x + (2y - 3)$
13. $3x + 4 - (2a - b)$
14. $y - (2x - 6)$
15. $(2x + 3) - (5y + 4z)$
16. $(3a - 5b) - (4x - 3y)$

2.1 Applications of the Distributive Law

For each of the following expressions, factor out the indicated quantity:

17. $5x^2 + 15x - 10$; 5
18. $4y^2 + 12y - 16$; 4
19. $ax + bx$; x
20. $6u + 8$; 2
21. $3x^2 + x$; x
22. $4y + 4z$; 4
23. $12x^2 + 8x + 4$; 4
24. $2m^3 + 6m^2 - 2m$; $2m$

Simplify each expression by combining like terms, if possible:

25. $2x + y + 3x$
26. $a + 3b + 4b$
27. $3a + b + 2a + 6b$
28. $5m + 2p + 3m + p$
29. $3m - 4p + m + p$
30. $2x - 4y - 2x - y$
31. $4u + 3 - u - 3$
32. $4x + y - x + 2y - 3x$
33. $x^2 - 3x + 4 + x^2 + 12x - 8$
34. $2y^2 + 6y - 11 - y^2 - 2y + 9$
35. $3m^2 - 3m + 9 + 6m - 4$
36. $3x + 4xy - y + x + xy$

Simplify each of following expressions by removing parentheses and combining like terms:

37. $(3a + 5) + (a - 4)$
38. $(4m + 2) + (3m - 5)$
39. $(x^2 - 2x) + (3x - 8)$
40. $(a^2 + 3a) + (a + 6)$
41. $(4p^2 + 3p) - (2p - 3)$
42. $(6x^2 - 4x) - (4x + 3)$
43. $(x^2 + 3x - 5) - (2x^2 - 6x - 11)$
44. $(4y^2 - 6y + 1) - (y^2 - 11y + 3)$
45. $(2m^2 - 3m - 8) - (m^2 - 3m + 13)$
46. $(4y^2 + y - 9) - (4y^2 - 3y - 9)$
47. $2(3p - 5) - 2(p + 4)$
48. $3(p + 6) - 3(2p + 1)$
49. $x(x + 3) - x(x - 6)$
50. $a(a - 4) - 2(a + 3)$
51. $x^2 - 3(x + 9) + 2(x - 3)$
52. $2a + 3(a - 5) - 2(2a + 4)$
53. $4(x + 2) - x(x + 3)$
54. $6(y + 1) - y(y + 5)$

55. The formula for the surface area (S) of a square-topped box (Figure 2.1-2) is given by

$$S = 2x^2 + 4xy$$

Simplify this formula by writing the right side in factored form and removing the factor $2x$.

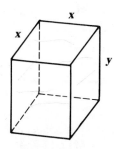

FIGURE 2.1-2

56. The formula for the surface area (S) of a cylinder (Figure 2.1-3) is given by

$$S = 2\pi r^2 + 2\pi rh$$

Simplify this formula by writing the right side in factored form and removing the factor $2\pi r$.

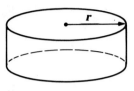

FIGURE 2.1-3

57. Translate each of the following phrases into an algebraic expression.

   a. Three times x plus two times y.

   b. Three times the sum of x and y.

58. Translate each of the following phrases into an algebraic expression.

   a. Four times m minus three times p.

   b. Four times the difference of m and p.

Review. Beginning with this section, each exercise set will include a few review problems. The problems will be either of a random nature designed to refresh your memory regarding topics not used for several sections, or to review procedures that will be used in subsequent sections.

59. Name the property of real numbers illustrated by each of the following statements. Assume that all variables represent real numbers.

   a. $B + 2A$ is a real number

   b. $2(A + 3) = 2 \cdot A + 2 \cdot 3 = 2A + 6$

   c. $x + 5 = 5 + x$

   d. $1(x - 2) = x - 2$

60. Evaluate each quotient (written in fractional form):

   a. $\dfrac{12}{3}$ b. $\dfrac{-18}{2}$ c. $\dfrac{25}{5}$ d. $\dfrac{-120}{6}$

61. Evaluate each expression for $x = 3$ and $y = -5$:

   a. $x(y + 3)$ b. $x^2 - 2y$ c. $\dfrac{y + 2}{x - 3}$ d. $\dfrac{2y + 10}{5x}$

2.2 Preliminaries

In further preparation for the task of solving equations, we review some concepts from arithmetic that will be needed: least common multiple (LCM) and least common denominator (LCD).

2.2 Preliminaries

Multiples of a natural number are the product of that number and the natural numbers 1, 2, 3, ... For example, some multiples of 2 are $2 \cdot 1 = 2$, $2 \cdot 2 = 4$, $2 \cdot 3 = 6$, $2 \cdot 4 = 8$, and so on. Some multiples of 3 are 3, 6, 9, 12, 15, ... By definition, the multiples of a number are exactly divisible by that number. The following are several multiples of 2, 3, and 4.

2: 2, 4, 6, 8, 10, 12, 14, 16, 18, 20, 22, 24, 26, ...

3: 3, 6, 9, 12, 15, 18, 21, 24, 27, 30, ...

4: 4, 8, 12, 16, 20, 24, 28, ...

A number appearing in all of the listings is called a **common multiple**. The common multiples of 2, 3, and 4 are 12, 24, 36, ... The smallest of these is called the **least common multiple** (**LCM**) of the numbers. In this example, the LCM of 2, 3, and 4 is 12, written LCM(2, 3, 4) = 12. That is, 12 is the smallest number exactly divisible by 2, 3, and 4. In general, we define:

> The *least common multiple* (**LCM**) of two or more numbers is the smallest number exactly divisible by all of the numbers.

For relatively simple numbers, we find the LCM by the following procedure:

1. Examine the largest of the numbers. If this number is exactly divisible by the others, it is the LCM.

2. If the largest of the numbers is not exactly divisible by all of the others, list multiples of the largest number until one is found that is exactly divisible by all others. This number is the LCM.

Example 1

Find the LCM of each set of numbers:

A. 2, 4, and 8
The largest of these is 8, and 8 is exactly divisible by 2 and 4. So 8 is the LCM, and we write:

LCM(2, 4, 8) = 8

B. 3, 6, and 4
The largest of these is 6, but 6 is not exactly divisible by 3 and 4. So we begin by listing multiples of 6 (often done mentally) until we find a number exactly divisible by 3 and 4:

Multiples of 6: 6, *12*, ...

12 is exactly divisible by 3 and 4, so it is the LCM.

LCM(3, 6, 4) = 12

C. 10, 4, and 5

The largest number, 10, is not exactly divisible by 4 and 5, so we list its multiples:

Multiples of 10: 10, *20*, . . .

The number 20 is exactly divisible by 4 and 5 so it is the LCM.

LCM(10, 4, 5) = 20

D. 12 and 9

The largest is 12, but it is not exactly divisible by 9, so we list its multiples:

Multiples of 12: 12, 24, *36*, . . .

LCM(12, 9) = 36 ◄

In Chapter 6 we will show another method for finding the LCM for numbers or expressions of any level of difficulty.

When working with fractions, we often need to find the least common multiple of all of the denominators. We call the least common multiple of the denominators the **least common denominator**, or LCD.

Example 2

For the given fractions, find the LCD:

A. $\dfrac{1}{2}, \dfrac{2}{3}, \dfrac{3}{4}$

The denominators are 2, 3, and 4.

LCD = LCM(2, 3, 4) = 12

B. $\dfrac{5}{8}, \dfrac{1}{6}, \dfrac{11}{12}$

The denominators are 8, 6, and 12. To find the LCM, list multiples of 12 until one is found that is exactly divisible by 8 and 6:

Multiples of 12: 12, *24*, 36, 48, . . .

LCD = LCM(8, 6, 12) = 24 ◄

To find the least common multiple of expressions containing variables, the variables must also be factors of the LCM.

Example 3

Find the LCM of each set of expressions:

A. 2, x, and $4x$

Find the LCM of the numerical coefficients

LCM(2, 1, 4) = 4

Two of the expressions have a factor of x, so the LCM must also have a factor of x.

LCM(2, x, $4x$) = $4x$

2.2 Preliminaries

B. $6x, 2y, 4$

Find the LCM of the numerical coefficients

$$\text{LCM}(6, 2, 4) = 12$$

One expression has a factor of x, one has a factor of y, so the LCM must also have x and y as factors.

$$\text{LCM}(6x, 2y, 4) = 12xy$$

C. $5, (x + y), 2$

$$\text{LCM}[5, (x + y), 2] = 10(x + y) \qquad \blacktriangleleft$$

Multiplying Fractions

To multiply a fraction by a whole number, multiply that number times the numerator. In symbols:

$$a \cdot \frac{b}{c} = \frac{a}{1} \cdot \frac{b}{c} = \frac{ab}{c}$$

If the number a is divisible by the denominator c, we divide. (Informally, we call this canceling.)

Example 4

Multiply:

A. $12 \cdot \dfrac{1}{3} = \dfrac{\overset{4}{\cancel{12}} \cdot 1}{\cancel{3}} = 4$

B. $16 \cdot \dfrac{3}{4} = \dfrac{\overset{4}{\cancel{16}} \cdot 3}{\cancel{4}} = 12$

C. $20 \cdot \dfrac{7}{4} = \dfrac{\overset{5}{\cancel{20}} \cdot 7}{\cancel{4}} = 35$

D. $8 \cdot \dfrac{(x + 2)}{4} = \dfrac{\overset{2}{\cancel{8}}(x + 2)}{\cancel{4}} = 2(x + 2) \qquad \blacktriangleleft$

To multiply a sum or difference of two or more fractions by an expression, we use the distributive law.

Example 5

Multiply:

A. $12\left(\dfrac{2}{3} + \dfrac{1}{4}\right)$

$$12\left(\dfrac{2}{3} + \dfrac{1}{4}\right) = \overset{4}{\cancel{12}} \cdot \dfrac{2}{\cancel{3}} + \overset{3}{\cancel{12}} \cdot \dfrac{1}{\cancel{4}}$$

$$= 4 \cdot 2 + 3 \cdot 1$$

$$= 8 + 3 = 11$$

B. $20\left(\dfrac{3}{4} + \dfrac{x}{5} - \dfrac{7}{10}\right)$

$$20\left(\dfrac{3}{4} + \dfrac{x}{5} - \dfrac{7}{10}\right) = \overset{5}{\cancel{20}} \cdot \dfrac{3}{\cancel{4}} + \overset{4}{\cancel{20}} \cdot \dfrac{x}{\cancel{5}} - \overset{2}{\cancel{20}} \dfrac{7}{\cancel{10}}$$

$$= 15 + 4x - 14$$

$$= 1 + 4x \qquad \blacktriangleleft$$

By definition, the LCD is the smallest number exactly divisible by all of the denominators of the fractions in an expression. So if an expression of one or more terms is multiplied by the LCD, all denominators will cancel, resulting in an expression free of fractions.

Example 6

For each expression, find the LCD and then multiply the expression by the LCD:

A. $\dfrac{3}{4} + \dfrac{1}{6}$

$\text{LCD} = \text{LCM}(4, 6) = 12$

Multiply:

$$\text{LCD} \cdot \left(\dfrac{3}{4} + \dfrac{1}{6}\right) = 12 \cdot \left(\dfrac{3}{4} + \dfrac{1}{6}\right) = \overset{3}{\cancel{12}} \cdot \dfrac{3}{\cancel{4}} + \overset{2}{\cancel{12}} \cdot \dfrac{1}{\cancel{6}}$$

$$= 9 + 2$$

$$= 11$$

B. $\dfrac{5}{8} - \dfrac{x}{6}$

$\text{LCD} = \text{LCM}(8, 6) = 24$

Multiply:

$$\text{LCD} \cdot \left(\dfrac{5}{8} - \dfrac{x}{6}\right) = 24 \cdot \left(\dfrac{5}{8} - \dfrac{x}{6}\right) = \overset{3}{\cancel{24}} \cdot \dfrac{5}{\cancel{8}} - \overset{4}{\cancel{24}} \cdot \dfrac{x}{\cancel{6}}$$

$$= 15 - 4x$$

C. $\dfrac{5}{4x} - \dfrac{1}{2y}$

$\text{LCD} = \text{LCM}(4x, 2y) = 4xy$

Multiply:

$$\text{LCD} \cdot \left(\dfrac{5}{4x} - \dfrac{1}{2y}\right) = 4xy \cdot \left(\dfrac{5}{4x} - \dfrac{1}{2y}\right) = \cancel{4xy} \cdot \dfrac{5}{\cancel{4x}} - \overset{2}{\cancel{4xy}} \cdot \dfrac{1}{\cancel{2y}}$$

$$= 5y - 2x$$

2.2 Preliminaries

D. $\dfrac{x}{6} - \dfrac{1}{x+y}$

$\text{LCD} = \text{LCM}[6, (x+y)] = 6(x+y)$

Multiply:

$$\text{LCD} \cdot \left(\dfrac{x}{6} - \dfrac{1}{x+y}\right) = 6(x+y) \cdot \left(\dfrac{x}{6} - \dfrac{1}{x+y}\right)$$

$$= \cancel{6}(x+y) \cdot \dfrac{x}{\cancel{6}} - 6\cancel{(x+y)} \cdot \dfrac{1}{\cancel{x+y}}$$

$$= (x+y)x - 6$$

$$= x^2 + xy - 6$$

Multiplying Decimals

In arithmetic, recall that when a decimal number is multiplied by 10, the decimal point is moved one position to the right. When a decimal number is multiplied by 100, the decimal point is moved two positions to the right. In general, when a decimal number is multiplied by 10, 100, 1000, ..., the number of positions that the decimal point moves to the right is the same as the number of zeros in the multiplier.

Example 7

Multiply as indicated:

A. 10(6.3)

$10(6.3) = 6{\scriptstyle\times}3. = 63$

B. 100(1.08)

$100(1.08) = 1{\scriptstyle\times}08. = 108$

C. 100(4.2)

$100(4.2) = 4{\scriptstyle\times}20. = 420$

D. 10(9.042)

$10(9.042) = (9{\scriptstyle\times}0.42) = 90.42$

E. 1000(4.2x − 1.096)

$1000(4.2x - 1.096)$

$= 1000(4.2x) - 1000(1.096)$ Distributive property

$= 4{\scriptstyle\times}200x - 1{\scriptstyle\times}096$

$= 4200x - 1096$

Example 8

Multiply each expression by the smallest appropriate number (10, 100, 1000, . . .) so that the product will have no decimals:

A. 7.06

This number has 2 decimal places, so we multiply by 100.

$$100(7.06) = 7.06 = 706$$

B. $x + 0.08x$

$x + 0.08x$
— 2 decimal places
— No decimal places, coefficient is 1

We multiply by 100
— 2 zeros

$$100(x + 0.08x) = 100x + 100(0.08x) \quad \text{Distributive property}$$
$$= 100x + 0.08x$$
$$= 100x + 8x$$
$$= 108x \quad \text{Combine like terms}$$

C. $7.3x - 4.668$

— 3 decimal places
— 1 decimal place

We multiply by 1000
— 3 zeros

$$1000(7.3x - 4.668) = 7.300x - 4.668$$
$$= 7300x - 4668$$

◀

Exercise Set 2.2

Find the LCM of each group of numbers or expressions:

1. 3, 6	**2.** 5, 10	**3.** 2, 6, 3	**4.** 2, 8, 4
5. 6, 9	**6.** 4, 6	**7.** 8, 10, 15	**8.** 3, 4, 8
9. 4, 6, 10	**10.** 6, 10, 12	**11.** $2x$, 8	**12.** $3y$, 15
13. $4x$, $6y$	**14.** $6x$, $3y$	**15.** $6x$, 8, $12y$	**16.** $9a$, $6b$
17. 15, $3x$, y	**18.** $4(x + y)$, $6x$	**19.** x, y, $(x + y)$	**20.** a, $2b$, $6c$

2.2 Preliminaries

Multiply as indicated:

21. $6 \cdot \left(\dfrac{2}{3}\right)$
22. $4 \cdot \left(\dfrac{1}{4}\right)$
23. $8 \cdot \left(\dfrac{3}{4}\right)$
24. $12 \cdot \left(\dfrac{5}{6}\right)$
25. $12 \cdot \left(\dfrac{1}{2} + \dfrac{3}{4}\right)$
26. $16 \cdot \left(\dfrac{5}{8} + \dfrac{1}{2}\right)$
27. $14 \cdot \left(\dfrac{5}{7} - \dfrac{1}{2}\right)$
28. $15 \cdot \left(\dfrac{4}{5} - \dfrac{2}{3}\right)$
29. $24 \cdot \left(\dfrac{x}{2} - \dfrac{y}{6} + \dfrac{3}{8}\right)$
30. $20 \cdot \left(\dfrac{x}{10} + \dfrac{y}{4} - \dfrac{3}{5}\right)$
31. $12 \cdot \left(\dfrac{x + 2}{4}\right)$
32. $8 \cdot \left(\dfrac{2x - 1}{4}\right)$

Determine the LCD for each of the expressions, and then complete the indicated multiplication:

33. $\text{LCD} \cdot \left(\dfrac{x}{3} + \dfrac{1}{4}\right)$
34. $\text{LCD} \cdot \left(\dfrac{a}{6} + \dfrac{1}{9}\right)$
35. $\text{LCD} \cdot \left(\dfrac{3}{8} - \dfrac{x}{6}\right)$
36. $\text{LCD} \cdot \left(\dfrac{y}{6} + \dfrac{x}{12}\right)$
37. $\text{LCD} \cdot \left(\dfrac{u}{2} + \dfrac{2}{u}\right)$
38. $\text{LCD} \cdot \left(\dfrac{m}{3} - \dfrac{3}{m}\right)$
39. $\text{LCD} \cdot \left(\dfrac{x}{y} + \dfrac{y}{x}\right)$
40. $\text{LCD} \cdot \left(\dfrac{2}{u} + \dfrac{u}{5}\right)$
41. $\text{LCD} \cdot \left(\dfrac{4}{x} - \dfrac{3}{5}\right)$
42. $\text{LCD} \cdot \left(\dfrac{5}{6} + \dfrac{2}{y}\right)$

Multiply as indicated:

43. $10(4.2)$
44. $10(2.6)$
45. $10(23)$
46. $100(16)$
47. $100(4.06)$
48. $100(3.19)$
49. $100(x + 1.92)$
50. $100(2.98 - 4x)$
51. $1000(2.406 - y)$
52. $1000(9.216x + 3)$

Multiply each expression by the smallest appropriate number, 10, 100, ..., such that the product will have no decimals:

53. 3.1
54. 4.7
55. 2.06
56. 1.93
57. $0.05x + 0.007$
58. $0.4 - .07y$
59. $3.015x - 0.06$
60. $0.06y + 0.1$
61. $0.7b + 1.0004c$
62. $2.16m - 0.0002u$

63. Explain how determining the smallest appropriate number, 10, 100, ..., needed to multiply the expression $0.05x + 6.381$ so it will have no decimals is like finding a least common denominator (LCD).

64. How is the notion of LCM related to the LCD of a collection of fractions?

Review

65. Complete the indicated operations:

   a. $(-6)(-4)$

   b. $\dfrac{-24}{-3}$

   c. $6 - 2 + 8 - 11$

   d. $\dfrac{2(-3) + 6}{4(-2) + 5}$

   e. $6 \cdot \left(\dfrac{1}{2} + \dfrac{2}{3}\right)$

   f. $100(x - 1.08)$

66. Set $A = \{1, 2, 3, 4, 5\}$, $B = \{2, 4\}$, $Q = $ the set of rational numbers. Indicate each statement below as true or false.

   a. $A \subseteq B$

   b. $B \subset A$

   c. $A \subset Q$

   d. $Q \subseteq B$

   e. $2.3 \in A$

   f. $\dfrac{3}{4} \notin B$

   g. $1.6 \subseteq Q$

   h. $\{1.6\} \subseteq Q$

2.3 Basic Definitions

An **algebraic equation** is a symbolic statement that two algebraic expressions are equal, indicated by the familiar symbol "=." Further, the statement of equality must contain at least one variable. For example, the statement $3 + 4 = 7$ is an equation, but it is not an *algebraic* equation since it does not contain a variable. On the other hand, $3 + x = 7$ is an algebraic equation. The two algebraic expressions separated by the "=" sign are called the **members**, or **sides**, of the equation.

The following are examples of algebraic equations:

Example 1

A. $2x^2 - 3xy + 5y^2 = 0$

B. $(x - 3)(x + 3) = x^2 - 9$

C. $\dfrac{2x + 3}{x - 1} = 8$

D. $\dfrac{(x - 2)}{2} + 3 = \dfrac{3(5 - 2x)}{4} - 3$

◀

2.3 Basic Definitions

Four important properties of equality are the reflexive, symmetric, transitive, and substitution properties. These properties are basic and are probably already familiar to you, although perhaps not by these names.

Properties of equality

1. Reflexive Property: If x is any expression, then $x = x$.
2. Symmetric Property: If x and y are any expressions and if $x = y$, then $y = x$.
3. Transitive Property: If x, y, and z are any expressions and if $x = y$ and $y = z$, then $x = z$.
4. Substitution Property: If x and y are any expressions and if $x = y$, then either expression may be substituted for the other in any statement without changing the truth or falseness of the statement.

Identities and Conditional Equations

An **identity** is an algebraic equation that is true for any value of the variable or variables. The properties of real numbers provide examples of identities since they are true for all values of the variables.

Example 2

Each of the following equations is an identity:

A. $a(b + c) = ab + ac$

B. $(a + b) + c = a + (b + c)$

C. $ab = ba$

D. $x + 3 = 3 + x$

E. $(x + 3) + 5 = x + 8$

(Can you name the preceding properties?) ◀

A **conditional** equation is a statement which may be true for some but not all values of the variable(s). To *solve* an equation means to find all possible values of the variables that make the equation true. For an equation in one variable, a value of the variable that makes the equation true is called a **solution**, or **root**, of the equation.

Example 3

Each of the following equations is a conditional equation:

A. $x^2 - 5x = -6$ is a conditional equation with the two roots: $x = 2$ and $x = 3$. Try these roots as well as other values.

B. $2x + 3 = -7 - 3x$ is a conditional equation with only one root, $x = -2$. Try it and others.

C. $2y + 5 = 13$ is a conditional equation with only one solution. Can you find it?

D. $2x + y = 10$ is a conditional equation in two variables and has many solutions. Some of the combinations of x and y that are solutions include $x = 2$ and $y = 6$; $x = 0$ and $y = 10$; $x = 5$ and $y = 0$; $x = 7$ and $y = -4$. Can you find other combinations of x and y that are solutions? Can you find combinations of values that are *not* solutions? Equations that contain more than one variable will be discussed in more detail later. ◀

Exercise Set 2.3

Substitute several values for the variable, and state whether each equation is a conditional equation or an identity:

1. $x + 1 = 1 + x$
2. $x(x + 1) = x^2 + x$
3. $-(x - y) = y - x$
4. $x + 3 = 5$
5. $2x^2 - 8 = 0$
6. $x^2 = -4$
7. $x + 4 = 7 + x$
8. $x + 3 = 4 - (1 - x)$

State three pairs of values of x and y that solve each equation:

9. $x + y = 5$
10. $x - 3 = y$
11. $2x - y = 4$
12. $x - 3y = 2$
13. $x + 1 = y - 1$
14. $2x - 1 = y$

Show that the number given is a root of the equation:

15. $3x - 5 = 7 + x$; 6
16. $9 - 4x = x - 6$; 3
17. $(x + 5)(x - 3) = x^2 + 2x - 15$; 15
18. $(2x + 1)(x + 3) = 2 + x(5 + x)$; -1
19. $3x - 2(x + 1) = 5 + 2(x - 3)$; -1
20. $4x - 3(x - 2) = 7 - 2(x + 2)$; -1
21. $x^2 - x - 12 = 0$; 4
22. $x^2 - x - 12 = 0$; -3

Each of the following equations has only one solution. Try to find it by guessing.

23. $3x + 2 = 5$
24. $6 - x = 2$
25. $3 + x = -2$
26. $5 + y = -6$
27. $2x + 3 = x - 7$
28. $2x - 1 = 8 - x$

Review

29. Evaluate each expression for $x = 6$.

    a. $5(x - 2)$
    b. $x^2 - 3x + 5$
    c. $2x + 3(x + 5)$
    d. $x(x - 1) + 3(x + 2)$

30. Write the negative, the absolute value, and the reciprocal of each number.

	Number	Negative	Absolute Value	Reciprocal
a.	-6			
b.	7			
c.	$-\dfrac{2}{3}$			
d.	0			
e.	1.4			

31. Clear grouping symbols and combine like terms:

    a. $2(y + 3) - 3(y - 4)$
    b. $(a^2 + 3a - 5) - (a^2 - 3a - 7)$
    c. $6b - 3[2 + 3(b - 1)]$

2.4 Solving Linear Equations in One Variable

In the preceding exercise set you were asked to solve some simple equations by guessing. In the sections that follow, we develop a systematic approach to solving equations that does not rely on guessing.

Equivalent Equations

Two or more equations are called **equivalent** if they have exactly the same solutions. For example, the equations

$$x = 6 - 1$$
$$x + 2 = 7$$
$$2x + 3 = 13$$

and $4x - 1 = 2x + 9$

are equivalent equations since they all have the same solution, $x = 5$.

To solve an equation, we perform operations on that equation that result in successively simpler equivalent equations. This process is continued until a final equation is obtained in the form $x = k$ (k a constant), whose solution is obvious.

We now state three properties that are used to rewrite an equation in simpler equivalent form:

PROPERTIES OF EQUATIONS		
	Given any algebraic equation, an equivalent equation results if:	
Addition Property	1.	The same quantity is added to or subtracted from both sides of the equation.
Multiplication Property	2.	Both sides are multiplied or divided by the same non-zero quantity.
Substitution Property	3.	Any quantity in the equation is replaced by a quantity to which it is equal.

In this chapter, the equations we solve are called linear equations in one variable. In such equations, the variable has an exponent of 1. All terms in the equation are either a constant, the variable, or a product of a constant and the variable. In general, a **linear equation** in the single variable x is any equation that can be written in the form: $ax = b$ where a and b are real numbers and $a \neq 0$.

Example 1

Each of the following is a linear equation in one variable:

A. $2x - 3 = 5$

B. $4y + 3 = 2y + 11$

C. $5(m + 1) = 3(2m - 1) + 8$

Each of the following is *not* a linear equation in one variable (why?):

D. $x^3 - 9 = 0$

E. $x^2 + 3x + 2 = 0$

F. $2x - y = 5$

G. $\dfrac{2}{x} = \dfrac{x}{3}$

2.4 Solving Linear Equations in One Variable

Solving an equation requires a game plan or general procedure to follow. To solve a linear equation in one variable, use the following steps:

Solving a linear equation in one variable

1. Express each side of the equation without parentheses or fractions (if present).

2. Add or subtract appropriate quantities to both sides of the equation so that all terms containing the variable are on one side, and all constant terms are on the other side.

3. Simplify each side of the equation by combining like terms.

   If Steps 1–3 are carried out properly, the resulting equation will be in the form $ax = b$ where a and b are constants and x is the variable.

4. If the coefficient of the variable (a) is other than positive one (1), divide both sides of the equation by this number. The resulting equation will be of the form $x = \frac{b}{a}$ and represents the solution to the original equation.

5. Check.

To check (Step 5), the solution is substituted for the variable in the original equation, and each side is evaluated. If the two sides are not equal, you must check your work. You have either made a mistake in solving the equation or in checking the solution.

Example 2

Solve and check each of the following equations:

A. $x + 3 = 5$

$$x + 3 = 5$$
$$x + 3 - 3 = 5 - 3 \quad \text{Add } -3 \text{ to both sides}$$
$$x = 2 \quad \text{Solution}$$

Check:
$$x + 3 = 5$$
$$2 + 3 \stackrel{?}{=} 5 \quad \text{Substitute}$$
$$5 = 5 \checkmark$$

B. $2x - 3 = 7$

$$2x - 3 = 7$$
$$2x - 3 + 3 = 7 + 3 \quad \text{Add } +3 \text{ to both sides}$$
$$2x = 10 \quad \text{Combine terms}$$
$$x = 5 \quad \text{Divide both sides by 2}$$

Check:

$$2x - 3 = 7$$
$$2(5) - 3 \stackrel{?}{=} 7 \quad \text{Substitute}$$
$$10 - 3 = 7$$
$$7 = 7 \checkmark$$

C. $x + 4 = 8 - x$

$$x + 4 = 8 - x$$
$$x + 4 - 4 = 8 - x - 4 \quad \text{Add } -4 \text{ to both sides}$$
$$x = 4 - x$$
$$x + x = 4 - x + x \quad \text{Add } x \text{ to both sides}$$
$$2x = 4$$
$$x = 2 \quad \text{Divide both sides by 2}$$

Check:

$$x + 4 = 8 - x$$
$$(2) + 4 \stackrel{?}{=} 8 - (2)$$
$$6 = 6 \checkmark$$

D. $2x - 5 = 15 - 3x$

$$2x - 5 = 15 - 3x$$
$$2x - 5 + 5 = 15 - 3x + 5 \quad \text{Add 5 to both sides}$$
$$2x = 20 - 3x$$
$$2x + 3x = 20 - 3x + 3x \quad \text{Add } 3x \text{ to both sides}$$
$$5x = 20$$
$$x = 4 \quad \text{Divide both sides by 5}$$

Check:

$$2x - 5 = 15 - 3x$$
$$2(4) - 5 \stackrel{?}{=} 15 - 3(4)$$
$$8 - 5 = 15 - 12$$
$$3 = 3 \checkmark$$

◀

Note the alignment that was used in Example 2. It is good form to keep the equal signs in a straight column. In addition, many of the steps used to solve an equation need not be written. For example, when adding the same quantity to both sides, the quantity is generally not written. Follow the next example step-by-step. The equations that you will solve should appear as those in Example 3.

2.4 Solving Linear Equations in One Variable

Example 3

Solve each equation for the indicated variable and check:

A. $2x - 5 = 3x + 6$

$2x - 5 = 3x + 6$

$2x = 3x + 11$ Add 5 to both sides

$-x = 11$ Subtract $3x$ from both sides

$x = -11$ Divide both sides by -1

Check:

$2(-11) - 5 \stackrel{?}{=} 3(-11) + 6$

$-22 - 5 = -33 + 6$

$-27 = -27$ ✓

B. $a + 5 = 9 - 3a$

$a + 5 = 9 - 3a$

$a = 4 - 3a$ Subtract 5 from both sides

$4a = 4$ Add $3a$ to both sides

$a = 1$ Divide both sides by 4

Check:

$1 + 5 \stackrel{?}{=} 9 - 3(1)$

$6 = 9 - 3$

$6 = 6$ ✓

C. $2r + 5 - 3r + 2 = 3 - 2r + 1$

$2r + 5 - 3r + 2 = 3 - 2r + 1$

$-r + 7 = 4 - 2r$ Combine like terms

$r + 7 = 4$ Add $2r$ to both sides

$r = -3$ Subtract 7 from both sides

Check:

$2(-3) + 5 - 3(-3) + 2 \stackrel{?}{=} 3 - 2(-3) + 1$

$-6 + 5 + 9 + 2 = 3 + 6 + 1$

$-6 + 16 = 10$

$10 = 10$ ✓ ◀

Next, we solve equations that contain parentheses and other symbols of grouping. We combine the rules for removing parentheses (Section 2.1) and the rules for solving equations.

Example 4

Solve each equation and check the solution:

A. $3x + (x + 2) = 2x + 6$

$$3x + (x + 2) = 2x + 6$$
$$3x + x + 2 = 2x + 6 \qquad \text{Remove parentheses}$$
$$4x + 2 = 2x + 6 \qquad \text{Combine like terms}$$
$$2x + 2 = 6 \qquad \text{Add } -2x \text{ to both sides}$$
$$2x = 4 \qquad \text{Add } -2 \text{ to both sides}$$
$$x = 2 \qquad \text{Divide both sides by 2}$$

Check:
$$3(2) + (2 + 2) \stackrel{?}{=} 2(2) + 6$$
$$6 + 4 = 4 + 6$$
$$10 = 10 \quad \checkmark$$

B. $2(5m - 3) - 4 = 4(2m + 3)$

$$2(5m - 3) - 4 = 4(2m + 3)$$
$$10m - 6 - 4 = 8m + 12 \qquad \text{Remove parentheses}$$
$$10m - 10 = 8m + 12 \qquad \text{Combine like terms}$$
$$10m = 8m + 22 \qquad \text{Add 10 to both sides}$$
$$2m = 22 \qquad \text{Subtract } 8m \text{ from both sides}$$
$$m = 11 \qquad \text{Divide both sides by 2}$$

Check:
$$2[5(11) - 3] - 4 \stackrel{?}{=} 4[2(11) + 3]$$
$$2(55 - 3) - 4 = 4(22 + 3)$$
$$2(52) - 4 = 4(25)$$
$$104 - 4 = 100$$
$$100 = 100$$

C. $2y - 5(3y - 2) = -3y - 2(-y + 1)$

$$2y - 5(3y - 2) = -3y - 2(-y + 1)$$
$$2y - 15y + 10 = -3y + 2y - 2 \qquad \text{Remove parentheses}$$
$$-13y + 10 = -y - 2 \qquad \text{Combine like terms}$$
$$-12y + 10 = -2 \qquad \text{Add } y \text{ to both sides}$$
$$-12y = -12 \qquad \text{Add } -10 \text{ to both sides}$$
$$y = 1 \qquad \text{Divide by } -12$$

2.4 Solving Linear Equations in One Variable

Check:
$$2(1) - 5[3(1) - 2] \stackrel{?}{=} -3(1) - 2(-1 + 1)$$
$$2 - 5(1) = -3 - 2(0)$$
$$2 - 5 = -3 - 0$$
$$-3 = -3 \checkmark$$

D. $4(3x + 3) + 1 = 2 - [2x - (x - 2)]$

$$4(3x + 3) + 1 = 2 - [2x - (x - 2)]$$
$12x + 12 + 1 = 2 - [2x - x + 2]$	Remove parentheses
$12x + 13 = 2 - [x + 2]$	Combine like terms
$12x + 13 = 2 - x - 2$	Remove brackets
$12x + 13 = -x$	Combine like terms
$13 = -13x$	Subtract $12x$ from both sides
$-1 = x$	Divide both sides by -13
$x = -1$	

Check:
$$4[3(-1) + 3] + 1 \stackrel{?}{=} 2 - \{2(-1) - [(-1) - 2)]\}$$
$$4(-3 + 3) + 1 = 2 - [-2 - (-3)]$$
$$4(0) + 1 = 2 - [-2 + 3]$$
$$0 + 1 = 2 - 1$$
$$1 = 1 \checkmark$$

◀

In the equations in the preceding examples, the numerical coefficients are integers. In practical problems, the coefficients are often fractions or decimals. To solve an equation containing fractions, first multiply both sides of the equation by a number that is evenly divisible by all of the denominators in the equation, so that all denominators cancel. The smallest number evenly divisible by all of the denominators is the LCD (see Section 2.2). This multiplication will produce an equivalent equation without fractions, which may then be solved by methods learned earlier. We call this process _clearing fractions_ from the equation.

Example 5

Solve and check each of the following equations:

A. $\dfrac{x}{2} - 1 = 3$

$$\frac{x}{2} - 1 = 3$$

$$2\left(\frac{x}{2} - 1\right) = 2(3) \qquad \text{Multiply both sides by the LCD} = 2$$

$$2\left(\frac{x}{2}\right) - 2 \cdot 1 = 2 \cdot 3 \qquad \text{Distribute the 2 over each term on the left side}$$

$$x - 2 = 6$$

$$x = 8$$

Check:

$$\frac{x}{2} - 1 = 3$$

$$\frac{8}{2} - 1 \stackrel{?}{=} 3$$

$$4 - 1 = 3$$

$$3 = 3 \;\checkmark$$

B. $\dfrac{2}{3} \cdot y + \dfrac{3}{1} = \dfrac{5}{8}$

$$\frac{2}{3} \cdot y + \frac{3}{1} = \frac{5}{8}$$

$$24 \cdot \left(\frac{2}{3} \cdot y + \frac{3}{1}\right) = 24 \cdot \frac{5}{8} \qquad \text{Multiply both sides by the LCD} = 24$$

$$\frac{\overset{8}{\cancel{24}}}{1} \cdot \frac{2 \cdot y}{\cancel{3}} + \frac{24}{1} \cdot \frac{3}{1} = \frac{\overset{3}{\cancel{24}}}{1} \cdot \frac{5}{\cancel{8}} \qquad \text{Distribute 24 on the left side and cancel}$$

$$16y + 72 = 15$$

$$16y = -57$$

$$y = -\frac{57}{16}$$

Check:

$$\frac{2}{3} \cdot y + \frac{3}{1} = \frac{5}{8}$$

$$\frac{2}{3} \cdot \left(-\frac{57}{16}\right) + \frac{3}{1} \stackrel{?}{=} \frac{5}{8}$$

$$\frac{\cancel{2}}{\cancel{3}} \cdot \frac{-\overset{19}{\cancel{57}}}{\underset{8}{\cancel{16}}} + \frac{3}{1} = \frac{5}{8}$$

2.4 Solving Linear Equations in One Variable

$$\frac{-19}{8} + \frac{3}{1} = \frac{5}{8}$$

$$-\frac{19}{8} + \frac{24}{8} = \frac{5}{8}$$

$$\frac{5}{8} = \frac{5}{8} \checkmark$$

C. $\quad \dfrac{5x}{8} + \dfrac{3}{4} = \dfrac{7}{10} \qquad$ LCD $= 40$

$$\frac{5x}{8} + \frac{3}{4} = \frac{7}{10}$$

$$40 \cdot \left(\frac{5x}{8} + \frac{3}{4}\right) = 40 \cdot \frac{7}{10} \qquad \text{Multiply both sides by 40}$$

$$\overset{5}{\cancel{40}} \cdot \frac{5x}{\cancel{8}} + \overset{10}{\cancel{40}} \cdot \frac{3}{\cancel{4}} = \overset{4}{\cancel{40}} \cdot \frac{7}{\cancel{10}} \qquad \text{Cancel}$$

$$25x + 30 = 28$$

$$25x = -2$$

$$x = -\frac{2}{25}$$

Check:

$$\frac{5x}{8} + \frac{3}{4} = \frac{7}{10}$$

$$\frac{5}{8} \cdot \left(\frac{-2}{25}\right) + \frac{3}{4} \overset{?}{=} \frac{7}{10}$$

$$\frac{\overset{1}{\cancel{5}}}{\underset{4}{\cancel{8}}} \cdot \frac{\overset{1}{\cancel{-2}}}{\underset{5}{\cancel{25}}} + \frac{3}{4} = \frac{7}{10}$$

$$\frac{-1}{20} + \frac{3}{4} = \frac{7}{10}$$

$$\frac{-1}{20} + \frac{15}{20} = \frac{7}{10}$$

$$\frac{14}{20} = \frac{7}{10}$$

$$\frac{7}{10} = \frac{7}{10} \checkmark$$

◀

TWO   Linear Equations and Inequalities in One Variable

To solve an equation containing decimals, multiply both sides by 10, 100, ... to remove all decimals. Note the selected number has the number of zeros equal to the largest number of decimal places.

Example 6

Solve and check each of the following equations:

A. $2.5x + 0.37 = 17.62$

$$2.5x + 0.37 = 17.62$$

$$100(2.5x + 0.37) = 100(17.62)$$ Since 0.37 and 17.62 each have 2 decimal places, multiply by 100 to clear decimals

$$100(2.5x) + (100)(0.37) = 100(17.62)$$ Distribute the 100 on the left side

$$250x + 37 = 1762$$

$$250x = 1725$$

$$x = \frac{1725}{250} = 6.9$$

Note: Since the original problem contained decimal numbers, the answer is expressed as a decimal.

Check:

$$2.5x + 0.37 = 17.62$$

$$2.5(6.9) + 0.37 \stackrel{?}{=} 17.62$$

$$17.25 + 0.37 = 17.62$$

$$17.62 = 17.62 \;\checkmark$$

B. $0.5(x + 3) - 0.3x = 3.2$

$$0.5(x + 3) - 0.3x = 3.2$$

$$10[0.5(x + 3) - 0.3x] = 10(3.2)$$ Multiply by 10 to clear decimals

$$(10)(0.5)(x + 3) - 10(0.3)x = 32$$ Distribute the 10 on the left side

$$5(x + 3) - 3x = 32$$

$$5x + 15 - 3x = 32$$

$$2x + 15 = 32$$

$$2x = 17$$

$$x = \frac{17}{2} = 8.5$$

2.4 Solving Linear Equations in One Variable

Check:
$$0.5(x + 3) - 0.3x = 3.2$$
$$0.5(8.5 + 3) - 0.3(8.5) \stackrel{?}{=} 3.2$$
$$0.5(11.5) - 0.3(8.5) = 3.2$$
$$5.75 - 2.55 = 3.2$$
$$3.2 = 3.2 \checkmark$$

◀

We now summarize the steps for solving linear equations in one variable:

Summary: solving equations   KNOW   IMPortANt!

1. If the equation contains fractions, clear all denominators by multiplying both sides of the equation by the LCD. If the equation has decimal coefficients, multiply both sides by 10, 100, 1000, ..., whichever number is necessary to clear the equation of decimals.

2. Remove grouping symbols, innermost to outermost on both sides.

3. Simplify both sides by combining like terms.

4. Gather all terms containing the variable on one side of the equation, and all terms without the variable on the other side

   If Steps 1–4 are carried out properly, the result will be an equation of the form $ax = b$, in which a and b are constants and x is the variable.

5. Divide both sides of the equation by the coefficient of the variable.

6. Check.

Exercise Set 2.4

Solve each equation for the variable and check:

1. $x + 3 = 7$
2. $x + 2 = -6$
3. $y - 5 = 3$
4. $3 - y = 4$
5. $2u + 1 = 9$
6. $2u - 1 = 11$
7. $3a - 2 = 7$
8. $3a + 1 = -11$
9. $x + 3 = 5 - x$
10. $2x - 3 = 9 - x$
11. $3y + 1 = 5 - y$
12. $2u + 6 = 9 - u$
13. $3m + 5 = 7 + m$
14. $5r + 3 = 2r - 12$
15. $14 - s = 2 - 5s$
16. $17 - 3w = 23$
17. $4y + 1 = 9 - 2y$
18. $4 - 6x = -11x - 21$
19. $-11 - u = 2u - 4$
20. $a + 6 = 22 - 7a$
21. $2c - 3 = -5 + 2c$

22. $2 + w - 3 = -1 + w$
23. $2a + 5 = 3 + 2a + 2$
24. $7 - 2d = -3 - 2d$
25. $3x + 5 = 11 - 6x$
26. $4x + 2 = -8x + 11$
27. $5y - 1 = 11 - 15y$
28. $-2x + 4 = 3 - 4x$ prove
29. $5 - 4x = 6x - 10$
30. $3 - 2x = 6x - 17$
31. $2x + (x - 1) = 2$
32. $3y + (y - 2) = 2y$
33. $3m = 5 - (3 - 2m)$
34. $2x + (3 - x) = 6$
35. $3(2a - 1) - 3 = 2(5 - a)$
36. $4(2 - m) = 15 - 3(m - 1)$
37. $5(2y + 1) = 32 - 9(3 - y)$
38. $2(3y - 1) = 11 + (8 - y)$
39. $2x - 5 + x + 1 = 3(x + 2) - 10$
40. $2(m + 2) + 1 = 4m + 5 - 2m$
41. $-3(2x - 8) = 4(7 - x)$
42. $3(y - 4) = -2(2y - 1)$
43. $12 - 2(1 - u) = 3(2u - 2)$
44. $13(3 + c) + 1 = 8(5 - c)$
45. $3(7 + 2x) - (x + 2) = 2(5 + x) + 3(3x + 13)$
46. $4(2x - 3) + 2(5 - x) = 3(2 + 3x) - (5x + 2)$
47. $5(2x - 1) - 3 = 3 - [x - (x - 1)]$
48. $x - [(x - 1) - x] = 2(2x + 1) - 5$

Solve each of the following equations:

49. $\dfrac{x}{5} = \dfrac{7}{15}$
50. $\dfrac{5}{24} = \dfrac{y}{4}$
51. $\dfrac{3}{4} + 2x = \dfrac{5}{2}$
52. $\dfrac{2}{3} = 4x - \dfrac{5}{6}$
53. $\dfrac{y}{2} - 2 = \dfrac{3}{2}$
54. $\dfrac{a}{5} + 2 = \dfrac{3}{25}$
55. $0.6x = 8$
56. $3.1m = 5$
57. $y + 2.3 = 4.7$
58. $7.2 + y = 9.1$
59. $3.5a - 2 = 3.1$
60. $4.5 = 2.3a - 1$
61. $5.2m + 4.3 = 2.1 - 1.6m$
62. $2.7 - 3.6y = 9.1 + 2.5y$
63. $\dfrac{x}{12} - \dfrac{2x}{3} = 4$
64. $\dfrac{x}{3} + \dfrac{1}{2} = \dfrac{x}{6}$
65. $\dfrac{2x}{3} + 1 = x$
66. $\dfrac{2x}{5} - \dfrac{4}{5} = 3$
67. $\dfrac{3x}{4} + \dfrac{5}{8} = \dfrac{x}{2} - \dfrac{3}{4}$
68. $\dfrac{2x}{5} - \dfrac{3}{4} = \dfrac{1}{2} + \dfrac{x}{3}$
69. $3x + \dfrac{1}{2} = 5 - \dfrac{x}{2}$
70. $\dfrac{2x}{3} + 4 = \dfrac{1}{3} - 2x$

2.4 Solving Linear Equations in One Variable

71. $\dfrac{x+1}{2} = \dfrac{14-x}{3}$

72. $\dfrac{b-4}{3} = \dfrac{2+b}{4}$

73. $\dfrac{1(x-1)}{2} = \dfrac{3(x+2)}{4}$

74. $\dfrac{2(y+2)}{3} = \dfrac{4(y-1)}{5}$

75. $\dfrac{x+2}{3} = \dfrac{x-1}{4} - 2$

76. $\dfrac{m-3}{2} + 3 = \dfrac{m+5}{5}$

77. $2.3(2x - 3) = 4.1(x + 2)$

78. $5.7(3m + 1) = 3.2(5m + 2)$

79. $1.4(3.2y - 0.7) = 2.5(1.2y + 1.1)$

80. $5.5(0.1m - 2.1) = 1.7(1.3m + 3.9)$

Translate each of the following sentences to an equation. Do not solve.

81. Two times a number x is the same as 5.

82. Five times a number y is the same as 7.

83. Three times a number z added to 6 is the same as 12.

84. Four times the sum of a number x and 3 is the same as three times the number.

85. Two times the sum of a number x and 3 is the same as three times the number.

86. Four times the difference of a number y and 2 is the same as two times the number.

Review

87. The formula for the area A and perimeter P of a rectangle are:

    $$A = LW \qquad P = 2L + 2W$$

    a. Find the area and perimeter of a rectangle with $L = 14$ and $W = 9$.

    b. A rectangle has a perimeter of 72 and a width of 7. Find its length.

    c. A rectangle has an area of 176 and a length of 16. Find its width.

    d. Identify the constants and variables in each of the formulas.

88. Evaluate the following expressions for $x = 8$:

    a. $3x - 7$

    b. $x^2 - 6x - 9$

    c. $\dfrac{3x + 2}{x + 5}$

    d. $x(x - 1) + 3x - 5$

89. Write $2 \cdot a \cdot a \cdot a \cdot b \cdot c \cdot c$ in exponential form.

2.5 Literal Equations

A **literal equation** is an equation that contains more than one letter (or literal, or variable). A **formula** is a literal equation that represents a relation between physical quantities.

To solve a literal equation for one of its variables means to find an expression for that variable in terms of the other variables in the equation. Solving a literal equation is done in the same manner as solving any other equation. However, since a literal equation contains more than one variable, the operations on these variables cannot be completed, only indicated.

Before working examples, we repeat the rules for solving an equation:

1. Remove fractions by multiplying by the LCD or remove decimals by multiplying by the appropriate number 10, 100, ...

2. Clear symbols of grouping.

3. Simplify both sides.

4. Gather terms with the variable on one side of the equation and the constant terms on the other side.

5. Divide both sides of the equation by the coefficient of the variable.

Example 1

Solve each equation for x. (Compare the steps in parts A and B.)

A. $2x + 7 = 11$

$$2x + 7 = 11$$
$$2x = 11 - 7 \quad \text{Subtract 7 from both sides}$$
$$2x = 4$$
$$x = \frac{4}{2} = 2 \quad \text{Divide both sides by 2}$$

B. $ax + b = c$

$$ax + b = c$$
$$ax = c - b \quad \text{Subtract } b \text{ from both sides}$$
$$x = \frac{c - b}{a} \quad \text{Divide both sides by } a \quad \blacktriangleleft$$

Example 2

Solve each equation for x. (Compare the steps in parts A and B.)

A. $2x + 3 = 5x - 7$

$$2x + 3 = 5x - 7$$

2.5 Literal Equations

$$2x - 5x = -7 - 3 \quad \text{Subtract } 5x \text{ and } 3 \text{ from both sides}$$
$$-3x = -10 \quad \text{Combine like terms}$$
$$x = \frac{-10}{-3} = \frac{10}{3}$$

B. $ax + b = cx - d$

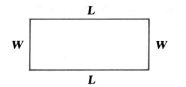

$$ax + b = cx - d$$
$$ax - cx = -d - b \quad \text{Subtract } cx \text{ and } b \text{ from both sides}$$

Isolate x by factoring out the x. (See Section 2.1.)

$$(a - c)x = -d - b$$
$$x = \frac{-d - b}{a - c} \quad \text{Divide each side by } a - c \quad \blacktriangleleft$$

Handwritten note: "can be done another way so that $x = \frac{b+d}{c-a}$"

Example 3

Write the formula for finding the area and the perimeter of a rectangle in terms of the length and width. Then solve each of these literal equations for the width.

[Diagram of rectangle with length L on top and bottom, width W on left and right sides]

$L = \text{length}, W = \text{width}$
$A = \text{area}, P = \text{perimeter}$

Area: $A = L \cdot W$ Solve for W.

$$A = L \cdot W$$
$$\frac{A}{L} = W \quad \text{Divide both sides by } L$$
$$\text{or} \quad W = \frac{A}{L} \quad \text{Symmetric property of equality}$$

Perimeter: $P = 2L + 2W$ Solve for W.

$$P = 2L + 2W$$
$$P - 2L = 2W \quad \text{Subtract } 2L \text{ from both sides}$$
$$\frac{P - 2L}{2} = W \quad \text{Divide both sides by } 2$$
$$\text{or} \quad W = \frac{P - 2L}{2} \quad \text{Symmetric property of equality} \quad \blacktriangleleft$$

Example 4

The amount of money (A) in a savings account is equal to the original principal (P) plus the interest, $I = PRT$. That is,

$$A = P + PRT$$

Solve this equation (A) for P; (B) for R.

A. $A = P + PRT$ Solve for P.

In this example, there are two terms containing P. Isolate P by factoring (Section 2.1) and then dividing both sides of the equation by the coefficient of P.

$$A = P + PRT$$
$$A = (1 + RT)P \qquad \text{Factor}$$
$$\frac{A}{1 + RT} = P \qquad \text{Divide both sides by } 1 + RT$$
$$\text{or} \quad P = \frac{A}{1 + RT}$$

B. $A = P + PRT$ Solve for R.

$$A = P + PRT$$
$$A - P = PRT \qquad \text{Subtract } P \text{ from both sides}$$
$$\frac{A - P}{PT} = R \qquad \text{Divide both sides by } PT$$
$$\text{or} \quad R = \frac{A - P}{PT}$$

◀

Example 5

The force due to gravitational attraction between two particles having masses M_1 and M_2 and separated by a distance d is given by the formula:

$$F = \frac{kM_1M_2}{d^2}$$

Solve this formula for k.

Note: M_1 and M_2 are called **subscripted variables**. The subscripts 1 and 2 should not be confused with exponents. The subscripts are used to indicate that M_1 and M_2 represent different quantities just as x and y do. M_1 is read as "M sub 1."

$$F = \frac{kM_1M_2}{d^2} \qquad \text{Multiply both sides by } d^2 \text{ to remove fractions}$$

2.5 Literal Equations

$$d^2 \cdot F = \frac{\cancel{d^2} \cdot kM_1M_2}{\cancel{d^2}} \qquad \text{Cancel } d^2$$

$$d^2F = kM_1M_2$$

$$\frac{d^2F}{M_1M_2} = k \qquad \text{Divide both sides by } M_1M_2$$

$$\text{or} \quad k = \frac{d^2F}{M_1M_2} \qquad \blacktriangleleft$$

Exercise Set 2.5

Solve each of the equations for the indicated variable:

1. $I = PRT$ for R (Simple interest)
2. $R = r_1 + r_2 + r_3$ for r_1 (Series resistors)
3. $I = \dfrac{E}{R}$ for R (Series circuit)
4. $C = 2\pi r$ for r (Circumference of a circle)
5. $d = rt$ for r (Distance/rate/time)
6. $d = rt$ for t (Distance/rate/time)
7. $ax + by + c = 0$ for x (Equation of a straight line)
8. $IQ = \dfrac{100\overline{MA}}{\overline{CA}}$ for $\overline{MA}$ (Intelligence quotient)
9. $F = \dfrac{kM_1M_2}{d^2}$ for M_1 (Gravitational attraction)
10. $A = \dfrac{1}{2}bh$ for h (Area of a triangle)
11. $F = \dfrac{9}{5} \cdot C + 32$ for C (Celsius to Fahrenheit conversion)
12. $C = \dfrac{5}{9}(F - 32)$ for F (Fahrenheit to Celsius conversion)
13. $P = 2L + 2W$ for L (Perimeter of a rectangle)
14. $V = LWH$ for H (Volume of a rectangular solid)
15. $V = \dfrac{1}{3}\pi r^2 h$ for h (Volume of a cone)

16. $I = \dfrac{E}{R_1 + R_2}$ for R_1 (Current in a series circuit)

17. $\dfrac{W_1}{W_2} = \dfrac{L_1}{L_2}$ for L_2 (Lever and fulcrum)

18. $S = \dfrac{N(A + L)}{2}$ for L (Arithmetic series)

19. $A = 2\pi r(h + r)$ for h (Surface area of cylinder)

20. $A = \dfrac{h}{2}(B + b)$ for B (Area of a trapezoid)

21. $a^2 = b^2 + c^2 - 2bcx$ for x (Cosine law)

22. $\dfrac{a}{x} = \dfrac{b}{y}$ for x (Proportion)

Review

23. Graph the following numbers on a number line:

    a. 7

    b. $\sqrt{7}$ (approximately 2.6)

    c. -4 d. $4\tfrac{1}{2}$ e. -1.8

24. Find the distance indicated:

    a. from 2 to 7   b. from 5 to -6

    c. from -1 to -5 d. from -9 to 0

25. Remove parentheses and combine like terms:

    a. $(9x + 3) - (x - 4)$ b. $2(3m - 5) - 4(m - 8)$

    c. $2(x^2 + x) - 3(x - 4)$

2.6 Solving Inequalities

The preceding sections of this chapter considered only the equality relation, indicated by the "=" sign. At this time we review other relations between expressions called **inequalities** (Section 1.3).

The expression "a is less than b" is written symbolically $a < b$. The symbol "$<$" is read "is less than." The statement "a is greater than b" is written $a > b$. Recall that in writing these symbols, the smaller end of the inequality symbol points to the smaller number.

2.6 Solving Inequalities

> **DEFINITION**
>
> $a < b$ if and only if there exists some positive number p such that $a + p = b$ or, equivalently, $b - a = p$. Further, if $a < b$, then $b > a$.

We can interpret inequalities geometrically using the number line. That is, $a < b$ if and only if a is to the left of b on the number line. Other inequality relations and symbols are listed in Figure 2.6-1.

FIGURE 2.6-1
Inequality Symbols

Symbolic Form	Verbal Form	Examples
$a < b$	a is less than b	$4 < 6,\ -2 < 0,\ -6 < -4$
$a > b$	a is greater than b	$6 > 4,\ 0 > -2,\ -4 > -6$
$a \leq b$	a is less than or equal to b	$4 \leq 6,\ -4 \leq -4,\ -6 \leq -4$
$a \geq b$	a is greater than or equal to b	$-4 \geq -6,\ -4 \geq -4,\ 6 \geq 4$
$a \neq b$	a is not equal to b	$5 \neq -5,\ 0 \neq -2,\ 7 \neq 3$

There are several properties that real numbers possess relative to the inequality relations we have discussed.

Properties of inequalities

1. Trichotomy Property: Given any two real numbers a and b, only one of three possible relations exists:

   a. $a = b$ **b.** $a > b$ **c.** $a < b$

2. Transitive Property: Given any three real numbers a, b, and c, if a is less than b and b is less than c, then a is less than c. In symbols: If $a < b$ and $b < c$, then $a < c$.

Any set of numbers in which both the trichotomy property and the transitive property hold is called an **ordered** set of numbers. The set of real numbers is an ordered set.

Graphs of Inequalities

When an inequality statement contains a variable, the solution is the set of all replacement values for the variable that make the statement true. The statement

$x \leq 5$ is true for all values of x that are 5 or less. Similarly, the solution to $x < 5$ is the set of all values of x that are strictly less than 5.

The graphs of the solutions to the statements $x \leq 5$ and $x < 5$ are shown by shading all points of the number line to the left of 5. If the number 5 is included in the solution ($x \leq 5$), a dot is drawn at 5 on the line; but if the number 5 is not included in the solution ($x < 5$), a hole is drawn at 5 on the line. (See Figure 2.6-2.)

FIGURE 2.6-2
Graphs of $x \leq 5$ and $x < 5$

Example 1

Graph each inequality:

A. $x > 3$

3 is not part of the solution

B. $c \leq 4$

4 is part of the solution

C. $m \geq -2$

-2 is part of the solution

D. $p < -5$

-5 is not part of the solution

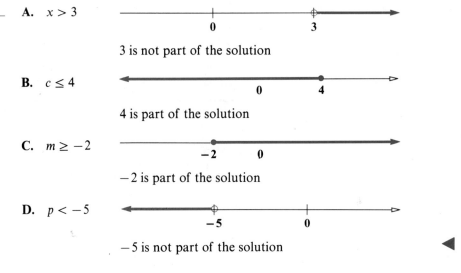

The inequality symbol is also convenient for representing positive or negative numbers. To indicate that x is some positive number, write $x > 0$; and to indicate that x is some negative number, write $x < 0$.

Solving Inequalities in One Variable

In a manner similar to equations, inequalities that have the same solution set are called **equivalent inequalities**. By substituting several values you will see that any value that satisfies the inequality $2x - 3 < 7 - 3x$ also solves the statement $x < 2$. These inequalities are equivalent. To *solve* an inequality means to replace it with successively simpler equivalent inequalities until the solution is obvious. This is done in a manner similar to that used to solve equations. To simplify an inequality, the following properties are used.

2.6 Solving Inequalities

Additional properties of inequalities

1. The sense (direction) of the inequality symbol is *unchanged* whenever:
   a. The same quantity is added to or subtracted from each member.
   b. Each member is multiplied or divided by the same *positive* quantity.
2. The sense of the inequality is *reversed* whenever:
   a. The members are interchanged.
   b. Each member is multiplied or divided by the same *negative* quantity.

Example 2 Consider the inequality $-2 < 4$. Show that a true statement results if we *leave the sense unchanged* and:

A. Add 7 to each member.
B. Subtract 6 from each member.
C. Multiply each member by 3.
D. Divide each member by 2.

A. $-2 < 4$
$-2 + 7 < 4 + 7$
$5 < 11$

B. $-2 < 4$
$-2 - 6 < 4 - 6$
$-8 < -2$

C. $-2 < 4$
$3(-2) < 3(4)$
$-6 < 12$

D. $-2 < 4$
$\dfrac{-2}{2} < \dfrac{4}{2}$
$-1 < 2$ ◀

Example 3 Consider the inequality $-2 < 4$. Show that a true statement results if we *reverse the sense* and:

A. Multiply each member by -3.
B. Divide each member by -2.
C. Interchange the members.

A. $-2 < 4$
$(-3)(-2) > (-3)4$
$6 > -12$

B. $-2 < 4$
$\dfrac{-2}{-2} > \dfrac{4}{-2}$
$1 > -2$

C. $-2 < 4$
$4 > -2$ ◀

To solve an inequality means to isolate the variable. We follow basically the same sequence of steps that was used to solve an equation. (See Section 2.4.)

Example 4

Solve and graph each inequality:

A. $2x + 3 \geq 9 - 4x$

$2x + 3 \geq 9 - 4x$
$6x + 3 \geq 9$ Add $4x$ to both members
$6x \geq 6$ Subtract 3 from both members
$x \geq 1$ Divide by 6

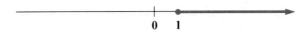

B. $2(x - 3) > 4x - (7 - 3x)$

$2(x - 3) > 4x - (7 - 3x)$
$2x - 6 > 4x - 7 + 3x$ Clear parentheses
$2x - 6 > 7x - 7$ Combine like terms
$-5x - 6 > -7$ Add $-7x$ to both members
$-5x > -1$ Add 6 to both members
$x < \dfrac{1}{5}$ Divide by -5 (*reverse sense*)

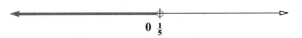

C. $\dfrac{x}{3} + \dfrac{5}{2} > \dfrac{3}{4} - \dfrac{2x}{3}$

$\dfrac{x}{3} + \dfrac{5}{2} > \dfrac{3}{4} - \dfrac{2x}{3}$

$12 \cdot \dfrac{x}{3} + 12 \cdot \dfrac{5}{2} > 12 \cdot \dfrac{3}{4} - 12 \cdot \dfrac{2x}{3}$ Multiply both members by the LCD = 12

$4x + 30 > 9 - 8x$
$12x + 30 > 9$ Add $8x$ to both members
$12x > -21$ Add -30 to both members
$x > \dfrac{-21}{12}$ Divide
$x > \dfrac{-7}{4}$ Reduce fraction ◀

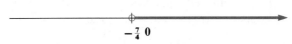

Double Inequalities

A symbolic statement such as $a < x < b$ (read "a is less than x, and x is less than b") is called a *double inequality* and means that both of the statements $a < x$ and $x < b$ must be true. For this double inequality $a < x < b$ to be valid, it must be true that $a < b$ by the transitive property.

In the statement $-1 < x \leq 4$, the values of the variable that make the statement true are those values of x that are both greater than -1 *and* less than or equal to 4. These values of x are between -1 and 4, and the solution is shown by shading that part of the number line. The graph is shown in Figure 2.6-3.

FIGURE 2.6-3
The Graph of
$-1 < x \leq 4$

Notice the use of the hole and dot to represent excluded and included numbers, respectively, in Figure 2.6-3 and in the examples that follow.

Example 5

Graph each of the following inequalities:

A. $-2 < x \leq 5$

B. $-10 \leq x < 40$

C. $4 \leq x \leq 10$

D. $12 < x < 14$

◀

Solving Double Inequalities

The same method as that of the preceding examples may be used to solve double inequalities such as $-5 \leq 2x + 7 < 9$. We rewrite the inequality so that the middle member is simply the variable, using two familiar steps:

1. Add or subtract the same quantity to all *three* members.

2. Divide all three members by the same non-zero coefficient of the variable. Reverse the sense of the inequalities if the divisor is negative.

Example 6

Solve and graph each inequality:

A. $-5 \leq 2x + 7 < 9$

$$-5 \leq 2x + 7 < 9$$
$$-12 \leq 2x < 2 \quad \text{Add } -7 \text{ to each member}$$
$$-6 \leq x < 1 \quad \text{Divide each member by 2}$$

B. $-2 < 5 - 3x < 11$

$$-2 < 5 - 3x < 11$$
$$-7 < -3x < 6 \quad \text{Add } -5 \text{ to each member}$$
$$\frac{7}{3} > x > -2 \quad \text{Divide each member by } -3 \text{ (reverse sense)}$$
$$-2 < x < \frac{7}{3} \quad \text{Rewrite}$$

Exercise Set 2.6

1. In the box, write =, <, or >, whichever is correct:
   - **a.** $2 \square 7 - 5$
   - **b.** $12 \square 3 \cdot 5$
   - **c.** $4 \square 7$
   - **d.** $-3 \square 4$
   - **e.** $3 \square -4$
   - **f.** $-3 \square -4$

2. In the box, write =, <, or >, whichever is correct:
   - **a.** $3 + x \square x + 3$
   - **b.** $2 \cdot 3 + 4 \square 2 \cdot 3 + 2 \cdot 4$
   - **c.** $-5 \square -7$
   - **d.** $0 \square -4$
   - **e.** $-7 \square -2$
   - **f.** $7 \square -7$

3. Express verbally:
   - **a.** $3 < x$
   - **b.** $x < 8$
   - **c.** $2 \leq 3 \cdot 8$
   - **d.** $x + 3 \geq 15$
   - **e.** $x - 2 \leq 15$
   - **f.** $2(3 + x) \neq x - 3$

2.6 Solving Inequalities

4. Express verbally:
   a. $-2 < 5$
   b. $-4 > -5$
   c. $x + 3 \leq 8$
   d. $5 \geq 5$
   e. $x - 3 < 7$
   f. $2(x + 7) \neq -2$

5. Write symbolically:
   a. -5 is less than zero
   b. $3 + 4$ is less than 8
   c. $x + 2$ is greater than 9
   d. x is not equal to 5
   e. $x + 3$ is greater than or equal to -4
   f. -5 is less than or equal to $x - 9$

6. Write symbolically:
   a. 7 is greater than -4
   b. x is not equal to -9
   c. $x - 2$ is less than $2x + 5$
   d. 5 is greater than $3x$
   e. $2x + 3$ is less than or equal to 15
   f. 5 is greater than or equal to $3y + 6$

7. Indicate each statement as true or false:
   a. $4 \geq 5$
   b. $4 < 5$
   c. $-7 \leq 5$
   d. $-5 > 0$
   e. $-7 \leq 7$
   f. $-3 \geq -3$

8. Indicate each statement as true or false:
   a. $-5 > 5$
   b. $-7 \geq -3$
   c. $5 < 5$
   d. $-3 \leq 0$
   e. $-8 < -3$
   f. $-3 \neq -5$

9. Graph the solution on the number line:
   a. $x < 4$
   b. $x > 3$
   c. $x < -2$
   d. $x > -6$
   e. $x \leq \dfrac{-1}{2}$
   f. $x \geq 12$
   g. $x \geq -6$
   h. $x \leq 9$

10. Graph the solution on the number line:
    a. $x < 4$
    b. $x > 8$
    c. $x < -5$
    d. $x > -1$
    e. $x \leq 0$
    f. $x \geq -6$
    g. $x \geq 0$
    h. $x \leq 10$

11. Graph the solution on the number line:
    a. $-2 < x < 6$
    b. $0 < x < 5$
    c. $-\dfrac{1}{2} < x < 0$
    d. $4 \leq x < 7$
    e. $-9 < x \leq -1$
    f. $-4 \leq x \leq 3$

12. Graph each set on the number line:

   a. $0 < x < 6$ b. $-3 < x \leq 0$ c. $2 \leq x < 8$

   d. $-7 \leq x \leq -2$ e. $-1 < x \leq 5$ f. $3 < x < \dfrac{14}{3}$

Solve and graph the solutions of each inequality on the real number line:

13. $x + 3 < 4$
14. $x - 4 < 5$
15. $x + 2 < 4 - x$
16. $x + 3 > 7 - x$
17. $2x + 3 \leq 9$
18. $3x - 1 \leq 11$
19. $x + 5 \leq 3x - 9$
20. $4x + 7 \geq x - 5$
21. $2(x - 3) \leq x - (2 - 3x)$
22. $3(x + 1) > x - 11$
23. $3(2x - 1) > 2x - 2(2 - x)$
24. $x - 3 < x + 2(x - 3)$
25. $\dfrac{x}{2} + 5 < 3$
26. $\dfrac{x}{3} - 2 > 2$
27. $\dfrac{x}{2} + \dfrac{5}{3} > \dfrac{x}{3}$
28. $\dfrac{x}{4} + \dfrac{3}{5} < 2 - x$
29. $-2 < x + 3 \leq 5$
30. $0 < x + 2 < 7$
31. $-2 \leq 2x + 1 < 9$
32. $3 \leq 3x + 2 \leq 11$
33. $-5 \leq 3 - 2x < 11$
34. $-3 < 3 - 4x < 19$
35. $-13 \leq 5 - 2x < -3$
36. $3 < 2 - 3x \leq 14$

37. A boy wants to buy a cassette player for $90. He has $46 in his bank account. If he adds $8 each week to his account, how long will it be until he can buy the cassette player (has at least $90)? Write an inequality and solve. *Hint*: Let $n =$ number of weeks until he has $90 or more in his account.

38. Ms. Smith has invested a principal of $2000 in an account earning 7 percent simple interest ($I = PRT$). How long must she leave her money invested to earn interest of at least $280? (That is, we want $I \geq \$280$.) Substitute into the inequality $I \geq 280$ and solve.

Review

39. Solve each equation for x:

   a. $3x - 5 = x + 7$

   b. $3(x - 2) + x = 2(x + 5)$

   c. $2x + 3y = 12$

40. Solve each equation:

   a. $\dfrac{a}{3} + 2 = \dfrac{a}{2}$

   b. $1.6p - 8 = 0.3p + 18$

2.7 Applications

We have seen how to solve some simple first-degree equations and inequalities in one variable. This skill enables us to solve a wide variety of practical problems. In practice, most problems are in stated form, rather than in the form of an equation. Our first task in solving such problems is to translate the problem into equation or inequality form.

In preparation for this task, we present the following table in Figure 2.7-1, which lists some of the more common words used to represent the four arithmetic operations.

FIGURE 2.7-1
Symbols and Words for Operations

Operation, Symbol(s)	Name of Result	Some Common Words
Addition, $a + b$	*Sum* of a and b	More, plus, increased, taller, older, heavier
Subtraction, $a - b$	*Difference* of a and b	Less, decreased, shorter, minus, younger
Multiplication, $(a)(b)$, $a \cdot b$, ab	*Product* of a and b	Times, multiplied, doubled, tripled
Division, $a \div b$, a/b, $\frac{a}{b}$, $b\overline{)a}$	*Quotient* of a divided by b	Divided by, half, into, how many

Representing Unknown Quantities with a Variable

We usually begin problem solving by assigning a variable to one of the unknown quantities from the problem. The assignment of a variable is generally done by a statement like "Let ____ = ____" or by labeling an appropriate sketch. As examples, we might begin solving a problem with a statement such as:

Let d = the number of dimes

Let L = the length of the side

Let N = the smaller number

Let A = Andy's present age

Let g = the weight of the gear

Let t = the number of 25-cent stamps

TWO   Linear Equations and Inequalities in One Variable

After the first variable has been selected, use the key words such as those suggested in Figure 2.7-1 to express all *other* unknown quantities of the problem in terms of that *same* variable. We begin with a simple example using ages.

Example 1

If John's age is known, then answer the following questions:

A. How old was he two years ago?
B. How old will he be when he is twice as old as now?
C. His father is three years younger than twice John's age. How old is John's father?
D. How old is John's sister, who is three years younger than John?
E. John's mother is 21 years older than John. How old is she?
F. His uncle is five years older than his mother. How old is his uncle?
G. How much older than John's mother is his father? ◀

In the following chart, we solve the problem in general using x to represent John's present age and also with a specific age of 26.

Age of:	When John's Present Age Is:	
	x	26
A. John, two years ago	$x - 2$	$26 - 2 = 24$
B. When he is twice as old	$2x$	$2 \cdot 26 = 52$
C. His father	$2x - 3$	$52 - 3 = 49$
D. His sister	$x - 3$	$26 - 3 = 23$
E. His mother	$x + 21$	$26 + 21 = 47$
F. His uncle	$(x + 21) + 5$	$47 + 5 = 52$
G. The difference between his father's and mother's age	$(2x - 3) - (x + 21)$	$49 - 47 = 2$

Money Problems

Problems involving money often express the *value* (or cost) of coins, postage stamps, or tickets. The value should be clearly expressed as either dollars or cents (do not mix them). Recall that each dollar has a value of 100 cents. In written money problems, the variable often represents the *number* of coins, tickets, or stamps. A convenient equation for you to remember when working money problems is:

> **Value = (Number of items) · (Unit value)**

For example, express the value of three quarters:

Value = (Number of items) · (Unit Value)

$= 3 \cdot 25 = 75$ cents

Example 2 Express the value of:

A. x number of basketball tickets at $10.00 each.
We must decide to express the value in either dollars or cents. If dollars are selected, then

Value = (Number of tickets) · (Cost of each ticket)

$= x \cdot 10 = 10x$ dollars

If cents are selected, then

Value = (Number of tickets) · (Cost of each ticket)

$= x \cdot 1000 = 1000x$ cents

B. x number of 25-cent stamps plus $(x - 5)$ number of 15-cent stamps.
In this example, we have the sum of two values, so they must both be expressed in the same units. In this case the obvious choice is cents.

Value of 25-cent stamps $= x \cdot 25 = 25x$ cents

Value of 15-cent stamps $= (x - 5) \cdot 15 = 15(x - 5)$ cents

Total value (in cents) $= 25x + 15(x - 5)$ cents ◀

Time-Rate-Distance Problems

In problems of time, rate, and distance, we have the basic relationship that the distance traveled is found by multiplying the average rate times the time traveled. In symbols,

> $d = r \cdot t$
>
> **distance = rate · time**

Example 3

How far could you travel if:

A. You drove for 5 hours at an average rate of 45 miles per hour?

$$d = r \cdot t$$
$$= 45 \cdot 5$$
$$= 225 \text{ miles}$$

B. You drove for $(t - 3)$ hours at an average rate of 60 kilometers per hour (kph)?

$$d = r \cdot t$$
$$= 60(t - 3) \text{ km} \qquad \blacktriangleleft$$

Consecutive Integer Problems

Consecutive integers follow each other in counting, as in 21, 22, 23, 24, ... Each integer is obtained by adding 1 to the preceding integer. If the first integer is identified as x, then the second would be $x + 1$, the third would be $(x + 1) + 1 = x + 2$, the fourth $x + 3$, and so on.

Consecutive *even* integers, such as 24, 26, 28, 30, ..., and consecutive *odd* integers, such as 35, 37, 39, ..., are obtained by repeatedly adding 2 to the preceding integer. So if the first of several even (or odd) consecutive integers is identified as x, then the next integers in sequence are $x + 2, x + 4, x + 6, \ldots$.

Example 4

A. Express the sum of three consecutive integers.
If the first integer is x, then the next two integers are $x + 1$ and $x + 2$, and their sum is $x + (x + 1) + (x + 2)$.

In the preceding example, the parentheses are unnecessary, but they help to show the *sum* of *three* integers. If the sum were written without the parentheses, it would appear $x + x + 1 + x + 2$, which is mathematically correct, but does not represent the stated problem nearly as well.

B. Express the product of two consecutive odd integers.
Let $x =$ the first odd integer.
Then $x + 2 =$ the next consecutive odd integer.
The product is $x(x + 2)$

C. Find the difference between two consecutive even integers.
Let $N =$ the first even integer.
Then $N + 2 =$ the next consecutive even integer.
Their difference is $(N + 2) - N$. $\qquad \blacktriangleleft$

Solving General Word Problems

It is difficult to overemphasize the importance of learning to solve stated word problems. In a typical job situation, a person is seldom asked by his or her

2.7 Applications

supervisor to solve a particular *equation*. Rather, the requirement is usually to solve a particular *problem*, such as to determine how much material is required to build a part, to compute the maximum amount of current that will flow in a circuit, or to determine how much a beam will expand as the temperature changes, and so on. In other words, the employee must understand the problem, be able to determine precisely what is to be found, write an equation representing the problem, and solve it. The solution must then be checked to see if it is correct.

In this section, you are asked to solve stated word problems. Some of the problems are simplified and do not necessarily represent job-related applications. The primary goal of solving such contrived problems is to give you practice in reading a problem, translating it to an equation, solving the equation, and checking your solution.

You are free to select any variable you wish to represent any of the unknowns. It is often the case that two people solving the same problem (both correctly, of course) will write equations looking quite different. Since the emphasis here is in learning to write an equation from a stated problem, you *must* practice writing and solving an equation rather than attempting to solve the problem by guessing the solution or by using arithmetic solutions. Some of the problems of this section may be fairly easy to solve by the guess method, but doing so will only slow the learning process. The student who solves simple problems by guessing will find he or she is unable to solve more challenging problems. It is the task of learning to express a stated problem as an equation that should be your first concern.

Each of the problems to be solved may be a bit different from the others, but a few generalities apply. Carefully read the following steps and notice how they are applied in the example problems in this section.

Steps in solving word problems

1. Read the problem <u>carefully</u>! Be sure that you understand what facts are given and exactly what you are to find.

2. Jot down all the given facts and/or draw a sketch.

3. Identify all the unknowns that you are to find in the problem. Let one of the unknowns be represented by a variable. In setting up a problem, you will probably find that there is often a best choice of which unknown is assigned the variable.

4. Express the other unknowns *in terms of this same variable*.

5. Write an equation from the facts of the problem. Often a form of the verb "to be" (is, are, were, was, etc.) is in the problem and is replaced by the "=" sign.

6. Solve the equation for the variable, then express the values of all unknowns.

7. Check the solution by verifying that the original problem is solved. It is not enough to check the solution in the equation you wrote, since this equation itself may not correctly represent the problem.

In the remaining examples of this section, several problems have been worked out in relative detail. Study them carefully before beginning the exercises.

Example 5

$x = \text{Sam} \qquad 2x = \text{Sam's father} \qquad x + 2x = 63$

Sam's father is twice as old as Sam. The sum of their ages is 63. How old is each?

Let S = Sam's age.

Then $2S$ = Sam's father's age.

$S + 2S = 63$ Form the equation

$3S = 63$

$S = 21$ Sam's age

$2S = 42$ His father's age ◀

Example 6

A play was attended by 120 people. If student tickets cost $0.75, and adult tickets cost $1.25, and the total collected from ticket sales was $113.50, how many adults and students were there?

Let s = the number of student tickets.

Then $120 - s$ = the number of adult tickets.

Since *each* student paid 75 cents, the total collected from all the students was $75s$ cents. Similarly, the total collected from all of the adults was

$125(120 - s)$ cents

The total amount of money collected was $113.50 (or 11,350 cents). Write and solve the equation:

$75s + 125(120 - s) = 11{,}350$

$75s + 15{,}000 - 125s = 11{,}350$ Distribute the 125

$15{,}000 - 50s = 11{,}350$ Combine like terms

$-50s = -3650$ Subtract 15,000 from both sides

$s = 73$ students   Divide both sides by -50

$120 - s = 47$ adults

Check:

   Students: 73   73 @ $ .75 = $54.75

   Adults: + 47   47 @ $1.25 = $58.75

                 120                  $113.50 total ✓ ◀

2.7 Applications

Example 7

Your father leaves on a trip at 10:00 A.M. You know that he generally averages 40 mph. One hour later you see that he has left without his briefcase. If you attempt to catch him by driving at 50 mph, at what time can you expect to catch him? How far will you have traveled?

Let x = the number of hours traveled by your father.

$x - 1$ = the number of hours you travel (why?).

$40x$ = distance traveled by your father ($d = rt$).

$50(x - 1)$ = your distance traveled.

Since you will both have gone the same distance when you finally catch him, write the equation:

$$40x = 50(x - 1)$$

Then solve:

$40x = 50x - 50$ Distribute the 50

$-10x = -50$ Subtract $50x$ from both sides

$x = 5$ hours traveled by your father

Five hours after 10:00 A.M., it will be 3:00 P.M. The distance traveled will be father's rate times the time, which is $(40) \cdot 5 = 200$ miles. (Note you may also multiply *your* rate times *your* time, giving the same distance.) ◀

Example 8

Find three consecutive integers whose sum is 54.

Let x = the first integer.

$x + 1$ = the second integer.

$x + 2$ = the third integer.

Write the equation:

$$x + (x + 1) + (x + 2) = 54$$

Solve:

$x + x + 1 + x + 2 = 54$ Remove parentheses

$3x + 3 = 54$ Combine like terms

$3x = 51$

$\left. \begin{array}{l} x = 17 \\ x + 1 = 18 \\ x + 2 = \underline{19} \end{array} \right\}$ Solution

Check:

Total = 54 ✓ ◀

Example 9

A 12-foot pipe is cut so that the long piece is 6 inches shorter than twice the length of the short piece. How long is each piece?

Let x = the length of the short piece in *inches*.

$2x - 6$ = the length of the long piece in inches.

Write and solve the equation expressing all lengths in inches. The equation will state that the length of the short piece plus the length of the long piece equals the total length.

$x + (2x - 6) = 144$	(Note: 12 feet = 144 inches)
$x + 2x - 6 = 144$	Remove parentheses
$3x = 150$	Combine like terms
$x = 50$ in. = 4 ft. 2 in.	Short piece
$2x - 6 = 94$ in. = 7 ft. 2 in.	Long piece ◀

Example 10

A boy is 20 years younger than his father. In three years the father will be three times as old as the boy. How old is each now?

Let a = the son's age *now*.

$a + 20$ = the father's age *now*.

$a + 3$ = the son's age *three years from now*.

$(a + 20) + 3$ = the father's age *three years from now*.

Father's age in three years = 3 · (son's age in three years)

$(a + 20) + 3 = 3(a + 3)$	
$a + 20 + 3 = 3a + 9$	Remove parentheses
$a + 23 = 3a + 9$	Combine like terms
$23 = 2a + 9$	Subtract a from both sides
$14 = 2a$	Subtract 9 from both sides
$a = 7$	Son's age *now*
$a + 20 = 27$	Father's age *now*

Check: In three years they will be 10 and 30 years old, and $30 = 3 \times 10$. ◀

In these examples, much care is taken in assigning the variable to one of the unknowns in the problem. Then all other unknowns in the problem are conveniently represented using the same variable. This care makes it easier to write the corresponding equation.

The solution to some word problems may be more appropriately determined by writing and solving a first-degree inequality.

2.7 Applications

Example 11

Mr. Jones has invested $5000 at 8 percent simple interest. How long must he leave his money invested to earn interest of <u>at least</u> $1600?

The simple interest formula is given by $I = PRT$, and in this problem $P = \$5000$ and $R = 8$ percent $= 0.08$.

To earn interest of at least $1600, we want $I \geq \$1600$. That is,

$$PRT \geq 1600$$

We now substitute and solve this inequality.

$$PRT \geq 1600$$
$$(5000)(.08)T \geq 1600 \quad \text{Substitute}$$
$$400T \geq 1600 \quad \text{Multiply}$$
$$T \geq \frac{1600}{400}$$
$$T \geq 4 \quad \text{Divide}$$

Mr. Jones must leave his money invested for at least 4 years to earn the desired interest. ◀

Exercise Set 2.7

1. Given some number N, express the number that is:
   a. 3 more than N
   b. 4 less than N
   c. twice N
   d. 3 less than twice N

2. Given some number A, express the number that is:
   a. A increased by 5
   b. smaller than A by 37
   c. half of A
   d. 8 divided by A

3. The height of a building is given as T feet. Express the height of another building that is:
   a. 20 feet taller
   b. 10 feet shorter
   c. 3 times as tall
   d. only one-third as tall

4. The height of a student is given as H inches. Express the height of another student who is:
   a. 4 inches taller
   b. 6 inches shorter
   c. 1 foot taller
   d. 2 feet shorter

TWO   Linear Equations and Inequalities in One Variable

5. Express the value *in cents* of:
   a. 3 dimes
   b. 6 nickels
   c. 2 dollars
   d. $28.45
   e. d number of dimes
   f. n number of nickels
   g. $(20 - c)$ dollars

6. Express the value *in cents* of:
   a. 13 dimes
   b. 21 quarters
   c. $(25 - x)$ number of dimes
   d. $2x$ quarters
   e. $25.90
   f. $x + 2$ number of dollars

7. Express the value *in dollars* of:
   a. 5 percent of $500.00
   b. 72 quarters
   c. 8 tickets @ $3.75
   d. 5 tickets at $4.25
   e. 4 percent of x dollars
   f. P dollars plus 6 percent of P

8. Express the value *in dollars* of:
   a. 45 cents
   b. x number of cents
   c. $4\frac{1}{2}$ percent of 4000 dollars
   d. 5 percent of $300
   e. N tickets @ $4.50
   f. T tickets at 75 cents

9. How far will a car travel if its average speed is 40 miles per hour and it travels for:
   a. 3 hours
   b. $4\frac{1}{2}$ hours
   c. 90 minutes
   d. 22 minutes
   e. x number of hours
   f. z number of minutes

10. How far will a jet plane travel if its average speed is 550 miles per hour and it travels for:
    a. 6 hours
    b. $3\frac{1}{2}$ hours
    c. 80 minutes
    d. 22 minutes
    e. x number of hours
    f. z number of minutes

11. Three years ago, how old was a person whose present age is:
    a. 12 years
    b. 22 years
    c. $(12 - x)$ years
    d. $(x - 12)$ years
    e. $(A + 5)$ years
    f. $(2A - 6)$ years

2.7 Applications

12. In five years, what will be the age of a person whose present age is:
    a. 15 years
    b. 22 years
    c. Y years
    d. $x + 5$ years
    e. $x - 3$ years
    f. $2F + 4$ years

13. Dan's age is d years. Express the age of a person:
    a. 5 years older than Dan
    b. 3 years younger than Dan
    c. 2 years less than twice Dan's age
    d. 4 years more than half Dan's age

14. Tom's age is given as T years. Express the ages of his family members whose ages are:
    a. 2 years older
    b. half as old
    c. three times as old
    d. 3 years less than double Tom's

15. Given an even number E, express:
    a. the next two even numbers
    b. triple E
    c. the next consecutive number
    d. the difference between E and 7
    e. 5 more than half of E
    f. E increased by $(x + 7)$

16. Given an odd number N, express:
    a. the next consecutive number
    b. the next consecutive odd number
    c. double N
    d. 6 more than N
    e. 3 less than four times N

17. A person has 20 coins, some of them are nickels, and the rest are dimes. If n represents the number of nickels, then express:
    a. the number of dimes

b. the value of the nickels (in cents)

c. the value of the dimes (in dollars)

18. A person purchased some 25- and some 30-cent stamps. If 40 stamps were purchased and N was the number of 25-cent stamps, then express:

    a. the number of 30-cent stamps

    b. the value of the 25-cent stamps (in cents)

    c. the value of the 30-cent stamps (in dollars)

19. A play was attended by some adults, whose number was 15 less than twice as many students. The cost of admission was $.75 per student and $2.00 per adult. Let S represent the number of students. Then express:

    a. the number of adults

    b. the amount (in cents) received from the students

    c. the amount (in dollars) received from the adults

    d. the total amount collected (in dollars)

    e. the total amount collected (in cents)

20. A ball game was attended by 336 people. Let n represent the number of children at $1.50. If adults each paid $3.50, then express:

    a. the number of adults

    b. the amount (in cents) collected from the children

    c. the amount (in dollars) collected from the adults

    d. the total amount collected (in dollars)

    e. the total amount collected (in cents)

21. One number is 6 more than a second number and their sum is 38. What are the two numbers?

22. Find three consecutive odd numbers whose sum is 57.

23. The home team outscored the visiting team by 6 points in a certain football game. If the two teams scored a total of 76 points, what was the final score of the game?

24. A sport coat and a pair of slacks cost $75. If the coat cost $5 more than triple the cost of the slacks, how much was each?

25. A basketball team scored 67 points in a game. If the number of field goals was one more than twice the number of free throws, then how many of each were scored? (Each field goal is worth 2 points: each free throw worth 1.)

2.7 Applications

26. In a certain basketball game, a team scored 75 times for a total of 139 points. How many field goals and how many free throws were scored?

27. A calculator and a trigonometry book together cost $43. If the cost of the calculator was $5 more than the text, how much did each cost?

28. A family was preparing to leave on a three-day trip of 1350 miles. How far would they travel each day if they decided that each day they would go 50 miles less than on the preceding day?

29. A rectangular field requires 460 feet of fencing. If the length is 20 feet longer than the width, what are the dimensions of the field? Use $P = 2L + 2W$.

30. Part of $10,000 is invested at 6 percent and the rest at 8 percent. If the total annual interest is $740, how much is invested at each rate? Use $I = PRT$; for 1 year, $I = PR$.

31. Part of $3300 is invested at 5 percent and the rest at 6 percent. If the annual interest is the same from the investments, then how much is invested at each rate?

32. Part of $5000 is invested at 4 percent and the rest at 5 percent. The annual interest on the 5 percent was 124 dollars more than the interest on the 4 percent investment. How much was invested at each rate?

33. The sum of the ages of a boy and his sister, who is three years older, is the same as their mother's age, 39. How old is the boy?

34. A man is three times as old as his son. Four years ago he was four times as old. How old is each now?

35. A woman is 25 years older than her daughter, and the sum of their ages is 41. How old is each?

36. A board 39 meters long is to be cut into three pieces, each 1 meter longer than the previous one. How long will each piece be?

37. A woman's salary was increased by 10 percent to $1430 per month. What was her monthly salary before the increase?

38. A total of 20 nickels and dimes has a value of $1.75. How many coins of each type are there?

39. Some 15-cent and 25-cent stamps were purchased for a total cost of $11.50. If 50 stamps were purchased, how many of each type were there?

40. How many dimes and quarters would you have if you had four more quarters than dimes and their total value was $3.80?

41. Mr. Smith begins a trip at 8:00 A.M. One hour later, Mr. Jones leaves the same place traveling the same route. If Mr. Smith averages 40 mph and Mr. Jones averages 55 mph, then at what time would Mr. Jones pass Mr. Smith?

42. Towns A and B are 905 miles apart. Mr. Zimm leaves town A, heading toward town B at 8:00 A.M. Mr. Yam leaves town B at 10:00 A.M., heading toward town A. If Mr. Zimm averages 40 mph and Mr. Yam averages 35 mph, then at what time will they pass each other?

43. Mr. Poff bought some 25-cent and 30-cent stamps. The number of 25-cent stamps was two more than the number of 30-cent stamps, and the total cost was $8.20. How many of each kind did he buy?

44. A girl collected $9.95 worth of nickels, dimes, and quarters. She had two more dimes than nickels and twice as many quarters as nickels. How many coins of each type did she have?

45. A school play was attended by 10 more students than adults. The student admission was 50 cents. The adult admission was $1.00 and the total collected was $62.00. How many adults were there?

46. A school play was attended by 100 persons: adults at $1.75 and students at $.75. If the total collected was $102.00, then how many students and how many adults attended?

47. Find three consecutive even numbers with the property that the third is $\frac{4}{3}$ times the first.

48. Find three consecutive odd integers whose sum is 39.

49. Donna invests $10,000 into an account paying 9 percent simple interest. How long must she leave her money invested to earn interest of at least $2700?

50. Meg invests $1000 into an account paying 5 percent simple interest. How long must she leave her money invested to earn interest of at least $250?

51. Mr. Williams must drive from Detroit to Cincinnati, a total of 320 miles. How fast must Mr. Williams drive (on average) to make the trip in no more than 5 hours?

52. Melanie must drive from Cleveland to Fort Lauderdale, a total of 1450 miles. How fast must she drive (on average) to make the trip in no more than 30 hours?

53. The sum of three consecutive odd integers is greater than 75. What are the possible values for the smallest integer? What is the least value the smallest integer can have?

54. The sum of three consecutive even integers is less than 85. What are the possible values for the largest integer? What is the greatest value the largest integer can have?

Review

55. Solve each equation for y:

    a. $3x + y = 5$ b. $x + 3y = 5$ c. $3x - 2y = 9$

56. Solve each of the following equations for y, given $x = 2$:

   a. $2x + y = 9$ **b.** $3x + 2y = 12$ **c.** $\dfrac{x}{2} + \dfrac{y}{3} = 1$

57. Name the property of real numbers illustrated. Assume x represents a real number.

   a. $2 + x = x + 2$ **b.** $x \cdot 3 = 3x$
   c. $(x + 2) + 5 = x + (2 + 5)$ **d.** $3(2x) = (3 \cdot 2)x$

2.8 Review and Chapter Test

(2.1) The distributive law $a(b + c) = ab + ac$ is used to remove parentheses (and other symbols of grouping) in an expression. Mechanically, we remove parentheses as follows:

Parentheses preceded by a "+" are removed by writing each term inside with its sign unchanged.

Parentheses preceded by "−" are removed by writing each term inside with its sign reversed.

Parentheses preceded by a factor are removed by multiplying each term (including its sign) by the factor (including its sign).

When symbols of grouping are contained within other symbols of grouping, the innermost symbols are removed first, continuing from the innermost to the outermost.

Example 1

Remove symbols of grouping:

$$2 - 3[x^2 + (2x - y) - (3a - 5b)] = 2 - 3[x^2 + 2x - y - 3a + 5b]$$
$$= 2 - 3x^2 - 6x + 3y + 9a - 15b \blacktriangleleft$$

The distributive law is used in the form $ba + ca = (b + c)a$ to remove a common factor.

Example 2

Factor:

A. $6x + 6y = 6(x + y)$

B. $ax - 7x = (a - 7)x$ $\blacktriangleleft$

The *terms* of an expression are the parts that are added together. Each term is generally the product of a constant called the *numerical coefficient* and a letter or product of letters called the *literal part*. A term with no literal part is called a *constant term*.

Example 3

Discuss the terms and their parts in the expression $x^3 + 5x^2y - xy^2 + 11$.

Term No.	Term	Numerical Coefficient	Literal Part
1st	x^3	1	x^3
2nd	$5x^2y$	5	x^2y
3rd	$-xy^2$	-1	xy^2
4th	11	11	none

◀

Like terms, also called *similar terms*, of an expression are those terms that have *identical literal parts*. Constant terms are like terms. *Combining like terms* is done by adding the numerical coefficients. The literal part remains unchanged.

Example 4

Combine like terms:

A. $2x - 6y + 3x + 2y = 5x - 4y$

B. $y^2 - 4y + 3 + 2y^2 - y + 7 = 3y^2 - 5y + 10$ ◀

Expressions containing symbols of grouping are simplified by removing grouping symbols and then combining like terms (if there are any).

Example 5

Remove parentheses and combine like terms:

A. $x^2 - (2x + 3) + (3x^2 + x - 4)$

$= x^2 - 2x - 3 + 3x^2 + x - 4$

$= 4x^2 - x - 7$

B. $-3(2a - 3b) + 2(a - 4b)$

$= -6a + 9b + 2a - 8b$

$= -4a + b$ ◀

(2.3) An algebraic equation is a symbolic statement that two algebraic expressions are equal. Further, a variable is contained in at least one of the expressions.

Example 6

A. $\dfrac{x + 3}{2x - 1} = 7$ is an algebraic equation.

B. $7 - 3 = 4$ is *not* an algebraic equation (no variable). ◀

Properties of equality

1. Reflexive Property: If x is any expression, then $x = x$.
2. Symmetric Property: If x and y are any expressions and if $x = y$, then $y = x$.
3. Transitive Property: If x, y, and z are any expressions and if $x = y$ and $y = z$, then $x = z$.
4. Substitution Property: If x and y are any expressions and if $x = y$, then either expression may be substituted for the other in any statement without changing the truth or falseness of the statement.

An equation is an *identity* if it is true for every value of the variable(s) in the equation.

Example 7 $2x - 8 = 2(x - 4)$ is an identity. ◀

A *conditional equation* is a statement that is true for some but not all values of the variable.

Example 8 $3x + 2 = 8$ is a conditional equation since it is true only for $x = 2$. ◀

The values of the variable(s) that make an equation true are the *solutions* or *roots* of the equation.

Example 9   Since the equation $3x + 2 = 8$ is true for $x = 2$, 2 is a root of the equation. ◀

(2.4)   Two or more equations are called *equivalent equations* if they have the same solution(s).

Example 10   The equations $y - 3 = 7$ and $2 + y = 12$ are equivalent since they have the same solution, $y = 10$. (Try it.) ◀

Axioms for equations

Given any algebraic equation, an equivalent equation results if:

1. The same quantity is added to or subtracted from both sides.
2. Both sides are multiplied or divided by the same non-zero quantity.
3. Any quantity is replaced by a quantity to which it is equal.

A linear equation in a single variable x is any equation that can be written in the form $ax + b = 0$ where a and b are real numbers and $a \neq 0$. The steps used to solve a linear equation in one variable use the axioms for equations to rewrite the equation in a simpler equivalent form.

Summary: solving equations

1. Clear fractions by multiplying both sides of the equation by the LCD. If the equation has decimal coefficients, multiply both sides by 10, 100, 1000 to clear the equation of decimals.
2. Remove symbols of grouping, innermost to outermost.
3. Simplify both sides by combining like terms.
4. Gather all terms containing the variable on one side of the equation and all terms without the variable on the other side.

   Note: If Steps 1–4 are carried out properly, the result will be an equation of the form $ax = b$, in which a and b are constants and x is the variable.

5. Divide both sides of the equation by the coefficient of the variable.
6. Check.

Example 11

Solve and check each equation:

A. $3x + 2(2x - 3) = 7 - (3 - 2x)$

$$3x + 2(2x - 3) = 7 - (3 - 2x)$$
$$3x + 4x - 6 = 7 - 3 + 2x \qquad \text{Remove parentheses}$$
$$7x - 6 = 4 + 2x \qquad \text{Combine like terms}$$
$$5x - 6 = 4 \qquad \text{Subtract } 2x \text{ from both sides}$$
$$5x = 10 \qquad \text{Add 6 to both sides}$$
$$x = 2 \qquad \text{Divide both sides by 2}$$

Check:
$$3x + 2(2x - 3) = 7 - (3 - 2x)$$
$$3(2) + 2[2(2) - 3] \stackrel{?}{=} 7 - [3 - 2(2)]$$
$$6 + 2(4 - 3) = 7 - (3 - 4)$$
$$6 + 2(1) = 7 - (-1)$$
$$6 + 2 = 7 + 1$$
$$8 = 8 \checkmark$$

B. $\dfrac{x}{3} + \dfrac{1}{2} = \dfrac{x - 2}{2}$

$$\dfrac{x}{3} + \dfrac{1}{2} = \dfrac{x - 2}{2}$$

$$6 \cdot \left(\dfrac{x}{3} + \dfrac{1}{2}\right) = 6 \cdot \left(\dfrac{x - 2}{2}\right) \qquad \text{Multiply both sides by 6 (LCD)}$$

$$\cancel{6}^2 \cdot \frac{x}{\cancel{3}} + \cancel{6}^3 \cdot \frac{1}{\cancel{2}} = \cancel{6}^3 \cdot \frac{(x-2)}{\cancel{2}} \qquad \text{Distribute 6 on the left side}$$

$$2x + 3 = 3(x - 2)$$

$$2x + 3 = 3x - 6 \qquad \text{Remove parentheses}$$

$$9 = x$$

$$x = 9$$

Check:

$$\frac{x}{3} + \frac{1}{2} = \frac{x-2}{2}$$

$$\frac{9}{3} + \frac{1}{2} \stackrel{?}{=} \frac{9-2}{2}$$

$$3 + \frac{1}{2} = \frac{7}{2}$$

$$3\frac{1}{2} = 3\frac{1}{2} \checkmark$$

C. $4.5x - 0.13 = 2.11x + 5.1$

$$4.5x - 0.13 = 2.11x + 5.1$$

$$100(4.5x - 0.13) = 100(2.11x + 5.1) \qquad \text{Multiply by 100 to clear decimals}$$

$$450x - 13 = 211x + 510$$

$$239x = 523$$

$$x = \frac{523}{239}$$

$$x = 2.19 \qquad \blacktriangleleft$$

(2.5) A *literal equation* is an equation that contains more than one letter (or literal or variable). To solve a literal equation for one of its variables means to find an expression for that variable in terms of the other variables in the equation. A *formula* is a literal equation that represents a relation between physical quantities.

Example 12

Solve the equation $A = \frac{h}{2}(B + b)$ for b:

$$A = \frac{h}{2}(B + b)$$

$$2A = 2\left(\frac{h}{2}\right)(B + b) \qquad \text{Multiply both sides by 2 (LCD)}$$

$$2A = h(B + b)$$
$$2A = hB + hb \qquad \text{Distribute the } h \text{ on the right side}$$
$$2A - hB = hb \qquad \text{Subtract } hB \text{ from both sides}$$
$$\frac{2A - hB}{h} = \frac{\cancel{h}b}{\cancel{h}} \qquad \text{Divide both sides by } h$$
$$b = \frac{2A - hB}{h} \qquad \text{Rewrite}$$
$$\text{or} \quad b = \frac{2A}{h} - B$$

◀

(2.6) Definition of less than (<): $a < b$ if and only if there exists some positive real number p such that $a + p = b$. On the number line, $a < b$ if and only if a is to the left of b.

Property of trichotomy

If a and b are any two real numbers, then one and only one of the following is true:

1. $a < b$,
2. $a = b$, or
3. $a > b$

Transitive Property

If a, b, and c are any real numbers and if $a < b$ and $b < c$, then $a < c$. A statement such as $a < x < b$ means that both $a < x$ and $x < b$. Such a statement is a double inequality. To solve an inequality means to replace it with successively simpler equivalent inequalities until the solution is obvious.

Properties used to solve inequalities

1. The sense of an inequality is *unchanged* whenever:
    a. The same quantity is added to or subtracted from each member.
    b. Each member is multiplied or divided by the same positive quantity.
2. The sense of an inequality is *reversed* whenever:
    a. The members are interchanged.
    b. Each member is multiplied or divided by the same negative quantity.

2.8 Review and Chapter Test

Example 13

Solve and graph each inequality:

A. $3(2x - 1) < -4(3 - x)$

$$3(2x - 1) < -4(3 - x)$$
$$6x - 3 < -12 + 4x \quad \text{Remove parentheses}$$
$$2x - 3 < -12 \quad \text{Subtract } 4x \text{ from each member}$$
$$2x < -9 \quad \text{Add 3 to each member}$$
$$x < -\frac{9}{2} \quad \text{Divide each member by 2}$$

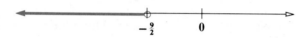

B. $-4 \leq 3x - 7 < 2$

$$-4 \leq 3x - 7 < 2$$
$$3 \leq 3x < 9 \quad \text{Add 7 to each member}$$
$$1 \leq x < 3 \quad \text{Divide each member by 3}$$

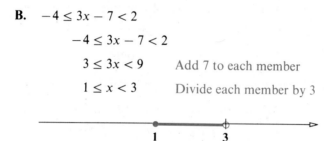

(2.7) Steps in solving applied problems:

1. Read the problem carefully! Be sure that you understand what facts are given and exactly what you are to find.

2. Jot down all the given facts and/or draw a sketch.

3. Identify all the unknowns in the problem that you are to find. Let one of the unknowns be represented by a variable. In setting up a problem, you will probably find that there is often a best choice of which unknown is assigned the variable.

4. Express the other unknowns *in terms of this same variable*.

5. Write an equation or inequality from the facts of the problem. Often a form of the verb "to be" (is, are, were, was, etc.) is in the problem and is replaced by the "=" sign.

6. Solve the equation for the variable; then express the values of all unknowns.

7. Check the solution by verifying that the original problem is solved. It is not enough to check the solution in the equation you wrote, since this equation itself may not correctly represent the problem.

Example 14

A concert is attended by 350 people. Let s represent the number of students attending at a ticket price of $2.50 each. If non-students each paid $4.50, then express:

A. The number of non-students:

$350 - s$ non-students

B. The amount collected (in dollars) from the students:

Amount = (Number of students) · (Price per student)

$= s(\$2.50)$

$= 2.50s$ dollars

C. The amount collected (in dollars) from the non-students:

Amount = (Number of non-students) · (Price per person)

$= (350 - s)(4.50)$

$= 4.50(350 - s)$ dollars

D. The total amount collected (in dollars):

Amount = (Amount from students) + (Amount from non-students)

$= 2.50s + 4.50(350 - s)$ dollars

$= 2.50s + 2025 - 4.50s$ dollars

$= 2025 - 2s$ dollars ◀

Example 15

Find three consecutive even integers whose sum is 72.

Let $\quad x =$ first even integer.
$x + 2 =$ second consecutive even integer.
$x + 4 =$ third consecutive even integer.

Then,

$x + (x + 2) + (x + 4) = 72$

$x + x + 2 + x + 4 = 72$

$3x + 6 = 72 \quad$ Combine like terms

$3x = 66 \quad$ Subtract 6 from both sides

$\left. \begin{array}{l} x = 22 \\ x + 2 = 24 \\ x + 4 = 26 \end{array} \right\}$
Divide both sides by 3

Solution

Check: $\quad 72 \checkmark \quad$ Add ◀

Exercise Set 2.8

Factor each expression, removing the factor indicated:

1. $2x^2 + yx^2$; x^2
2. $6a - ax$; a
3. $5m^2 + m$; m
4. $xy - y$; y

Remove parentheses and combine like terms:

5. $x^2 - x(x + 3)$
6. $2x + x(x - 4)$
7. $3x + 2(x - 1)$
8. $4y - 2[y - 3(y + 2) + 3]$

Show that the number given is a root of the equation:

9. $(x + 1)(2x - 3) = 4x - 5$; 2
10. $x^2 - 3x - 10 = 0$; -2

State three pairs of values of x and y that solve each equation:

11. $x + 2y = 7$
12. $x - 1 = y + 1$

Substitute several values for the variable and state whether each equation is a conditional equation or an identity:

13. $2x^2 - 18 = 0$
14. $2x - 5 = x - 2 - (3 - x)$

Solve each of the following equations:

15. $3x + 2 = 7 - x$
16. $5 - 5y = 2y - 6$
17. $5(3m - 2) = -3(2m + 3)$
18. $2y - (4 - 3y) = 5 - 3y$
19. $4 - [3m - (2 - m)] = -4(2m + 3)$
20. $\dfrac{x}{3} + 2 = \dfrac{2x}{5}$
21. $\dfrac{3a}{5} - \dfrac{1}{2} = \dfrac{a}{3}$
22. $3.2 - 4.6y = 4.2(5 - 3y)$
23. $\dfrac{(a - 2)}{4} - 3 = \dfrac{(a + 1)}{3}$
24. $7.3(1.4 - 0.5x) = 2.2(3.1x - 2.5)$

Solve each of the following literal equations for the indicated variable:

25. $ax + by + c = 0$ for y
26. $I = PRT$ for P
27. $S = \dfrac{N(A + L)}{2}$ for A
28. $I = \dfrac{E}{R_1 + R_2}$ for R_2
29. $\dfrac{a}{b} = \dfrac{c}{d}$ for b

Solve and graph each of the following inequalities:

30. $3x - 5 < 8$

31. $4 - 2x \geq 5$

32. $3x + 7 \leq -2x - 3$

33. $\dfrac{x}{5} - \dfrac{2}{3} \leq \dfrac{x}{3}$

34. $-4 < 3x - 2 < 6$

35. $-3 < 4 - 3x \leq 3$

36. A jacket and pair of trousers together cost $70. If the cost of the jacket was $14 more than the trousers, how much did each cost?

37. A woman is 25 years older than her son and the sum of their ages is 61. How old is each?

38. A total of 18 quarters and nickels has a value of $3.10. How many coins of each type are there?

39. Part of $10,000 is invested at 8 percent and the rest at 11 percent. The annual interest on the 11 percent investment was $435 more than the annual interest on the 8 percent investment. How much was invested at each rate?

40. The sum of four consecutive integers is at least 82. What are the possible values for the smallest of these integers? What is the least value the smallest integer can have?

Chapter 2 Review Test

1. Factor by removing the indicated factor:

    a. $15x + 25y$; 5
    b. $2x^2 - 8x$; $2x$
    c. $4x^2 + x$; x

2. Remove symbols of grouping and combine like terms:

    a. $(6b + 3) + (b - 4) - (2b - 5)$
    b. $(y^2 - 5y + 3) - 3(y^2 + 2)$
    c. $4m - 2[m - 3(m + 4)]$

3. a. Find LCM (8, 12, 6)

    b. Multiply and simplify $24\left(\dfrac{5}{8} - \dfrac{x}{12} + \dfrac{1}{6}\right)$

    c. Multiply $100(2.7 - 4.01y)$

4. Solve each of the following equations:

    a. $3x - 4 = 11$
    b. $4x - 3 = 17 - x$

2.8 Review and Chapter Test

    c. $3(m - 2) - 1 = 2m - 3$ **d.** $\dfrac{x}{2} + 3 = \dfrac{x}{4} + 1$

    e. $p + 0.08p = 486$

5. Solve each equation for the indicated variable:

    a. $3x - 5y = 12$ for y

    b. $A = P + PRT$ for P

    c. $3(m - 2) = 2(a + 3)$ for a

6. Graph each inequality:

    a. $u \leq 6$ **b.** $y > -8$ **c.** $-4 \leq x < \dfrac{5}{2}$

7. Solve and graph each inequality:

    a. $2m - 3 \leq 12 - m$

    b. $3(u - 5) + 4 > 2(u - 1) - 8u$

    c. $\dfrac{t}{3} + 1 \geq 2t - \dfrac{5}{6}$

    d. $-8 < 5x + 2 < 17$

    e. $2 \leq 3 - \dfrac{x}{2} \leq \dfrac{19}{2}$

8. If P dollars are invested for 1 year at 8 percent annual interest, interest earned will be 8 percent of P, or $0.08P$.

    a. Write the equation expressing the value of the account, A, at the end of 1 year.

    b. If $2000 is invested, how much is in the account at the end of 1 year?

    c. Find the amount invested if the balance at the end of 1 year is $4104.

9. The length of a certain rectangle is 2 feet less than triple the width. Find the length and width, given the perimeter is 68 feet. (Hint: $P = 2L + 2W$)

10. Some 25-cent and 15-cent stamps were purchased for $6.20. If 28 stamps were purchased, how many of each denomination were there?

CHAPTER THREE

LINEAR EQUATIONS AND INEQUALITIES IN TWO VARIABLES

The preceding chapter discussed solving equations and inequalities that could be written with one variable. There are many applications that express a relation between *two* quantities, such as relating the distance an object falls to the time it falls, relating the time required to stop an automobile to its speed, relating the price of an object to the demand for that object, and relating the pressure of a gas in a fixed volume to its temperature. This chapter discusses drawing graphs that show linear relationships between two quantities.

3.1 The Cartesian Plane

Suppose you were given the problem of finding two numbers such that the second number subtracted from the first gives 3. You could probably find several pairs of numbers that work, such as 5 and 2, 10 and 7, 0 and -3, and so on. If we use variables to represent the two unknown quantities, the preceding problem can be expressed as an equation. If we let x represent the first number and let y represent the second number, then the problem can be represented by the equation $x - y = 3$. A *solution* of an equation with two variables is any pair of numbers that satisfies the equation.

Each solution may be shown as an *ordered pair*, (a, b). The number a, called the *first coordinate*, represents the first number, and b, called the *second coordinate*, represents the second number. When the variables used are x and y, it

3.1 The Cartesian Plane

is common practice to assume that x is the first and y is the second number. Some solutions of the equation $x - y = 3$ shown as ordered pairs are (5, 2), (10, 7), (0, −3), etc.

Ordered pairs of numbers can be shown as points in a plane. Such a plane is called the **Cartesian coordinate system** and is constructed as follows. Draw a horizontal number line and a vertical number line that intersect at their origins. The point of intersection is called the **origin** of the system. On the horizontal number line (called the **horizontal axis**), draw uniformly spaced marks to the right of the origin to represent positive numbers and to the left to represent negative numbers. Then draw marks upward on the **vertical axis** to represent positive numbers and downward to represent negative numbers. These two axes are collectively called the **coordinate axes**. The length of segment used to represent one unit determines the **scale**, which is generally (although not always) the same on the horizontal and vertical axes.

The coordinate axes divide the plane into four regions called **quadrants**. The quadrants are designated I, II, III, and IV as shown in Figure 3.1-1.

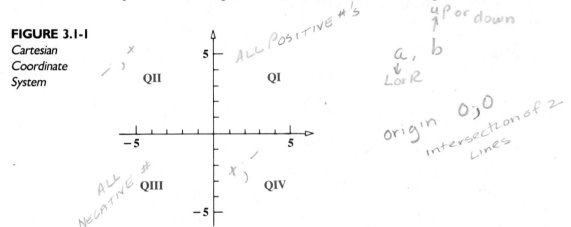

FIGURE 3.1-1
Cartesian Coordinate System

To graph the ordered pair (a, b) in this system, count a number of units along the horizontal axis (to the right if positive, to the left if negative), and then count b number of units vertically (up if positive, down if negative). Place a dot at that location and write the ordered pair or a letter near the dot to name the point.

Note that we are using this system to represent pairs of numbers (an algebra concept) with points in the plane (a geometry concept). The dot corresponding to the ordered pair of numbers is called the **plot** or the **graph** of the ordered pair. The numbers of the ordered pair corresponding to the point are called the **coordinates** of the point. The first number of the ordered pair is the **abscissa**, and the second number is the **ordinate**.

144   THREE   Linear Equations and Inequalities in Two Variables

Example 1

Graph each ordered pair:

A. (3, 0)
B. (−1, −3)
C. (0, −2)
D. (2, 2)
E. (0, 1)
F. (2, −1)
G. (−3, 0)
H. (−3, 2)

Solution:

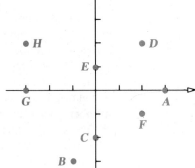

Example 2

List the coordinates of the points A, B, C, D, E, F, G, and H of Figure 3.1-2.

FIGURE 3.1-2

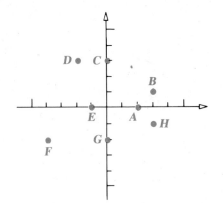

Solution:

A. (2, 0)
B. (3, 1)
C. (0, 3)
D. (−2, 3)
E. (−1, 0)
F. (−4, −2)
G. (0, −2)
H. (3, −1)

When the Cartesian coordinate system is used to graph ordered pairs of numbers that are solutions to an equation in x and y, the horizontal axis is used to

FIGURE 3.1-3
x-y Plane

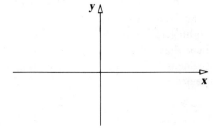

3.1 The Cartesian Plane

represent the x-values and is called the *x-axis*. Similarly, the vertical axis is used to represent the corresponding y-values, and is called the *y-axis*. When this notation is used, the Cartesian coordinate system is called an *x-y plane*. See Figure 3.1-3. When variables other than x and y are used, there must be some agreement as to which of the axes is assigned to each variable. Regardless of which variables are used, the first coordinate of an ordered pair (the abscissa) always corresponds to the value on the horizontal axis; the second component (the ordinate) corresponds to the value on the vertical axis.

The Cartesian coordinate system is also called the *rectangular coordinate system*, the *rectangular coordinate plane*, the *Cartesian plane*, and even combinations of these names.

Exercise Set 3.1

1. On the same coordinate system, graph each ordered pair:
   a. $(2, -3)$ b. $(-2, 3)$ c. $(-2, -3)$ d. $(2, 3)$ e. $(3, 0)$

2. On the same coordinate system, graph each ordered pair:
   a. $(1, 5)$ b. $(1, -5)$ c. $(-1, 5)$ d. $(-1, -5)$ e. $(-1, 0)$

3. On the same coordinate system, graph each ordered pair:
   a. $(2, 5)$ b. $(-3, 1)$ c. $(-2, -4)$ d. $(0, 5)$ e. $(3, -2)$

4. On the same coordinate system, graph each ordered pair:
   a. $(5, 2)$ b. $(1, -3)$ c. $(-4, -1)$ d. $(-2, 0)$ e. $(-6, 1)$

5. Find the coordinates of each point shown:

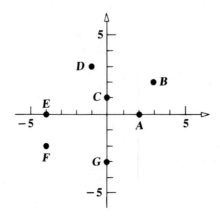

6. Find the coordinates of each point shown:

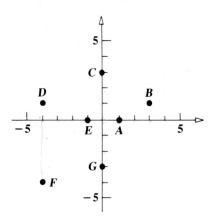

7. In which quadrant would the graph of each of the following ordered pairs be plotted?

   a. (4, 2)   b. (4, −2)   c. (−4, −2)   d. (−4, 2)

8. In which quadrant would the graph of each of the following ordered pairs be plotted?

   a. (5, 3)   b. (−5, −3)   c. (−5, 3)   d. (5, −3)

9. What are the signs of the first and second coordinates of ordered pairs whose graph is in quadrant:

   a. I   b. II   c. III   d. IV

For each of the following equations, find five solutions and plot them in the Cartesian plane:

10. $2x + y = 9$
11. $x + 2y = 12$
12. $3x - 2y = 12$
13. $2x + 3y = 6$

14. What sort of points, if any, do not belong in any of the four quadrants?

15. How many solutions exist for the equation $x + y = 2$?

16. In what quadrants will the abscissa and ordinate of an ordered pair have the same sign? Have opposite signs?

Review

17. Solve the equation $2x + y = 5$ for y, given x has the indicated value:

    a. $x = 0$ b. $x = 3$ c. $x = -2$

18. Solve the equation $2x - 3y = 12$ for y, given x has the indicated value:

    a. $x = 0$ b. $x = 6$ c. $x = -3$

19. Draw a number line and plot the points:

    a. 0   b. -5 c. 3

20. Jerry has 14 coins with a value of $1.50. If his coins are nickels and quarters only, how many coins of each type does he have?

3.2 Graphs of Linear Equations

We saw in the preceding section that an equation in two variables has many solutions and that each solution can be written as an ordered pair. Each ordered pair can be graphed in the Cartesian plane. The *graph of an equation* is the figure determined by plotting all ordered pair solutions of that equation.

Any equation that can be written in the form $Ax + By = C$, where A, B, and C are real numbers, with not both A and B equal to zero and x and y as variables, is called a **linear equation in two variables**. It can be shown that the graph of a linear equation in two variables is a straight line. Let's discuss how to draw the graph of the equation $x - y = 3$, mentioned in the preceding section. Since this equation is already written in the form $Ax + By = C$ ($A = 1, B = -1, C = 3$), its graph is a straight line. Only two points are needed to determine a straight line, but we will plot three points representing solutions to the equation and verify that they lie in a straight line.

Rather than write ordered pairs, we prefer to show solutions in a table, as in Figure 3.2-1. To prepare such a table (other than simply by "guess") we select arbitrary values for one of the variables, then for each value selected, compute the corresponding value of the other variable. For example, if we select $x = 5$, then substituting into the equation gives:

FIGURE 3.2-1
Table of Some Solutions of $x - y = 3$

x	y
5	2
10	7
0	-3

$x - y = 3$
$5 - y = 3$ Substitute $x = 5$
$-y = -2$ Subtract 5
$y = 2$ Divide by -1

Therefore, (5, 2) is one solution. The other entries can be found in the same manner. Now each pair of table values corresponds to an ordered pair. Plot each of these in an x-y plane. See Figure 3.2-2.

FIGURE 3.2-2
Graph of
$x - y = 3$

x	y
5	2	$\longrightarrow (5, 2)$
10	7	$\longrightarrow (10, 7)$
0	−3	$\longrightarrow (0, -3)$

Plot these points

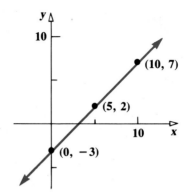

Now with a straight edge, draw the straight line that passes through these three points. Draw arrowheads on the ends of the line to indicate that it extends indefinitely in both directions. The process is summarized as follows.

To draw the graph of the linear equation in two variables $Ax + By = C$:

1. Prepare a table of three (or more) solutions. Select arbitrary values for x. For each value chosen, compute the corresponding value of y.

2. Draw a Cartesian plane. Assign x to the horizontal axis and y to the vertical axis.

3. Plot the ordered pairs determined in Step 1.

4. Draw the straight line through these points. Place an arrowhead at each end of the line.

It should be emphasized that the values selected for the first variable, usually x, are completely arbitrary. You may choose any values you find convenient. Since each equation of this type has a graph that is a straight line, it is only necessary to plot two points to draw the line. It is common practice, however, to graph a third point as a check.

Example 1

Draw the graph of the linear equation $2x + 3y = 9$.

Determine three x- and y-values; enter them in a table; plot these three points; then draw the straight line through them.

If $x = 0$:

$2 \cdot 0 + 3y = 9$

$3y = 9$

$y = 3$

If $x = -3$:

$2(-3) + 3y = 9$

$-6 + 3y = 9$

$3y = 15$

$y = 5$

If $x = 3$:

$2(3) + 3y = 9$

$6 + 3y = 9$

$3y = 3$

$y = 1$

3.2 Graphs of Linear Equations

x	y
0	3 → (0, 3)
−3	5 → (−3, 5)
3	1 → (3, 1)

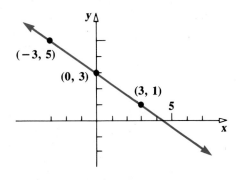

Some comments on graphs of linear equations:

1. If the variables chosen are x and y, the horizontal axis is used for x-values and is called the x-axis, and similarly the vertical axis is used for y-values and is called the y-axis. For simplicity, the Cartesian plane is then called the x-y plane.

2. Arrowheads are drawn at both ends of the graph of the line to indicate that the line extends without end in both directions.

3. A very important relation exists between the equation and its graph: *any point on the graph has coordinates that satisfy the equation, and any x- and y-values that satisfy the equation are the coordinates of a point on the graph.*

Example 2

Draw the graph of the equation $x - 2y = -8$.

x	y
0	4
−4	2
−8	0

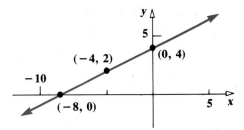

Example 3

Draw the graph of the equation $P = 3m + 8$.

In this equation, since P can easily be computed if a value of m is given, we select m as our first variable and prepare the following table.

m	P
0	8
2	14
4	20

Now, since the values of *P* are quite large, we select different scales for the horizontal (*m*) and vertical (*P*) axes, as shown in Figure 3.2-3.

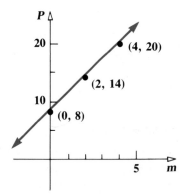

FIGURE 3.2-3
Graph of
$P = 3m + 8$

Graphs of Equations of the Type $x = k$ or $y = k$

If an equation contains only one variable, it may be written in the form $x = k$ or $y = k$, where *k* is a constant. We defined a linear equation as one that can be written in the form $Ax + By = C$ in which *not both A and B* are zero. Each of the equations $x = k$ and $y = k$ fits this definition, so each is linear and has a straight line graph.

Consider first the equation $x = 4$. How can the table of *x*- and *y*-values be prepared? The variable *y* isn't even in the equation! The equation $x = 4$ means that *x* has the value 4 regardless of the value of *y*. So we might fill out a table of values as follows:

x	y
4	0
4	−5
4	4
4	10

Of course any other values of *y* may be selected, but this will give enough points to plot and draw the graph. The points will all lie along a *vertical* line parallel to the *y*-axis. In other words, the graph is a picture of all of the points that have an *x*-coordinate of 4.

3.2 Graphs of Linear Equations

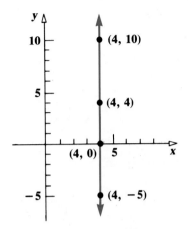

Similarly, an equation that may be written in the form $y = k$ would have as its graph the *horizontal* line made up of all points with the same y-value, k. The equation $y = -2$ states that each point has a y-value of -2 regardless of the value of x.

Example 4

Draw the graph of each of the equations:

A. $x = -3$

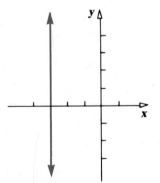

B. $2x - 5 = 0$

$$2x - 5 = 0$$
$$2x = 5$$
$$x = \frac{5}{2}$$

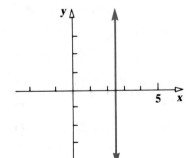

C. $y = 4$

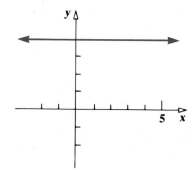

D. $3y + 7 = 0$

$3y + 7 = 0$
$3y = -7$
$y = \dfrac{-7}{3}$

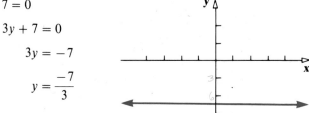

Note in parts B and D that it was necessary to isolate the variable to put the equation in the correct form.

In equations of this type, no table of x- and y-values should be needed.

Exercise Set 3.2

For each of the following equations, make a table of solutions containing three ordered pairs by selecting arbitrary values of x and computing the corresponding value of y for each. Draw the graph:

1. $2x + y = 6$
2. $3x + y = 6$
3. $2x - y = 8$
4. $3x - y = -6$
5. $x + 2y = 10$
6. $x - 3y = 9$
7. $2x + 3y = 12$
8. $3x - 2y = -12$
9. $5x - y = -10$
10. $6x + y = -6$
11. $3x - 5y = 15$
12. $4x - 2y = 3$
13. $3x - 4y = 24$
14. $x + y = 7$
15. $y = 2x + 4$
16. $y = 2x + 5$
17. $y = \dfrac{2}{3}x + 5$
18. $y = \dfrac{1}{2}x - \dfrac{7}{2}$
19. $2y = 5x - 12$
20. $5y = x - 15$

3.2 Graphs of Linear Equations

Draw the graph of each equation:

21. $x = -4$
22. $2x - 5 = 0$
23. $y = 6$
24. $y = -1$
25. $2x + 3 = 7$
26. $3y - 4 = 7$

27. Without graphing, tell which of the following points lie on the graph of $2x + 3y = 9$:

   a. (3, 1)   b. (0, −3)   c. (6, −1)   d. (−6, 5)

28. Without graphing, tell which of the following points lie on the graph of $y = 3x - 6$:

   a. (4, 13)   b. (−3, 0)   c. (0, −6)   d. (2, 0)

29. Which of the following are linear equations?

   a. $2x = \dfrac{1}{y}$ b. $xy = 3$
   c. $x = 2y$ d. $x - 2y - 5 = 3$

30. Which of the following are linear equations?

   a. $x^2 = y + 2$ b. $2xy + 4 = x$
   c. $x = 2y + 4$ d. $x - 3y = \dfrac{1}{2}$

31. How many different lines can be drawn through one point? How many different lines can be drawn through two fixed points? In general, how many different lines can be drawn through three arbitrary fixed points?

32. If the abscissa of an ordered pair solution to an equation is zero, where is this point in the Cartesian plane? If the ordinate is zero, where is the point?

Review

33. Given the equation $3x - 2y = 18$:

   a. solve for y b. find y when $x = 0$
   c. find x when $y = 0$

34. Solve the following equations for the variable:

   a. $6a - 3 = 18 - a$ b. $\dfrac{b}{2} + 3 = 1 - \dfrac{b}{3}$
   c. $x + 0.06x = 185.5$

35. Clear symbols of grouping and combine like terms:

   a. $3x - y + 2 - x + 5y - 11$ b. $3(x + 2) - 5(x + 3)$
   c. $2(3a + 5) - [2a - (a - 8)]$

3.3 Further Topics on Graphing

We restate a fundamental relationship that exists between an equation and the graph of the equation:

> Every pair of values that satisfies the equation determines the coordinates of a point on the graph, and
>
> Every point on the graph has coordinates that satisfy the equation.

The first of these statements permits us to use *any* ordered pairs of x and y that satisfy the equation to draw the graph. The second statement permits us to select any point on the line and be assured that its coordinates satisfy the equation.

Graphing by Solving the Equation for y

Any equation in the standard form $Ax + By = C$ with $B \neq 0$ may be rewritten to isolate y, giving an equation of the form $y = mx + b$ where m and b are constants. With the equation written in the form $y = mx + b$, the process of finding the corresponding y-value for a chosen x-value is simplified, and we can often make choices for x that will avoid fractional values of y.

Example 1

Draw the graph of $2x + 5y = 15$ by first rewriting the equation in the form $y = mx + b$.

$$2x + 5y = 15$$
$$5y = -2x + 15 \qquad \text{Add } -2x \text{ to both sides}$$
$$y = \left(-\frac{2}{5}\right)x + 3 \qquad \text{Divide both sides by 5}$$

With the equation in this form it can be seen that values of x that are divisible by 5 will give integer values of y.

x	y
0	3
5	1
−5	5

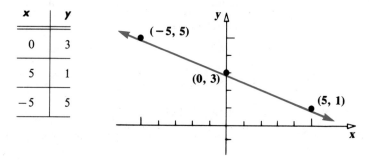

3.3 Further Topics on Graphing

The Intercept Method of Graphing

Drawing the graph of an equation by selecting values of x at random and determining the corresponding values of y can be a somewhat tedious task. Even rewriting the equation in the form $y = mx + b$ does not necessarily remove all of the difficulties.

A graphing method that is often efficient involves finding the points where the line crosses the x- and y-axes. From the preceding sections you have already discovered that a convenient value of x to select is $x = 0$. If this observation is extended to its next logical step, then we may also set $y = 0$ and solve for x. For convenience, when $x = 0$ we will designate the corresponding value of y as b; and when $y = 0$ we will designate the corresponding value of x as a. Thus, the points $(a, 0)$ and $(0, b)$ are the points where the graph crosses the axes, and these points are called the **x-** and **y-intercepts**, respectively. Although $(a, 0)$, and $(0, b)$ are technically the x- and y-intercepts, it is common practice to refer to the x-intercept as a and the y-intercept as b.

The values of a and b can generally be computed mentally if the equation is in the form $Ax + By = C$. Setting $x = 0$, the equation becomes $By = C$, and mentally dividing both sides by B, the equation becomes $y = \frac{C}{B}$, the y-intercept. Similarly, setting $y = 0$ and then dividing both sides by A gives $x = \frac{C}{A}$, the x-intercept. These two points are then plotted and the line is drawn through them.

Example 2

Use the intercept method to draw the graph of $2x + 3y = 12$.

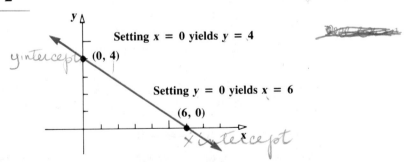

Setting $x = 0$ yields $y = 4$
$(0, 4)$

Setting $y = 0$ yields $x = 6$
$(6, 0)$

Despite the efficiency of this method, it does not work well for all equations. If the equation is of the type $Ax + By = 0$, then letting $x = 0$ and then $y = 0$ in turn gives only the one point $(0, 0)$. So another point must be selected before the graph can be drawn. Another situation in which you may wish to select a point different from the intercepts is one in which the two intercepts are very close together. In this case, another point should be selected to verify the graph.

Example 3

Use the intercept method to draw the graph of $3x - 4y = -2$.

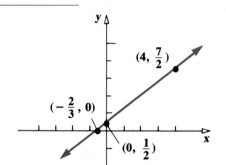

x	y
0	$\frac{1}{2}$
$-\frac{2}{3}$	0
4	$\frac{7}{2}$

Since these values are fractional and close together, we select another value for x, say $x = 4$, then compute $y = \frac{7}{2}$.

In the graphs we have presented, you may have noticed that more accurate graphs are drawn when the two points selected are not extremely close together. Therefore, in Example 3 we picked the value $x = 4$, which is farther from the intercepts.

Exercise Set 3.3

In Problems 1–8, proceed as follows. Select *two* values of x and compute the corresponding values of y. Plot the two points and draw the graph through them. Then select two *other* values of x, compute the corresponding values of y, and plot these two points. Verify that they are also on the graph. Finally, mark some arbitrary point on the graph. Read its coordinates as closely as your graph permits. Substitute the coordinates into the equation. Verify that they satisfy the equation.

1. $-2x + y = 7$
2. $3x - 4y = 12$
3. $4x - y = -8$
4. $x + 5y = 10$
5. $-3x - 5y = 30$
6. $-2x - 4y = -6$
7. $2x - 7y = 0$
8. $-4x + 10y = 0$

Graph each of the following equations by first rewriting in the form: $y = mx + b$.

9. $2x - 3y = 18$
10. $3x + y = 5$
11. $x - y = 5$
12. $x - y = -3$
13. $2x + 3y = 6$
14. $3x + 2y = 6$
15. $x - 3y = 15$
16. $x - 4y = 8$
17. $3x + 4y = 9$
18. $2x - 5y = 8$
19. $4x + 5y = -13$
20. $3x - 7y = 28$

3.3 Further Topics on Graphing

Draw the graph of each of the following equations by the intercept method:

21. $x + y = 3$
22. $x - y = 5$
23. $2x + 3y = 12$
24. $3x - y = 9$
25. $4x + 2y = 9$
26. $3x + 2y = 10$
27. $3x - 8y = 24$
28. $2x - 5y = 15$
29. $\dfrac{x}{2} + \dfrac{y}{3} = 1$
30. $\dfrac{x}{4} + \dfrac{y}{1} = 1$
31. $\dfrac{x}{-2} + \dfrac{y}{-3} = 1$
32. $\dfrac{x}{-3} + \dfrac{y}{4} = 1$

33. Which is the equation of the graph shown?

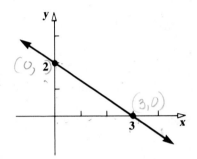

a. $2x - 3y = 1$
b. $3x + 2y = 0$
c. $y = 3x + 2$
d. $2x + 3y = 6$

34. Which is the equation of the graph shown?

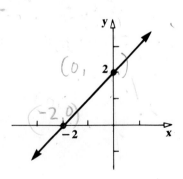

a. $-y = x + 2$
b. $-2x + 2y = 1$
c. $y = x + 2$
d. $x - y = 0$

35. Which is the equation of the following graph?

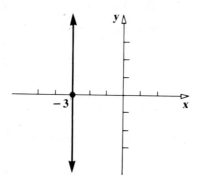

a. $x = -3y$ b. $x = -3$
c. $y = -3$ d. $x + y = -3$

36. Which is the equation of the following graph?

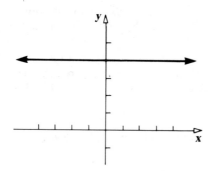

a. $x = 4y$ b. $x = 4$
c. $x + y = 4$ d. $y = 4$

Review

37. Complete the indicated divisions, if possible:

   a. $\dfrac{12}{2}$ b. $\dfrac{-12}{2}$ c. $\dfrac{12}{-2}$

   d. $\dfrac{-12}{-2}$ e. $\dfrac{12}{0}$ f. $\dfrac{0}{-2}$

38. Use the formula to find the distance D from A to B: $D = B - A$ to find the indicated distances:

   a. from -2 to 6 b. from 6 to -2

c. from 0 to -4
e. from -5 to -13
d. from 5 to 8
f. from 0 to 7

39. Let $A = \{4, 5, 6\}$, $B = \{1, 2, 3, 4, 5, 6, 7\}$. Indicate whether each of the following statements is true or false:

   a. $4 \in A$
   b. $4 \subseteq A$
   c. $\{4\} \subset A$
   d. $A \subset B$
   e. $B \subset A$
   f. $7 \notin B$

3.4 Slope and Slope-Intercept Form

This section introduces the concept of the *slope* of a line segment between two points, or the slope of a line. Informally, we say the slope of a line segment or line is the measure of its steepness.

Consider the two points P and Q as shown in Figure 3.4-1. As we move from P to Q along the line segment joining them, we encounter a certain vertical change and a corresponding horizontal change. We denote slope using the letter m, and define the slope of the line segment between P and Q as the ratio of the vertical change to the corresponding horizontal change. Informally, the vertical change and horizontal change are called the *rise* and *run*, respectively.

FIGURE 3.4-1
Slope of Line Segment PQ

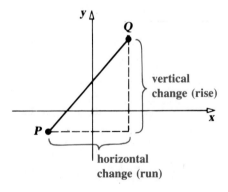

$$\text{slope} = m = \frac{\text{vertical change}}{\text{horizontal change}} = \frac{\text{rise}}{\text{run}}$$

Recall from Chapter 1 that the directed distance D from A to B is given by $D = B - A$. If we assign coordinates (x_1, y_1) to P and (x_2, y_2) to Q, then we see in Figure 3.4-2a that the vertical change is given by $y_2 - y_1$ and in Figure 3.4-2b that the horizontal change is given by $x_2 - x_1$.

FIGURE 3.4-2
Rise and Run

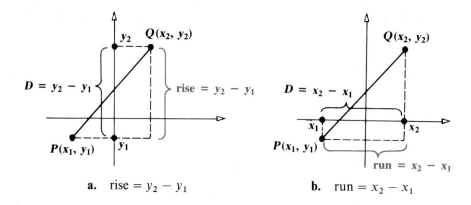

a. rise = $y_2 - y_1$ b. run = $x_2 - x_1$

Using this notation, we write

> **The slope of the line segment between $P(x_1, y_1)$ and $Q(x_2, y_2)$:**
>
> $$\text{slope} = m = \frac{\text{vertical change}}{\text{horizontal change}} = \frac{\text{rise}}{\text{run}} = \frac{y_2 - y_1}{x_2 - x_1}$$

Recall from arithmetic that the value of a fraction is unchanged if we multiply the numerator and denominator by the same quantity. We use this property of fractions to show that it makes no difference which point is labeled (x_1, y_1) and (x_2, y_2) since

$$\frac{y_2 - y_1}{x_2 - x_1} = \frac{(-1)(y_2 - y_1)}{(-1)(x_2 - x_1)} = \frac{-y_2 + y_1}{-x_2 + x_1} = \frac{y_1 - y_2}{x_1 - x_2}$$

Example 1

Find the slope of the line segment between the given pairs of points. Reduce fractions. Graph the line containing each pair of points.

A. $(-2, 3), (4, 7)$

$$\text{slope} = \frac{y_2 - y_1}{x_2 - x_1} = \frac{7 - 3}{4 - (-2)} = \frac{4}{6} = \frac{2}{3}$$

Note: The same slope is also obtained by taking the difference of respective coordinates in the reverse order.

3.4 Slope and Slope-Intercept Form

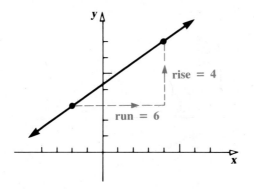

$$\frac{\text{rise}}{\text{run}} = \frac{4}{6} = \frac{2}{3}$$

B. $(5, 1), (4, 4)$

$$\text{slope} = \frac{y_2 - y_1}{x_2 - x_1} = \frac{4 - 1}{4 - 5} = \frac{3}{-1} = -3$$

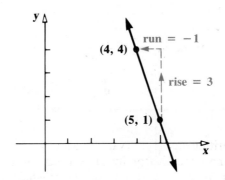

C. $(0, 4), (2, -3)$

$$\text{slope} = \frac{y_2 - y_1}{x_2 - x_1} = \frac{-3 - 4}{2 - 0} = \frac{-7}{2}$$

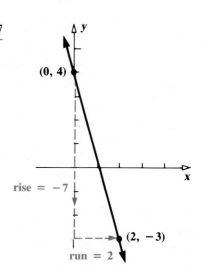

D. $(-7, -1), (3, -5)$

$$\text{slope} = \frac{y_2 - y_1}{x_2 - x_1} = \frac{-5 - (-1)}{3 - (-7)} = \frac{-5 + 1}{3 + 7} = \frac{-4}{10} = \frac{-2}{5}$$

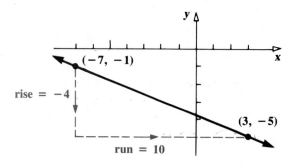

E. $(-3, 2), (4, 2)$

$$\text{slope} = \frac{y_2 - y_1}{x_2 - x_1} = \frac{2 - 2}{4 - (-3)} = \frac{0}{7} = 0$$

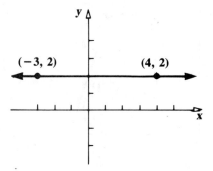

F. $(3, -1), (3, 4)$

$$\text{slope} = \frac{y_2 - y_1}{x_2 - x_1} = \frac{4 - (-1)}{3 - 3} = \frac{5}{0} = \text{undefined}$$

A Vertical Line Has Undefined Slope

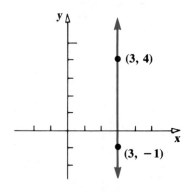

3.4 Slope and Slope-Intercept Form

Note in Example 1 that a line whose slope is positive rises to the right; a line whose slope is negative falls to the right; a horizontal line has a slope of zero; and a vertical line has an undefined slope. This is summarized in the following chart.

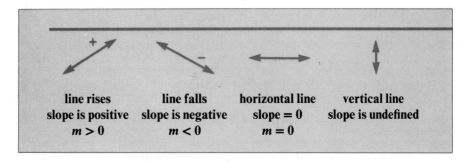

line rises	line falls	horizontal line	vertical line
slope is positive	slope is negative	slope = 0	slope is undefined
$m > 0$	$m < 0$	$m = 0$	

We have seen that an equation of the form $Ax + By = C$, $B \neq 0$, can be written in the form $y = mx + b$, where m and b are constants. We also see that when $x = 0$, $y = b$. So, the point $(0, b)$ is on the graph and is the point where the graph crosses the y-axis. Therefore, in the equation $y = mx + b$, b is the *y-intercept*.

To gain an understanding of the role of the constant m in the equation $y = mx + b$, consider the graph of the specific equation $y = 3x + 2$. Note that the points $(-2, -4)$ and $(1, 5)$ satisfy the equation and may be used to calculate the slope of the line segment between them.

$$m = \frac{y_2 - y_1}{x_2 - x_1} = \frac{5 - (-4)}{1 - (-2)} = \frac{9}{3} = 3$$

In geometry we show that the slope of a line is the same between *any* two points on the line, so we say 3 is the slope *of the line*. The value 3 (the slope) appears as the numerical coefficient of x in the equation $y = 3x + 2$. In general, when the equation of a line is written in the form $y = mx + b$, the graph of the line has slope m, the coefficient of x in the equation.

To summarize, we have seen that when an equation is in the form $y = mx + b$, b is the y-intercept and m is the slope.

$$y = \underset{\text{slope}}{m}x + \underset{y\text{-intercept}}{b}$$

This is called the **slope-intercept form** for the line.

THREE Linear Equations and Inequalities in Two Variables

Example 2

Determine the slope and y-intercept of the line whose equation is given by:

A. $y = -\dfrac{2}{3}x - 4$

This equation is written in slope-intercept form, so its slope is $m = -\dfrac{2}{3}$ (the coefficient of x) and the y-intercept is the constant, -4.

B. $3x - 5y = 25$

This equation can be written in the form $y = mx + b$.

$$3x - 5y = 25$$
$$-5y = -3x + 25 \qquad \text{Add } -3x$$
$$y = \dfrac{3}{5}x - 5 \qquad \text{Divide by } -5$$

The slope, $m = \dfrac{3}{5}$, the coefficient of x.
The y-intercept $= -5$, the constant. ◀

Graphing a Line Using the Slope-Intercept Form of the Line

The slope-intercept form of the equation of a line is useful for drawing the graph of the line. The process is outlined here and then illustrated by examples.

1. If necessary, write the equation in the form $y = mx + b$.
2. Plot the y-intercept (the point $(0, b)$) on the y-axis.
3. Consider the value of m expressed as a fraction. From the point $(0, b)$, move a certain number of units vertically (equal to the numerator of m) and then a corresponding number of units horizontally (equal to the denominator of m). That is, keep the ratio of vertical change to horizontal change $= m$. Mark the point.
4. Draw the line passing through the two points.

Example 3

Use the slope and y-intercept to draw the graph of $y = \dfrac{1}{2}x - 2$.

Locate the y-intercept $(0, -2)$.
The slope is $\dfrac{1}{2}$, so from the y-intercept, we move one unit up and two units to the right to obtain a second point.
Draw the line.

3.4 Slope and Slope-Intercept Form

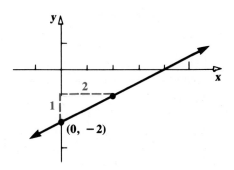

Note in Example 3, from the y-intercept we might move up 2, right 4; or up 5, right 10; or down 1, left 2; or any number of units whose *ratio* is $\frac{1}{2}$ and causes the line to *rise* to the right (since the slope is positive).

Example 4

Use the slope and y-intercept to draw the graph of $3x + 4y = 12$.

Rewrite the equation. Solve for y:

$$3x + 4y = 12$$
$$4y = -3x + 12 \qquad \text{Add } -3x$$
$$y = -\frac{3}{4}x + 3 \qquad \text{Divide by 4}$$

slope: $m = -\frac{3}{4}$
y-intercept: $b = 3$
Plot the y-intercept at (0, 3).
From the point (0, 3), move three units *down* (to move toward the origin—a good graphing technique) then four units to the *right*. The line falls to the right, since the slope is negative.

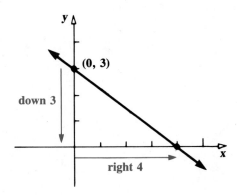

Exercise Set 3.4

In Problems 1–8, identify the slope and the y-intercept of the graph of each equation. *Hint:* If the equation is not in the form $y = mx + b$, it must first be placed in that form by solving for y.

1. $y = 2x - 5$
2. $y = -2x + 1$
3. $y = x + 4$
4. $y = -x - 3$
5. $2x + y = 6$
6. $3x + y = 5$ $y = -3x + 5$
7. $2x - 3y = 8$
8. $x - 4y = -12$ $-4y = -x - 12$

Find the slope of the line through the given points:

9. $(2, 3), (1, 0)$
10. $(0, -2), (3, 4)$
11. $(-2, 1), (1, -2)$
12. $(-3, 2), (4, 1)$
13. $(-5, -3), (-7, -5)$
14. $(2, 2), (-4, 0)$
15. $(-2, -1), (2, 4)$
16. $(-3, 1), (-4, 0)$
17. $(4, 2), (-1, 2)$ $y = 2$
18. $(6, -1), (1, -1)$ $y = -1$
19. $(0, 4), (0, -3)$
20. $(5, -1), (5, 2)$ $x = 5$

Use the slope and intercept to draw the graph of each of the following equations:

21. $2x + y = 3$
22. $2x - y = 5$
23. $x + 2y = 8$
24. $x - 2y = 4$
25. $2x + 3y = 6$
26. $3x + 2y = -12$
27. $5x + 2y = 8$
28. $3x - 4y = 16$
29. $4x - 2y = 5$
30. $5x + 2y = 10$
31. $6x - 3y = 9$
32. $8x - 4y = -6$
33. $3x - 2y = -11$
34. $4x + 5y = 21$

In Problems 35–40, (a) read the y-intercept b from the graph, (b) determine the slope m of the line by finding the rise and run required to move from the y-intercept to the second heavily graphed point shown, and (c) write the equation of the line in the form $y = mx + b$:

35.

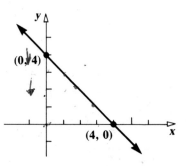

36.

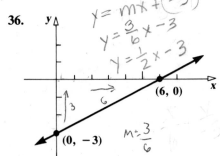

$y = mx + (-3)$
$y = \frac{3}{6}x - 3$
$y = \frac{1}{2}x - 3$

$M = \frac{3}{6}$

3.5 Linear Inequalities in Two Variables

37.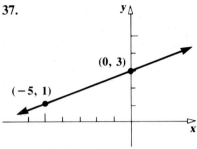

38.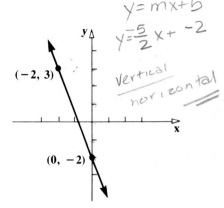

$y = mx + b$
$y = \frac{-5}{2}x + -2$
Vertical
horizontal

39.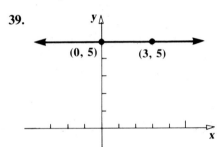

$y = mx + 5$

40.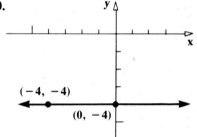

Review

41. Solve and graph the following inequalities:

   a. $3x - 4 < x + 6$
   b. $2(x + 3) - 4 \geq 5(x + 1)$
   c. $-1 \leq 5 - 2x < 13$

42. Plot the following points in a Cartesian plane:

   a. $(-2, 6)$
   b. $(-1, -3)$
   c. $(4, -2)$
   d. $(1, 3)$
   e. $(0, 2)$
   f. $(-6, 0)$
   g. $(0, 0)$

43. Draw the graph of the following equations:

   a. $3x - y = 6$
   b. $y = -\frac{1}{2}x + 3$
   c. $x = -2$
   d. $y = 4$

3.5 Linear Inequalities in Two Variables

Having shown how the Cartesian plane is used to display all solutions (the graph) of an equation of the type $Ax + By = C$, we now illustrate how to graph all solutions of a *linear inequality in two variables*, such as $Ax + By > C$, $Ax +$

$By \geq C$, $Ax + By < C$, or $Ax + By \leq C$. Here, as before, x and y are variables, and A, B, and C are constants with not both A and B equal to zero.

Consider the inequality $x + y \geq 5$. Let's select values of x and determine corresponding values of y that make the statement true.

If $x = 2$, the statement becomes

$$2 + y \geq 5$$

or $y \geq 3$

In the Cartesian plane, Figure 3.5-1, the line ① is all points in the plane for which $x = 2$ and $y \geq 3$.

FIGURE 3.5-1
Some Solutions of $x + y > 5$

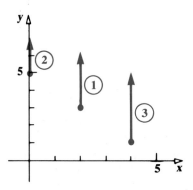

If $x = 0$, then

$$0 + y \geq 5$$

or $y \geq 5$

See line ② in Figure 3.5-1.

If $x = 4$, then $y \geq 1$. See line ③ in Figure 3.5-1. If we were to continue this process we would generate a graph as in Figure 3.5-2, in which the shaded area shows the complete solution of $x + y \geq 5$.

FIGURE 3.5-2
Complete Solution of $x + y \geq 5$

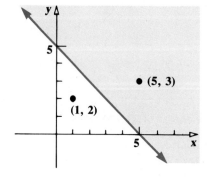

3.5 Linear Inequalities in Two Variables

We observe in this graph that the Cartesian plane is divided into two regions called **half-planes**, one on either side of the line $x + y = 5$. The graph of $x + y = 5$ is called the *boundary line*. The solution to the inequality $x + y \geq 5$ is all of the points in one of these half-planes.

Remember that a graph is simply a picture of all points (x, y) that make a statement true or false. In Figure 3.5-2 our graph illustrates that all points in the *shaded* region have coordinates that make the statement $x + y \geq 5$ true. Also, the points in the *unshaded* region make the statement $x + y \geq 5$ false.

Notice that the line $x + y = 5$ is also part of the solution since the "=" is part of the "$\geq$" symbol. To draw the graph of the similar inequality $x + y > 5$, we would draw the line $x + y = 5$ as a *dashed* line to show that points on the line are not in the solution. See Figure 3.5-3.

FIGURE 3.5-3
Graph of
$x + y > 5$

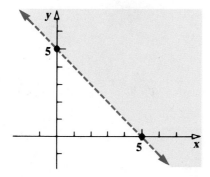

From these examples, we generalize the process of graphing a linear inequality in two variables. We use the symbol Ⓡ to indicate any of the relations $>$, $\geq$, $<$, or $\leq$.

To graph the solution of $Ax + By$ Ⓡ C:

1. Draw the graph of the boundary line $Ax + By = C$
   a. *Solid* if the "=" is included in Ⓡ.
   b. *Dashed* if the "=" is not included in Ⓡ.

2. Pick an arbitrary point in one of the half-planes determined by the boundary line. Substitute its coordinates into $Ax + By$ Ⓡ C.
   a. If the inequality is made *true* by the coordinates of the selected point, shade the half-plane containing that point.
   b. If the inequality is made *false* by the coordinates of the selected point, shade the half-plane on the opposite side of the boundary line from that point.

Example 1

Graph $2x - 3y < 6$:

1. Find the intercepts and sketch the graph of the equation $2x - 3y = 6$ as a *dashed* line, since the "=" is not included.

2. Test a point. The origin, (0, 0) is a good choice since it gives easy computations and is not on the boundary line.

$$2x - 3y < 6$$
$$2 \cdot 0 - 3 \cdot 0 < 6 \quad \text{Substitute } x = 0, y = 0$$
$$0 < 6 \quad \text{True}$$

Since the point (0, 0) makes the statement *true*, shade the half-plane containing (0, 0). See Figure 3.5-4.

FIGURE 3.5-4
Graph of
$2x - 3y < 6$

Intercepts

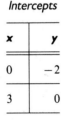

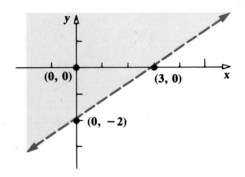

Example 2

Graph $y \geq -\frac{2}{3}x + 3$:

1. Sketch the graph of the equation $y = -\frac{2}{3}x + 3$ as a solid line. Since this equation is in slope-intercept form, we graph using the slope and y-intercept. See Figure 3.5-5.

2. Test the point (0, 0):

$$y \geq -\frac{2}{3}x + 3$$
$$0 \geq -\frac{2}{3} \cdot 0 + 3 \quad \text{Substitute } x = 0, y = 0$$
$$0 \geq 3 \quad \text{False}$$

Shade the half-plane opposite the point (0, 0). See Figure 3.5-5.

FIGURE 3.5-5
Graph of
$y \geq -\frac{2}{3}x + 3$

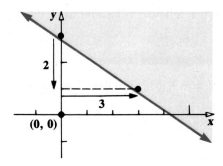

Example 3

Graph $4x > 3y$:

1. Sketch the graph of the equation $4x = 3y$. Write the equation in slope-intercept form.

$$4x = 3y$$
$$3y = 4x$$
$$y = \frac{4}{3}x \quad \text{(Note the } y\text{-intercept is 0)}$$

The line with y-intercept 0 and slope $\frac{4}{3}$ is sketched as a dashed line. See Figure 3.5-6.

FIGURE 3.5-6
Graph of
$4x > 3y$

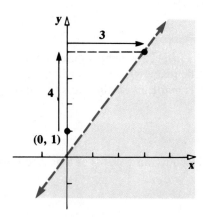

2. The point (0, 0) is on the line, so we select some other point, say (0, 1), as the test point.

$$4x > 3y$$
$$4 \cdot 0 > 3 \cdot 1$$
$$0 > 3 \quad \text{False}$$

Shade the half-plane opposite the point (0, 1). See Figure 3.5-6.

The two examples that follow may be done using the steps suggested earlier, or by inspection.

Example 4

Graph $x \geq 3$:

1. Sketch the graph of $x = 3$, a vertical line, drawn solid. See Figure 3.5-7.
2. Test the point $(0, 0)$:

   $x \geq 3$

   $0 \geq 3$ Substitute $x = 0$, $\underbrace{y = 0}_{\text{not in equation}}$

   False

   Shade the half-plane opposite $(0, 0)$. See Figure 3.5-7.

FIGURE 3.5-7
Graph of
$x \geq 3$

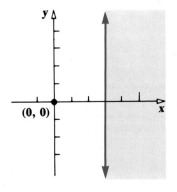

To draw the graph by inspection simply note that the inequality $x \geq 3$ has as its graph all points in the plane whose x-coordinate is greater than or equal to 3. That is, the solution is made up of all points to the right of and including the line $x = 3$.

Example 5

Graph $y < -2$:

1. Draw the graph of the equation $y = -2$ as a dashed horizontal line. See Figure 3.5-8.

FIGURE 3.5-8
Graph of
$x < -2$

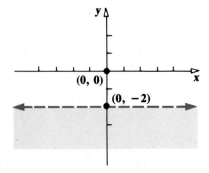

3.5 Linear Inequalities in Two Variables

2. Test the point (0, 0):

   $y < -2$

   $0 < -2$ False

   Shade the half-plane opposite (0, 0).

Exercise Set 3.5

Graph each of the following inequalities. Use intercepts to graph the boundary line.

1. $x + 3y \geq 6$
2. $3x + y \geq 12$
3. $3x - y > 3$
4. $4x - y > 4$
5. $2x - 5y \leq 10$
6. $3x - 4y \leq 12$
7. $4x + 2y < 7$
8. $5x + 3y < 9$

Graph each of the inequalities. Use the slope and y-intercept to graph the corresponding boundary line.

9. $y \leq 2x - 1$
10. $y \leq x + 2$
11. $y > \frac{2}{3}x - 3$
12. $y > \frac{1}{2}x - 2$
13. $y \geq -\frac{1}{3}x + 2$
14. $y \geq -\frac{2}{5}x + 4$
15. $y < \frac{3}{2}x + 4$
16. $y < \frac{5}{2}x - 1$

Graph each of the following inequalities:

17. $3x \geq 5y$
18. $2x \geq 5y$
19. $x < 3y$
20. $x < 2y$
21. $x \geq 2$
22. $x \geq -3$
23. $y < -4$
24. $y < 3$
25. $x < 4$
26. $x \leq -2$
27. $y \geq 0$
28. $x > 0$

Review

29. Solve $x + ax = b$ for x.
30. Solve each equation:

    a. $\dfrac{x}{2} = 8 - \dfrac{2x}{3}$

    b. $4.2y + 9 = 1.6y - 13$

31. Find the slope of the line through $(-1, 2)$ and $(5, -2)$.
32. Draw the graph of $4x - y = 6$

3.6 Point-Slope Form   y intercept

It is common in math and math-related courses to write the equation of a line whose slope is known and which passes through a known point. The slope of a line is defined as the ratio of the difference of *y*-coordinates (the rise) to the corresponding difference of *x*-coordinates (the run) between any two points on that line.

Consider a non-vertical line that passes through a fixed point $P(x_1, y_1)$. For emphasizing that this point is fixed, we show it fixed by a pin at P in Figure 3.6-1. Consider also any other point $Q(x, y)$ on the line. The given slope m is the ratio of rise to run between these two points.

$$m = \frac{\text{rise}}{\text{run}} = \frac{y - y_1}{x - x_1}$$

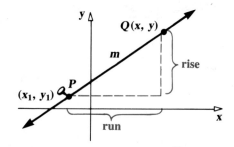

FIGURE 3.6-1

Since we are assuming the line is not vertical, then $x \neq x_1$ or equivalently $x - x_1 \neq 0$. So we may multiply both sides of this equation by the non-zero quantity $x - x_1$, giving

$$m(x - x_1) = y - y_1$$

or equivalently

$$y - y_1 = m(x - x_1)$$

In this equation x_1, y_1, and m all represent *constants*; x and y are *variables*. This equation is called the *point-slope form* of a linear equation.

Example 1

Write the equation in standard form, free of fractions, of the line that has a slope $\frac{2}{3}$ and contains the point $(3, -1)$.

Solution:

Write the point-slope form.

$$y - y_1 = m(x - x_1)$$

$$y - (-1) = \frac{2}{3}(x - 3) \qquad \text{Substitute } m = \frac{2}{3}, x_1 = 3, y_1 = -1$$

3.6 Point-Slope Form

$$y + 1 = \frac{2}{3}(x - 3)$$

$$3(y + 1) = 3\left[\frac{2}{3}(x - 3)\right] \quad \text{Multiply by 3 to clear fractions}$$

$$3y + 3 = 2(x - 3)$$

$$3y + 3 = 2x - 6$$

$$-2x + 3y + 3 = -6 \quad \text{Add } -2x \text{ to both sides}$$

$$-2x + 3y = -9 \quad \text{Add } -3 \text{ to both sides}$$

$$2x - 3y = 9 \quad \text{Divide both sides by } -1 \quad \text{why?}$$

We check that the graph of $2x - 3y = 9$ contains point $(3, -1)$ by substituting:

$$2x - 3y = 9$$
$$2(3) - 3(-1) \stackrel{?}{=} 9$$
$$6 + 3 = 9 \checkmark$$

◀

Example 2

Use the point-slope form to write the equation of the line whose slope is m and which contains point $(0, b)$:

$$y - y_1 = m(x - x_1)$$
$$y - b = m(x - 0) \quad \text{Substitute } x_1 = 0, y_1 = b$$
$$y - b = mx$$
$$y = mx + b$$

Observe that this result is the slope-intercept form!

◀

The point-slope form is also helpful for writing the equation of a line that passes through two fixed points. If the coordinates of two points are given, we first compute the slope m, then use the slope and the coordinates of *either* of the points in the point-slope formula.

Example 3

Write the equation (in standard form $Ax + By = C$, clear of fractions) of the line that contains points $(-3, 1)$ and $(-1, 7)$:

Solution:

We first find the slope

$$m = \frac{y_2 - y_1}{x_2 - x_1} = \frac{7 - 1}{-1 - (-3)} = \frac{6}{-1 + 3} = \frac{6}{2} = 3$$

Now substitute the slope $m = 3$ and coordinates of either point in the point-slope form; arbitrarily, we select $(-3, 1)$.

$$y - y_1 = m(x - x_1)$$
$$y - 1 = 3[x - (-3)] \quad \text{Substitute}$$
$$y - 1 = 3(x + 3)$$
$$y - 1 = 3x + 9$$
$$-3x + y = 10$$
$$\text{or} \quad 3x - y = -10 \quad \text{Divide by } -1$$

why

◀

You may check that this equation's graph contains both the given points by substituting their coordinates in the equation.

Exercise Set 3.6

Write the equation of the line with the indicated slope passing through the indicated point. Write the equation in standard form $Ax + By = C$, clear of fractions, with $A > 0$:

1. $m = 3; (4, 2)$
2. $m = 4; (2, 3)$
3. $m = -2; (3, 4)$
4. $m = -1; (1, 4)$
5. $m = 3; (-5, 1)$
6. $m = 2; (-1, -4)$
7. $m = \frac{2}{3}; (1, -4)$
8. $m = \frac{3}{4}; (2, -3)$
9. $m = -\frac{1}{3}; (2, 6)$
10. $m = -\frac{1}{2}; (4, -3)$
11. $m = -\frac{5}{4}; (2, 2)$
12. $m = -\frac{7}{2}; (1, -4)$
13. $m = -\frac{3}{8}; (4, 0)$
14. $m = -\frac{2}{5}; (6, 0)$

Write the equation of the line with the indicated slope containing the indicated point. Write the equation in slope-intercept form, $y = mx + b$:

15. $m = \frac{2}{3}; (1, -1)$
16. $m = \frac{3}{4}; (-2, 3)$
17. $m = 4; (-2, -2)$
18. $m = 2; (-3, 1)$
19. $m = \frac{3}{2}; (4, -1)$
20. $m = \frac{4}{3}; (-2, -5)$
21. $m = \frac{2}{5}; (1, 2)$
22. $m = -\frac{5}{3}; (-3, 7)$

3.6 Point-Slope Form

Write in standard form the equation of the line containing points:

23. $(1, 2), (3, 6)$
24. $(1, 3), (3, 9)$
25. $(2, -1), (5, 1)$
26. $(3, -1), (-2, 2)$
27. $(4, 1), (-3, -2)$
28. $(0, 2), (4, 0)$
29. $(0, -3), (2, 0)$
30. $(2, 0), (7, 1)$

Write in slope-intercept form the equation of the line containing points:

31. $(3, 1), (-1, -2)$
32. $(4, 1), (1, -4)$
33. $(-5, -2), (1, 2)$
34. $(4, -3), (0, 1)$
35. $(0, 3), (5, -1)$
36. $(-1, -3), (5, 2)$

Write in slope-intercept form the equation of the lines graphed in Problems 37–42:

37.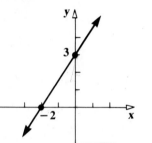

38.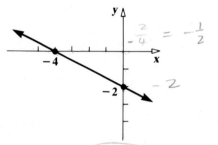

39.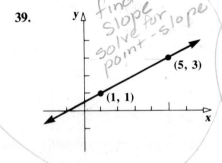

40.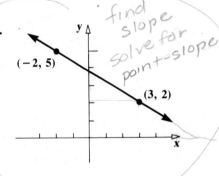

41.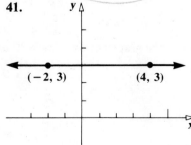

42.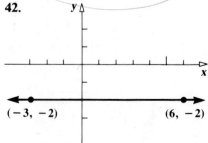

Review

43. Find the coordinates of each point in the Cartesian plane shown:

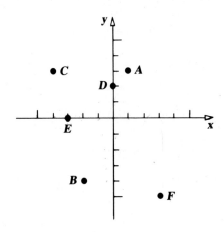

44. Solve the equation:

$$3y - 1 = 11 - y$$

45. Solve and graph the inequality:

$$3y - 1 \leq 11 - y$$

46. Write $3 \cdot x \cdot y \cdot y \cdot y \cdot y \cdot z \cdot z$ in exponential form.

47. Complete the indicated operations:

   a. $(-6)(8)$

   b. $\dfrac{27}{-3}$

   c. $8 - 11 + 2 - 6 - 5$

48. Indicate which of the following points are on the graph of the line $3x - 2y = 12$:

   a. $(2, -3)$ b. $(-2, 3)$ c. $(2, 3)$
   d. $(0, 6)$ e. $(4, 0)$ f. $(5, 1)$

49. In the same plane, draw the graphs of $3x - 2y = 12$ and $4x + y = 5$.

3.7 Review and Chapter Test

(3.1) The *Cartesian Plane* is formed by a horizontal and a vertical number line intersecting at their origins. The regions determined are called *quadrants* and are numbered as shown in Figure 3.7-1.

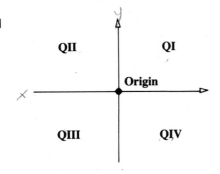

FIGURE 3.7-1
The Cartesian Plane

An ordered pair of numbers is written in the form (a, b). The first number is the *abscissa*, and the second number is the *ordinate*.

To graph or plot the ordered pair (a, b) in the Cartesian plane, count a units along the horizontal axis (to the right if a is positive, to the left if a is negative); then count b units vertically (up if b is positive, down if b is negative). Place a dot in the place indicated.

Example 1

Graph each of the indicated ordered pairs and indicate the quadrant.

A. $(2, -3)$
B. $(-3, -1)$
C. $(4, 4)$
D. $(-2, 5)$

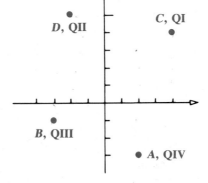

(3.2) An equation that can be written in the form $Ax + By = C$, where A, B, and C are real numbers, with A and B not both zero and x and y as variables, is called a *linear equation in two variables* and has a straight line as its graph.

One method to draw the graph of a linear equation in two variables is to find three ordered pairs of numbers that satisfy the equation, plot these points, and draw the line through them.

Example 2

Draw the graph of $2x - y = 3$. Shown are some ordered pair solutions.

x	y
0	−3
2	1
4	5

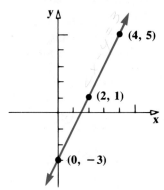

In the form $Ax + By = C$, if A or B equals zero, the equation can be rewritten in the form $x = k$ or $y = k$ (k is a constant) whose graph is a vertical or horizontal line, respectively.

Example 3

Shown are the graphs of the indicated equations:

A. $x = -2$

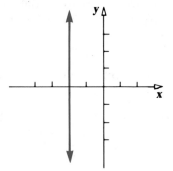

B. $3y - 8 = 0$
$3y = 8$
$y = \dfrac{8}{3}$

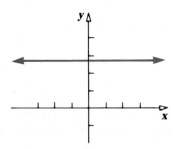

(3.3) One method of simplifying the process of drawing the graph of a line is to solve the equation for y.

Example 4

Solve for y; then find ordered pair solutions and draw the graph of the equation $3x - 2y = 12$:

$$3x - 2y = 12$$
$$-2y = -3x + 12$$
$$\frac{-2y}{-2} = \frac{-3x}{-2} + \frac{12}{-2}$$
$$y = \frac{3}{2}x - 6$$

x	y
0	-6
2	-3
4	0

In this case, even numbers are chosen for values of x to avoid fractions.

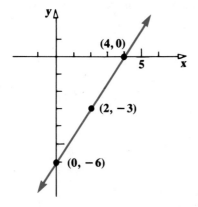

A simple method of graphing is to find the intercepts. This is done by setting $x = 0$, and then solving for y (to find the y-intercept); and setting $y = 0$, and then solving for x (to find the x-intercept). The entire process can generally be done mentally.

Example 5

Find the intercepts and draw the graph of $2x + 3y = 8$:

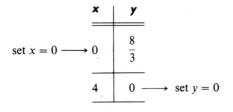

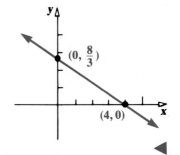

(3.4) The slope of the line through the two points $P(x_1, y_1)$ and $Q(x_2, y_2)$ is given as

$$\text{slope} = m = \frac{y_2 - y_1}{x_2 - x_1} = \frac{\text{vertical change}}{\text{horizontal change}} = \frac{\text{rise}}{\text{run}}$$

Example 6

Find the slope of the line through $(5, -1)$ and $(2, 3)$:

$$m = \frac{y_2 - y_1}{x_2 - x_1} = \frac{-1 - 3}{5 - 2} = \frac{-4}{3}$$

A linear equation $Ax + By = C$ may be solved for y, giving $y = mx + b$, where $m = $ slope and $b = y$-intercept. The slope and intercept may then be used to draw the graph.

Example 7

Graph $3x - 4y = 8$:

Solve for y:

$$3x - 4y = 8$$
$$-4y = -3x + 8$$
$$y = \frac{3}{4}x - 2$$

The slope is $\frac{3}{4}$; the y-intercept is -2.

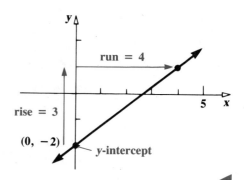

The relationship between a line and its slope is summarized in the following chart:

Slope:	Positive	Negative	Zero	Undefined
Line:	Rises	Falls	Horizontal	Vertical
Example:				

(3.5) A linear inequality in two variables is a statement that can be written in the form $Ax + By \;\text{\textcircled{R}}\; C$, where $\text{\textcircled{R}}$ is one of the symbols $<, \leq, >,$ or $\geq$.

To draw the graph of $Ax + By \;\text{\textcircled{R}}\; C$:

1. Draw the boundary line, the graph of the corresponding equation $Ax + By = C$
   a. Solid if the "$=$" is included in $\text{\textcircled{R}}$.
   b. Dashed if "$=$" is not included in $\text{\textcircled{R}}$.
2. Test a point <u>not on the boundary line</u> by substituting its coordinates into the original inequality.
   a. If the statement is made true by the coordinates of the point, shade the half-plane containing the test point.
   b. If the statement is made false, shade the half-plane on the opposite side of the boundary line from that point.

Example 8

Graph $3x - y > 6$:

1. Draw the graph of $3x - y = 6$ as a *dashed* line (since the "$=$" is not included).
2. Test a point. The point $(0, 0)$ is usually a good choice.

   $$3x - y > 6$$
   $$0 - 0 > 6 \quad \text{Substitute}$$
   $$0 > 6 \quad \text{False}$$

   Shade the half-plane opposite $(0, 0)$

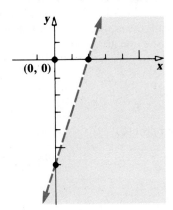

THREE Linear Equations and Inequalities in Two Variables

Example 9

Graph the inequalities:

A. $x \leq 2$

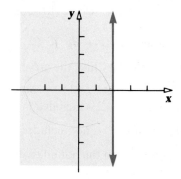

B. $y > -1$

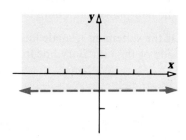

(3.6) The line that has a slope m and contains point $p(x_1, y_1)$ has an equation

$y - y_1 = m(x - x_1)$	Point-slope form

This form is useful for writing the equation of a line if its slope and a point are known, or if two points on the line are known.

Example 10

Write an equation in standard form $Ax + By = C$ that satisfies the following conditions:

A. Slope = $\frac{2}{3}$; containing point $(4, -1)$

$\qquad y - y_1 = m(x - x_1) \qquad$ Point-slope form

$\qquad y - (-1) = \frac{2}{3}(x - 4) \qquad$ Substitute given values

3.7 Review and Chapter Test

$$y + 1 = \frac{2}{3}(x - 4)$$

$3(y + 1) = 2(x - 4)$ Multiply both sides by 3

$3y + 3 = 2x - 8$

$-2x + 3y = -11$

$2x - 3y = 11$

B. Containing points $(4, -2), (1, 3)$ $y - y_1 = m(x - x_1)$
First find the slope:

$$m = \frac{y_2 - y_1}{x_2 - x_1} = \frac{3 - (-2)}{1 - 4} = \frac{3 + 2}{1 - 4} = \frac{5}{-3} = -\frac{5}{3}$$

Now substitute the slope and one point in the point-slope form. We will use $(4, -2)$.

$$y - y_1 = m(x - x_1)$$

$y - (-2) = \frac{-5}{3}(x - 4)$ Substitute values

$y + 2 = \frac{-5}{3}(x - 4)$

$3(y + 2) = -5(x - 4)$ Multiply both sides by 3

$3y + 6 = -5x + 20$

$5x + 3y = 14$

◀

Exercise Set 3.7

1. For each of the ordered pairs given:
   a. Indicate the abscissa and the ordinate
   b. Plot in the Cartesian plane
   c. Indicate the quadrant

   A. $(3, -2)$ B. $(-2, -3)$ C. $(-4, 1)$ D. $(4, 2)$

THREE   Linear Equations and Inequalities in Two Variables

2. Write the ordered pair associated with each point shown:

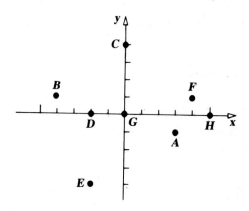

3. Draw the graph of each equation:

   a. $3x + y = 6$
   b. $2x - 5y = -10$
   c. $y = 3x + 2$
   d. $y = \dfrac{-1}{2}x + 3$
   e. $2x + 4 = 0$
   f. $3y - 2 = 0$

4. Find the slope of the line through the indicated pairs of points. Without graphing, state whether the line through them is horizontal, vertical, rises to the right, or falls to the right.

   a. $(2, 5), (-1, -1)$
   b. $(-3, 1), (2, -1)$
   c. $(4, 3), (4, -2)$
   d. $(5, -2), (0, -2)$

5. Write each of the following equations in the form $y = mx + b$, identify the slope and the y-intercept, and then use the slope and y-intercept to draw the graph:

   a. $3x - y = 2$
   b. $2x + 3y = 18$
   c. $y = 3$

6. Draw the graph of each of the following inequalities:

   a. $2x + 3y \geq 6$
   b. $3x + y < 6$
   c. $4x - 3y > 12$
   d. $y \geq -\dfrac{2}{5}x + 3$
   e. $2x < 6$
   f. $y > -2$

7. Write the equation in standard form of the line:

   a. With slope $-\tfrac{3}{4}$, containing point $(1, -5)$
   b. Containing points $(-4, -1), (-1, 6)$

Chapter 3 Review Test

1. Draw an x-y plane and plot the following points:

   a. $(3, -1)$ b. $(0, 4)$ c. $(-3, -2)$
   d. $(-2, 0)$ e. $(2, 3)$ f. $(-1, 4)$

2. Write the coordinates of each point in the x-y plane shown.

3. Find the slope of the line:

   a. containing points $(-4, -1), (0, 5)$
   b. whose equation is $y = -\frac{1}{2}x + 3$
   c. whose equation is $3x + 7y = -12$

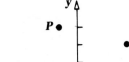

4. Draw an x-y plane and sketch the graphs of each of the following equations:

   a. $3x - 2y = 9$ b. $y = -\frac{1}{3}x + 2$
   c. $x = -3$ d. $2y - 5 = 0$

5. For each inequality, draw an x-y plane and then graph:

   a. $2x - 5y \leq 10$ b. $y > \frac{2}{3}x - 4$ c. $x \leq 6$

6. Without graphing, tell whether each of the ordered pairs *is* or *is not* a solution of $3x - 5y = 30$:

   a. $(5, -3)$ b. $(-5, -3)$
   c. $(10, 0)$ d. $(0, 6)$

7. Write the equation of each line in standard form, free of fractions:

   a. slope $= \frac{3}{5}$, y-intercept $(0, -4)$

   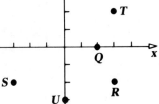

   b. slope $= -\frac{1}{2}$, containing point $(-3, 2)$
   c. containing points $(-1, 3)$ and $(4, -1)$
   d. vertical, containing point $(-3, -2)$

CHAPTER FOUR

SYSTEMS OF EQUATIONS AND INEQUALITIES

Systems of Equations

In Chapter 3, we determined that there are an unlimited number of ordered pairs that are solutions to equations of the type $Ax + By = C$, a linear equation in two variables. Further, we have seen that the graph of all the solutions to such an equation is a straight line.

In these sections we wish to find all pairs of numbers that are solutions to *two* equations of the type $Ax + By = C$. A set of two (or more) equations in two (or more) variables is called a **system of equations**. Although systems may contain any number of equations and variables, in this text we consider only systems of two linear equations in two variables. To *solve* a system of equations is to find all ordered pairs of numbers that satisfy all equations in the system.

Example 1

Show that (4, 1) is the solution of the system of equations:

$x - y = 3$

$x + y = 5$

Substitute $x = 4$, $y = 1$ into each equation.

$x - y = 3$	$x + y = 5$
$4 - 1 = 3$	$4 + 1 = 5$
$3 = 3$	$5 = 5$
Yes	Yes

(4, 1) is a solution to the system since it is a solution to *both* equations in the system. ◀

There are several methods used to obtain the solution(s) to such a system of equations. We consider three such methods: the graphing method, the elimination method, and the substitution method.

4.1 Solving Systems of Equations by Graphing

A system of two linear equations in two variables is of the form:

$$Ax + By = C$$
$$Dx + Ey = F$$

in which A, B, C, D, E, and F are all constants, and x and y are variables. Each equation has a graph that is a straight line (Chapter 3). These two lines will either intersect in a single point, be parallel, or be collinear (the same line). Recall that every point on the graph of an equation has coordinates that satisfy the equation. If the two lines intersect in a single point, the coordinates of that point will satisfy *both* equations. The coordinates of this point make up the *solution* of the system.

SOLVING A SYSTEM OF EQUATIONS BY GRAPHING

1. **Draw the graph of each equation from the system in the same coordinate plane.**
2. **If these two lines intersect at a unique point, the coordinates of that point represent the solution to the system.**

Example 1

Solve each of the following systems of equations:

A. $x - y = 3$
$\quad\; x = 2y$

Graph both equations on the same coordinate plane. The point of intersection of the two graphs is found by inspection to be the point (6, 3), which is the *solution* to the system. To check, the values $x = 6$ and $y = 3$ must be substituted into *both* equations

Check:

$x - y = 3$	$x = 2y$
$6 - 3 \stackrel{?}{=} 3$	$6 \stackrel{?}{=} 2(3)$
$3 = 3$	$6 = 6$
Yes	Yes

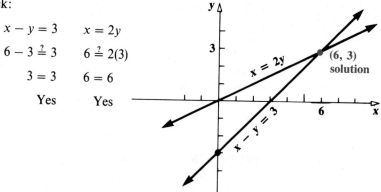

B. $2x + y = 8$
$x - y = 1$

The graphs of these equations are shown. By inspection, the coordinates of the point of intersection are (3, 2). Check by substituting into both equations.

Check:

$2x + y = 8$	$x - y = 1$
$2(3) + 2 \stackrel{?}{=} 8$	$3 - 2 \stackrel{?}{=} 1$
$6 + 2 = 8$	$1 = 1$
$8 = 8$	Yes
Yes	

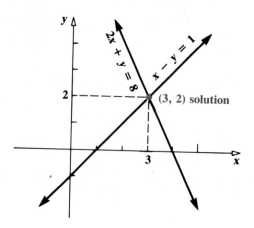

4.1 Solving Systems of Equations by Graphing

Example 2

Solve the following system by graphing:

$$2x + y = 10$$

$$x + \frac{1}{2}y = 4$$

y = mx + b
y = -2x + 10

The graphs of the two equations are shown at the right. Our graphs are a pair of parallel lines that will never meet! This system of equations has <u>*no solution*</u>.

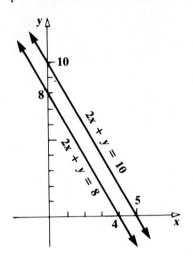

Note: If the second equation is multiplied by 2, it becomes $2x + y = 8$. Comparing this to the first equation, it appears that we are trying to find two numbers with the property that if we add twice the first number to the second number, we will get two different answers. This is impossible. ◀

Example 3

Solve the system by graphing:

$$y = 3x - 2$$

$$6x - 2y = 4$$

infinite # of solutions

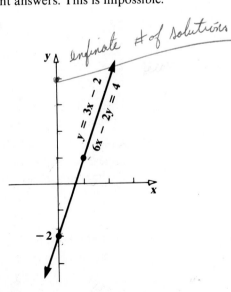

$-2y = -6x + 4$
$\overline{-2} \quad \overline{-2} \quad \overline{-2}$
$y = 3x - 2$

Attempting to solve this system by graphing, we see that graphs of the two equations are collinear (the same line). This system has an infinite number of solutions since the coordinates of every point on the line satisfy both equations. ◂

Solving systems of equations by graphing has the advantage that it gives a "picture" of the system. But, it has the disadvantage that even when the graphing is done carefully, the solution is often inaccurate, particularly when the solution contains fractional or irrational values. The x- and y-axes are generally not calibrated well enough for the exact solutions to be read. Methods are discussed in the next section for arriving at exact solutions to systems of equations.

Exercise Set 4.1

Solve each of the systems by graphing:

1. $x + y = 3$
   $x - y = 7$

2. $y = 2x$
   $3x + 2y = -21$

3. $x - y = -2$
   $x + y = -6$

4. $2x + 3y = 12$
   $-3x + y = 4$

5. $2y = -3x + 36$
   $y = 3x$

6. $x - 3y = 2$
   $2x + y = 11$

7. $2x - y = 3$
   $3x + 2y = 8$

8. $2x - y = -4$
   $-4x + 2y = 12$

9. $x - 3y = 7$
   $-2x + 6y = -14$

10. $x + 2y = -6$
    $3x - y = 3$

11. $2x + y = -4$
    $2y = -8 - 4x$

12. $2x - 4y = 8$
    $3x + y = 12$

13. $2x + y = 5$
    $4x - y = 7$

14. $4x + 3y = 9$
    $-3x + 8y = -17$

15. $2x + 4y = 12$
    $-x - 2y = 5$

16. $2x - 3y = 2$
    $5x - y = 5$

17. $x + 3y = -1$
    $x = 2$ vertical line

18. $3x - 2y = 7$
    $2y = 5$ horizontal line

19. Discuss and summarize the three possibilities that can occur when we solve a system of two equations in two variables.

20. Suppose that in solving a system by graphing we obtain two lines that appear to be parallel. Is there any way we can tell, for certain, whether the lines are parallel or whether they only appear to be parallel because of our graph?

Review

21. Simplify the following expressions by removing parentheses and combining like terms:

    a. $(3x - y) + (2x + y)$ **b.** $(2x - 3y) - (5x - 3y)$

c. $(6x + 8y) + (2x - 8y)$ d. $(4x - 6y) - (4x - 2y)$

22. For each of the following equations, clear fractions by multiplying both sides by the LCD:

    a. $\dfrac{x}{3} - \dfrac{y}{2} = 2$ b. $\dfrac{2x}{5} + \dfrac{y}{3} = 4$

    c. $\dfrac{x}{6} - \dfrac{3y}{4} = -2$

23. Solve each of the following equations:

    a. $2x = 12$ b. $-3x = -18$

    c. $4y = -8$ d. $5y = -27$

4.2 Solving Systems of Equations by the Elimination Method

In working with systems of equations we use the properties of equations. Remember that an equivalent equation results whenever:

1. the same quantity is added to both sides of the equation, or
2. both sides of the equation are multiplied or divided by the same non-zero constant.

Changing one or both of the equations in a system by one of these properties results in an **equivalent system**, one that has the same solution(s) as the original system.

Consider the system

$2x + y = 5$
$2x - y = 3$

We may add the left sides and the right sides of the two equations together. We say more simply that we add the two equations together. The sum is $4x = 8$, an equation in *one* variable whose solution is $x = 2$.

$\left. \begin{array}{l} 2x + y = 5 \\ 2x - y = 3 \end{array} \right\}$ Add the left members by combining like terms; add the right members

$\overline{}$

$4x + 0 = 8$

$4x = 8$

$x = 2$ Solve for x

We now substitute this value of x into *either* of the two original equations and solve for y, giving $y = 1$. The solution to the system is $(2, 1)$, which is checked by substituting these values into both of the original equations.

$$2x + y = 5 \qquad 2x - y = 3$$
$$2(2) + 1 \stackrel{?}{=} 5 \qquad 2(2) - 1 \stackrel{?}{=} 3$$
$$4 + 1 = 5 \qquad 4 - 1 = 3$$
$$5 = 5 \qquad 3 = 3$$
$$\text{Yes} \qquad \text{Yes}$$

Similarly, if we subtract the bottom equation from the top equation, the result is $2y = 2$. Solving this equation for y, we obtain $y = 1$. Substituting $y = 1$ into either of the original equations gives $x = 2$. As before, the solution to the system is $(2, 1)$.

$$\left. \begin{array}{r} 2x + y = 5 \\ 2x - y = 3 \end{array} \right\} \quad \text{Subtract the bottom equation}$$
$$\overline{0 + 2y = 2}$$
$$2y = 2$$
$$y = 1 \qquad \text{Solve for } y$$

The method for solving a system that we have just discussed worked nicely because the coefficients of the x-variable and of the y-variable in the two equations had the same absolute value. Consequently, either adding the equations together or subtracting one from the other resulted in an equation with only one variable, since one variable was eliminated.

To solve a system of equations in which the absolute values of the coefficients of the corresponding variables are not the same, we proceed as outlined in the following steps.

The elimination method

1. Write each equation in the form $Ax + By = C$.
2. Select a variable to be eliminated and, if necessary, multiply each equation in the system by an appropriate constant to make the coefficient of that variable the same in each equation, except possibly for the sign.
3. Add or subtract the equations, as appropriate, to eliminate one variable and create a single equation in the remaining variable.
4. Solve the resulting equation for its variable.
5. Substitute the result from Step 4 into one of the original equations. Solve to determine the value of the other variable.
6. Check the result by substituting the two values in both of the original equations.

4.2 Solving Systems of Equations by the Elimination Method

Example 1

Solve the system:

$$x - 2y = 5$$
$$3x + y = 8$$

We may eliminate x or y. Here we choose to eliminate y. Multiply both sides of the second equation by 2 to make the coefficients of y opposites, then add the two equations.

$$x - 2y = 5$$
$$2(3x + y) = 2 \cdot 8 \quad \text{Multiply by 2}$$

$$\left.\begin{array}{r} x - 2y = 5 \\ 6x + 2y = 16 \end{array}\right\} \quad \text{Add}$$
$$\overline{7x = 21}$$
$$x = 3$$

Substitute $x = 3$ into the first equation to find y.

$$x - 2y = 5$$
$$(3) - 2y = 5 \quad \text{Substitute}$$
$$-2y = 2$$
$$y = -1$$

Solution: $(3, -1)$

Check:

$$x - 2y = 5 \qquad\qquad 3x + y = 8$$
$$3 - 2(-1) \stackrel{?}{=} 5 \qquad 3(3) + (-1) \stackrel{?}{=} 8$$
$$3 + 2 = 5 \qquad\qquad 9 + (-1) = 8$$
$$5 = 5 \qquad\qquad\qquad 8 = 8$$
$$\text{Yes} \qquad\qquad\qquad \text{Yes}$$

◀

Example 2

Solve the system:

$$2x + 3y = 1$$
$$-3x + 5y = 27$$

In this example, we must modify both equations. We have two choices: either multiply the first equation by 3 and the second equation by 2 to make the coefficients of x opposites or multiply the first equation by 5 and the second equation by 3 to make the coefficients of y the same. Since the first choice gives smaller numbers, we will eliminate x.

$$3(2x + 3y) = 3 \cdot 1 \quad \text{Multiply by 3}$$
$$2(-3x + 5y) = 2 \cdot 27 \quad \text{Multiply by 2}$$
$$\left.\begin{array}{r}6x + 9y = 3 \\ -6x + 10y = 54\end{array}\right\} \quad \text{Add}$$
$$19y = 57$$
$$y = \frac{57}{19} = 3$$

Substitute $y = 3$ into the first equation.
$$2x + 3y = 1$$
$$2x + 3(3) = 1$$
$$2x + 9 = 1$$
$$2x = -8$$
$$x = -4$$

Solution: $(-4, 3)$

Check:

$$2x + 3y = 1 \qquad\qquad -3x + 5y = 27$$
$$2(-4) + 3(3) \stackrel{?}{=} 1 \qquad -3(-4) + 5(3) \stackrel{?}{=} 27$$
$$-8 + 9 = 1 \qquad\qquad 12 + 15 = 27$$
$$1 = 1 \qquad\qquad\qquad 27 = 27$$
$$\text{Yes} \qquad\qquad\qquad\quad \text{Yes}$$ ◀

Example 3

Solve the system:
$$3x - 4y = 5$$
$$5x + 3y = 9$$

Multiply the first equation by 5 and the second by 3 to eliminate x.
$$\left.\begin{array}{r}15x - 20y = 25 \\ 15x + 9y = 27\end{array}\right\} \quad \text{Subtract}$$
$$-29y = -2$$
$$y = \frac{2}{29}$$

Now, we can see that if we are to substitute this value into one of the original equations, we will have a difficult computation with fractions. As an alternative, we may return to the original system and solve for x by first eliminating y.

4.2 Solving Systems of Equations by the Elimination Method

Multiply the first equation by 3 and the second by 4, giving the system

$$\left.\begin{array}{r}9x - 12y = 15 \\ 20x + 12y = 36\end{array}\right\} \text{Add}$$

$$\overline{29x = 51}$$

$$x = \frac{51}{29}$$

Solution: $\left(\dfrac{51}{29}, \dfrac{2}{29}\right)$

Check:

$$3x - 4y = 5 \qquad\qquad 5x + 3y = 9$$

$$3\left(\frac{51}{29}\right) - 4\left(\frac{2}{29}\right) \stackrel{?}{=} 5 \qquad 5\left(\frac{51}{29}\right) + 3\left(\frac{2}{29}\right) \stackrel{?}{=} 9$$

$$\frac{153}{29} - \frac{8}{29} = 5 \qquad\qquad \frac{255}{29} + \frac{6}{29} = 9$$

$$\frac{145}{29} = 5 \qquad\qquad\qquad \frac{261}{29} = 9$$

$$5 = 5 \qquad\qquad\qquad 9 = 9$$

Yes Yes ◀

Example 4

Solve the system:

$$\frac{x}{3} + \frac{y}{2} = 0$$

$$\frac{x}{2} - \frac{y}{4} = 4$$

Note in this system, the equations contain fractional coefficients. In this case, we multiply both sides of each equation by the LCD to remove the fractions.

$$6 \cdot \left(\frac{x}{3} + \frac{y}{2}\right) = 6 \cdot 0$$

$$4 \cdot \left(\frac{x}{2} - \frac{y}{4}\right) = 4 \cdot 4$$

$$6 \cdot \frac{x}{3} + 6 \cdot \frac{y}{2} = 6 \cdot 0$$

$$4 \cdot \frac{x}{2} - 4 \cdot \frac{y}{4} = 4 \cdot 4$$

$$2x + 3y = 0$$
$$2x - y = 16$$

We now proceed to solve the system by elimination. Subtract the second equation from the first to eliminate x.

$$\left.\begin{array}{r} 2x + 3y = 0 \\ 2x - y = 16 \end{array}\right\} \quad \text{Subtract}$$
$$\overline{}$$
$$4y = -16$$
$$y = -4$$

Substitute $y = -4$ into the equation $2x + 3y = 0$ to find x.

$$2x + 3y = 0$$
$$2x + (-4) = 0$$
$$2x = 12$$
$$x = 6$$

Solution: $(6, -4)$
The solution may be checked by substitution into the original system. ◀

Example 5

Solve the system:
$$2x - y = 9$$
$$4x - 2y = 6$$

Multiply the first equation by 2; then subtract to eliminate x.

$$\left.\begin{array}{r} 4x - 2y = 18 \\ 4x - 2y = 6 \end{array}\right\} \quad \text{Subtract}$$
$$0 = 12$$

What has happened? Certainly 0 is not equal to 12. Therefore, the preceding system has *no solution*. The graph of such a system is two parallel lines. ◀

Example 6

Solve the system:
$$x - 2y = 5$$
$$-2x + 4y = -10$$

Multiply the first equation by 2; then add to eliminate x.

$$\left.\begin{array}{r} 2x - 4y = 10 \\ -2x + 4y = -10 \end{array}\right\} \quad \text{Add}$$
$$0 = 0$$

4.2 Solving Systems of Equations by the Elimination Method

There is nothing wrong mathematically with stating $0 = 0$, but it doesn't help us solve the system either. In such a system, every solution of one equation is also a solution to the other. Recall that the graph of such a system is two collinear lines.

Exercise Set 4.2

Solve each system by the elimination method:

1. $x + y = 5$
   $x - y = -3$

2. $2x + 3y = 16$
   $2x - y = 0$

3. $2x - y = 3$
   $3x + 2y = 8$

4. $x - 3y = 10$
   $x - 2y = 7$

5. $7x + 4y = 1$
   $7x - 5y = -80$

6. $3x - y = 4$
   $5x + 2y = 25$

7. $2x - 4y = -8$
   $3x + 5y = 10$

8. $4x - 3y = 11$
   $6x - 3y = 12$

9. $2x + 4y = 12$
   $-x - 2y = 5$

10. $3x + y = 6$
    $6x + 2y = 5$

11. $3x - 5y = 49$
    $4x - y = 3$

12. $5x - 4y = 5$
    $x - 2y = 5$

13. $2x - 4y = 8$
    $3x + y = 12$

14. $x - 3y = 8$
    $2x - 6y = 16$

15. $2x - y = 5$
    $-6x + 3y = -15$

16. $2x + 3y = -7$
    $3x + 5y = -13$

17. $2x - 5y = 6$
    $4x + 3y = 8$

18. $4x - 2y = 9$
    $6x - 5y = 11$

19. $8x + 3y = 6$
    $3x - 7y = -3$

20. $3x + 2y = -8$
    $7x + 4y = 11$

21. $\dfrac{x}{2} - \dfrac{3y}{2} = -1$
    $\dfrac{2x}{5} + \dfrac{3y}{4} = 7$

22. $\dfrac{2x}{3} - \dfrac{y}{3} = 1$
    $\dfrac{5x}{2} + y = 24$

23. $x - \dfrac{y}{2} = 17$
    $\dfrac{x}{4} + \dfrac{3y}{2} = 1$

24. $\dfrac{x}{6} + \dfrac{y}{2} = 0$
    $\dfrac{2x}{3} - \dfrac{3y}{4} = -11$

25. Multiply the first equation in the following system by 100 to eliminate the decimals and then solve the resulting system as usual:

    $0.05x + 0.20y = -0.25$
    $2x - 3y = 12$

26. Multiply the second equation in the following system by 100 to eliminate the decimals and then solve the resulting system as usual:

    $2x + y = 2$
    $0.03x - 0.05y = -0.23$

Solve the following systems of equations. Use the properties of equations to rewrite each equation in the system with the terms involving the variables x and y on the left side and constants on the right side of the equations.

27. $4x + y + 3 = x + 2y + 4$
    $x + 2y = 12$

28. $2x + y = 11$
    $2x + y - 1 = -x + 3y + 5$

29. $y = 2x + 8$
    $2y - 2 = -3x$

30. $3x - 1 = -4y$
    $x = 3 - 3y$

Review

31. Solve the following equations for the indicated variable:

    a. $3x + y = 6$, for y
    b. $-x + 3y = -2$, for x
    c. $2x - 5y = 12$, for x
    d. $2x - 5y = 12$, for y

32. Solve each of the following equations:

    a. $3x + 2(3x - 3) = 12$
    b. $2(3y - 10) + 5y = 2$

33. Solve each equation for y using the given value of x:

    a. $3x + 5y = -9, x = -8$
    b. $3x - 4y = -12, x = 4$
    c. $y = \dfrac{3x - 5}{2}, x = 7$

4.3 Solving Systems of Equations by the Substitution Method

Another method of solving systems of equations is the method of *substitution*. It is based on the property of equality that any quantity may be substituted for its equal in a statement without changing the truth or falseness of that statement. This method is outlined in the following steps.

Substitution method

1. Select *one of the equations* in the system and rewrite it to isolate one of the variables.
2. Substitute the resulting expression for the isolated variable into *the other equation*. This gives us a single equation in one variable.
3. Solve the equation in Step 2, if possible, for its variable.
4. Substitute the result from Step 3 into one of the original equations or into the rewritten equation of Step 1 to determine the value of the second variable.
5. Check the solution by substituting both values into the original equations.

4.3 Solving Systems of Equations by the Substitution Method

Using the substitution method, we have four choices for solving a system such as:

$$3x + y = 5$$
$$2x + 3y = 1$$

We could solve the first equation for either variable and substitute into the second equation; or solve the second equation for either variable and substitute into the first equation. The resulting equation is generally simpler if you solve for a variable that has a coefficient of 1 or -1.

Example 1

Solve the system by the substitution method:

$$3x + y = 5 \quad (1)$$
$$2x + 3y = 1 \quad (2)$$

Solve equation (1) for y since it has a coefficient of 1. Then substitute this expression for y in equation (2) and solve.

$y = -3x + 5 \quad (1)$ Solve equation (1) for y

$2x + 3(-3x + 5) = 1 \quad (2)$ Substitute

$2x - 9x + 15 = 1$ Solve for x

$-7x = -14$

$x = 2$

Now substitute $x = 2$ into the equation $y = -3x + 5$:

$y = -3x + 5$

$y = -3(2) + 5$

$y = -6 + 5$

$y = -1$

Solution: $(2, -1)$

The solution is checked by substituting both values into the two original equations.

$$3x + y = 5 \qquad\qquad 2x + 3y = 1$$
$$3(2) + (-1) \stackrel{?}{=} 5 \qquad 2(2) + 3(-1) \stackrel{?}{=} 1$$
$$6 + (-1) = 5 \qquad\qquad 4 + (-3) = 1$$
$$5 = 5 \qquad\qquad\qquad 1 = 1$$
$$\text{Yes} \qquad\qquad\qquad \text{Yes}$$

◀

Example 2

Solve the system by substitution:

$$2x + 5y = -4 \quad (1)$$
$$3x + 4y = 1 \quad (2)$$

Neither equation contains a variable whose coefficient is 1 or -1. So, arbitrarily, we solve equation (1) for x and substitute into equation (2).

$$2x + 5y = -4$$

$2x = -5y - 4$	Subtract $5y$ from both sides
$x = \dfrac{-5y - 4}{2}$	Divide both sides by 2
$3\left(\dfrac{-5y - 4}{2}\right) + 4y = 1$	Substitute into equation (2)
$3(-5y - 4) + 8y = 2$	Multiply by 2 to clear fractions
$-15y - 12 + 8y = 2$	Clear parentheses
$-7y = 14$	Solve for y
$y = -2$	

Substitute into the equation $x = \dfrac{-5y - 4}{2}$:

$$x = \dfrac{-5(-2) - 4}{2}$$

$$x = \dfrac{10 - 4}{2}$$

$$x = \dfrac{6}{2}$$

$$x = 3$$

Solution: $(3, -2)$
Check:

$2x + 5y = -4$	$3x + 4y = 1$
$2(3) + 5(-2) \stackrel{?}{=} -4$	$3(3) + 4(-2) \stackrel{?}{=} 1$
$6 - 10 = -4$	$9 - 8 = 1$
$-4 = -4$	$1 = 1$
Yes	Yes

◀

4.3 Solving Systems of Equations by the Substitution Method

Example 3

Solve the system by substitution:

$$4x - 2y = 10 \quad (1)$$
$$y = 2x - 5 \quad (2)$$

Since equation (2) is already solved for y, substitute this expression for y into equation (1).

$$4x - 2y = 10 \quad (1)$$
$$4x - 2(2x - 5) = 10 \qquad \text{Substitute}$$
$$4x - 4x + 10 = 10$$
$$10 = 10$$

The statement $10 = 10$ tells us that every solution of one equation is also a solution of the other. The system has infinitely many solutions. If graphed, both equations would yield the same line. ◀

Example 4

Solve the system by substitution:

$$2x - y = 5 \quad (1)$$
$$-4x + 2y = 7 \quad (2)$$

Solve equation (1) for y; then substitute into equation (2).

$$2x - y = 5 \quad (1)$$
$$-y = -2x + 5$$
$$y = 2x - 5 \qquad \text{Divide by } -1$$
$$-4x + 2y = 7 \quad (2)$$
$$-4x + 2(2x - 5) = 7 \qquad \text{Substitute into equation (2)}$$
$$-4x + 4x - 10 = 7$$
$$-10 = 7$$

Again, as with the elimination method, one number cannot be equal to a different number, so we conclude that this system has *no solution*. ◀

Selecting the Appropriate Method

In systems such as

$$4x + 5y = -2$$
$$3x - 2y = -13$$

where none of the variables has a coefficient of 1 or -1, the elimination method is preferred since it usually allows us to avoid the use of fractions.

Example 5

Solve by elimination:

$$4x + 5y = -2 \quad (1)$$
$$3x - 2y = -13 \quad (2)$$

Multiply both sides of the equation (1) by 2 and multiply both sides of equation (2) by 5 to make the coefficients of y opposites.

$$\left.\begin{array}{r} 8x + 10y = -4 \\ 15x - 10y = -65 \end{array}\right\} \quad \text{Add}$$

$$23x = -69$$

$$x = -3 \qquad \text{Solve}$$

Substitute $x = -3$ into equation (1) to find y.

$$4x + 5y = -2$$
$$4(-3) + 5y = -2$$
$$-12 + 5y = -2$$
$$5y = 10$$
$$y = 2$$

Solution: $(-3, 2)$ ◀

For a system such as

$$x = 4y + 5$$
$$5x - 7y = 12$$

where one of the variables has a coefficient of 1 or -1, or where one variable is already isolated, the substitution method is recommended.

Example 6

Solve by substitution:

$$x = 4y + 5 \quad (1)$$
$$5x - 7y = 12 \quad (2)$$

Substitute the expression for x from equation (1) into equation (2).

$$5x - 7y = 12 \quad (2)$$
$$5(4y + 5) - 7y = 12 \qquad \text{Substitute}$$
$$20y + 25 - 7y = 12 \qquad \text{Remove parentheses}$$
$$13y + 25 = 12$$
$$13y = -13$$
$$y = -1$$

4.3 Solving Systems of Equations by the Substitution Method

Substitute $y = -1$ into equation (1) to obtain x.

$x = 4y + 5$

$x = 4(-1) + 5$

$x = -4 + 5$

$x = 1$

Solution: $(1, -1)$ ◀

If a system of equations involves fractional coefficients, then, as in the previous section, we multiply both sides of each equation by its LCD to clear the equation of fractions. Then we proceed to solve the equation as usual (whether we use the elimination or substitution method).

Exercise Set 4.3

Solve each of the following systems by substitution:

1. $x + y = -6$
   $-x + y = 2$

2. $2y = -3x + 36$
   $y = 3x$

3. $x - 3y = 7$
   $2x - 6y = 14$

4. $y = 3x - 4$
   $5x + 2y = 25$

5. $3x - 5y = 49$
   $4x - y = 3$

6. $5x - 4y = 22$
   $x - 2y = 5$

7. $2x - 3y = 6$
   $x + 4y = -8$

8. $2x + y = 5$
   $3x - 2y = -10$

9. $3x + y = 12$
   $-2x + 3y = -8$

10. $2x + 5y = -9$
    $4x - y = 4$

11. $-2x + y = 7$
    $4x - 2y = 11$

12. $y = 3x + 5$
    $y = -2x - 10$

13. $4x - 3y = 11$
    $6x - 3y = 12$

14. $3x - 5y = -1$
    $7x + 4y = 29$

15. $2x + 3y = -7$
    $3x + 5y = -13$

16. $x + 5y = 8$
    $-2x - 10y = -16$

Solve each of the following systems by all three methods—graphing, elimination, and substitution:

17. $2x + 3y = 16$
    $x - 2y = 1$

18. $-2x + 3y = -14$
    $3x + 4y = 4$

19. $x - 2y = -5$
    $2x + y = 5$

20. $2x - 3y = 2$
    $5x - y = 5$

Solve each of the following systems by an appropriate method:

21. $5x - y = -8$
    $2x + y = 1$

22. $3x + y = -4$
    $-x - y = 0$

23. $2x + 3y = 18$
    $5x - 2y = 7$

24. $2x - 3y = -1$
    $7x + 5y = 43$

25. $14x + 21y = -5$
    $7x - 7y = 5$

26. $5x - 10y = -5$
    $x + 3y = 0$

27. $3y = -2x - 1$
    $x - y = 1$

28. $4y = -x + 2$
    $2x - y = -1$

29. $\dfrac{5x}{2} - y = 23$
    $\dfrac{3x}{4} + \dfrac{y}{3} = 5$

30. $\dfrac{2x}{5} - \dfrac{3y}{4} = 9$
    $\dfrac{x}{3} + \dfrac{y}{2} = 3$

31. Discuss the characteristics of a system of equations that suggests it should be solved using:

    a. the elimination method
    b. the substitution method

32. Multiply the second equation in the following system by 100 to remove decimals; then solve the resulting system as usual.

    $$y = 2x + 11$$
    $$-0.02x + 0.03y = 0.17$$

33. Multiply the first equation in the following system by 100 to remove decimals; then solve the resulting system as usual.

    $$0.03x - 0.04y = 0.07$$
    $$x = 2y + 1$$

Review

34. Express the value, in *cents*, of:

    a. 9 nickels
    b. 7 quarters
    c. x nickels
    d. y quarters

35. Express the following values in decimal form. (Evaluate where possible.)

    a. 8 percent of $3000
    b. $6\frac{1}{2}$ percent of $700
    c. 9 percent of x dollars
    d. $4\frac{1}{2}$ percent of y dollars

36. Find the slope of each indicated line:

    a. Passing through $(-3, 2)$ and $(3, -4)$
    b. Whose equation is $y = \dfrac{3}{2}x + 4$
    c. Whose equation is $4x - 5y = 40$

4.4 Solving Applications with Two Equations in Two Variables

This section contains a variety of applications similar to those of Chapter 2. These problems are solved by writing and solving a system of two equations in two variables. To solve each problem, the following guidelines are suggested.

Steps for solving applied problems

1. Read the problem carefully, until you understand the known facts and exactly what it is that you are to find.
2. Draw a sketch (where appropriate) or jot down the known facts.
3. Identify the unknowns. Let each unknown be represented by a variable.
4. From the stated facts of the problem, write a system with two equations, using the two variables.
5. Solve the system of equations by the elimination or substitution method.
6. Check the solution by returning to the original problem.

Example 1

A collection of 35 coins consists of nickels and quarters and has a value of $5.75. How many coins of each type are there in the collection?

Let n = the number of nickels. Assign variables to the unknowns

Let q = the number of quarters.

Now write two equations in n and q:

Quantity of coins: $\begin{pmatrix} \text{Number} \\ \text{of nickels} \end{pmatrix} + \begin{pmatrix} \text{Number} \\ \text{of quarters} \end{pmatrix} = \begin{pmatrix} \text{Number} \\ \text{of coins} \end{pmatrix}$

$$n \quad + \quad q \quad = \quad 35 \quad (1)$$

Value of coins: (Value of nickels) + (Value of quarters) = (Value of coins)

(in cents) $\quad\quad 5n \quad + \quad 25q \quad = \quad 575 \quad (2)$

These two equations make up the system:

$n + q = 35 \quad (1)$

$5n + 25q = 575 \quad (2)$

Multiply equation (1) by 5 to make the coefficients of n the same.

$\left.\begin{matrix} 5n + 5q = 175 \\ 5n + 25q = 575 \end{matrix}\right\}$ Subtract

$-20q = -400$

$q = 20$

Substituting $q = 20$ into equation (1)

$$n + q = 35$$
$$n + (20) = 35 \quad \text{Substitute}$$
$$n = 15$$

Solving the system gives $n = 15$ nickels and $q = 20$ quarters. ◀

Example 2   A woman is 25 years older than her son. Five years ago she was six times as old as her son. How old is each of them now?

$\quad$ Let $x =$ woman's age now.

$\quad$ Let $y =$ son's age now. $\qquad$ Assign variables

One equation in our system is:

$\quad$ (Woman's age now) = (Son's age now) + 25

$$x \quad = \quad y \quad + 25 \quad (1)$$

Then,

$\quad x - 5 =$ woman's age 5 years ago

$\quad y - 5 =$ son's age 5 years ago

The second equation in our system is:

$\quad$ Woman's age 5 years ago = $6 \cdot$ (Son's age 5 years ago)

$$x - 5 \quad = \quad 6(y - 5)$$

Or,

$$x - 5 = 6(y - 5)$$
$$x - 5 = 6y - 30$$
$$x - 6y = -25 \quad (2)$$

Therefore, the system for this problem is:

$$x = y + 25 \quad (1)$$
$$x - 6y = -25 \quad (2)$$

Since x is isolated in equation (1), we choose to solve this system by substitution. We substitute this expression for x into equation (2)

$$x - 6y = -25 \quad (2)$$
$$(y + 25) - 6y = -25 \qquad \text{Substitute}$$
$$-5y + 25 = -25 \qquad \text{Remove parentheses}$$

4.4 Solving Applications with Two Equations in Two Variables

$$-5y = -50$$
$$y = 10 \quad \text{Son's age now}$$

Substitute $y = 10$ into equation (1) to find x.

$$x = y + 25$$
$$x = (10) + 25$$
$$x = 35 \quad \text{Woman's age now} \quad \blacktriangleleft$$

Example 3

How many pounds of $1.55 per pound candy and how many pounds of $2.15 per pound candy should be combined to make 45 pounds of a mixture to sell for $1.79 per pound?

Let x = number of pounds of $1.55 candy.

Let y = number of pounds of $2.15 candy.

Then,

$$\begin{pmatrix}\text{Number of pounds} \\ \text{of \$1.55 candy}\end{pmatrix} + \begin{pmatrix}\text{Number of pounds} \\ \text{of \$2.15 candy}\end{pmatrix} = \begin{pmatrix}\text{Total pounds} \\ \text{of candy}\end{pmatrix}$$

$$x \quad + \quad y \quad = \quad 45 \quad (1)$$

The second equation comes from the observation:

$$\begin{pmatrix}\text{Value of} \\ \text{\$1.55/lb} \\ \text{candy}\end{pmatrix} + \begin{pmatrix}\text{Value of} \\ \text{\$2.15/lb} \\ \text{candy}\end{pmatrix} = \begin{pmatrix}\text{Value of} \\ \text{\$1.79/lb} \\ \text{mixture}\end{pmatrix}$$

$$1.55x \quad + \quad 2.15y \quad = \quad 1.79(45) \quad (2)$$

Note: The value in dollars is the product of the number of pounds of candy times the price in dollars per pound.

The system for this problem is:

$$x + y = 45 \quad (1)$$
$$1.55x + 2.15y = 80.55 \quad (2)$$

We now multiply equation (2) by 100 to clear decimals from the equation. Multiply equation (1) by 155 to make the coefficients of x the same in both equations; then subtract.

$$x + y = 45 \qquad \text{Multiply by 155}$$
$$1.55x + 2.15y = 80.55 \qquad \text{Multiply by 100}$$

$$155x + 155y = 6975$$
$$\underline{155x + 215y = 8055} \qquad \text{Subtract}$$
$$-60y = -1080$$

$$y = \frac{-1080}{-60}$$

$$y = 18 \text{ pounds of } \$2.15 \text{ candy}$$

Substitute $y = 18$ into equation (1) to find x.

$$x + y = 45 \quad (1)$$

$$x + (18) = 45$$

$$x = 27 \text{ pounds of } \$1.55 \text{ candy} \quad \blacktriangleleft$$

Example 4

The length of a rectangle is 4 cm less than twice the width. If the perimeter of the rectangle is 40 cm, what are the dimensions of the rectangle?

Let L = the length of the rectangle.

Let W = the width of the rectangle.

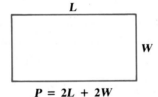

$$P = 2L + 2W$$

The system is

$$2L + 2W = 40 \quad (1) \qquad \text{Perimeter is 40}$$

$$L = 2W - 4 \quad (2)$$

Since L is isolated in equation (2), substitute for L into equation (1).

$$2L + 2W = 40$$

$$2(2W - 4) + 2W = 40 \qquad \text{Substitute}$$

$$4W - 8 + 2W = 40 \qquad \text{Remove parentheses}$$

$$6W - 8 = 40$$

$$6W = 48$$

$$W = 8 \text{ cm} \qquad \text{Width}$$

Substitute $W = 8$ into equation (2) to find L.

$$L = 2W - 4$$

$$L = 2(8) - 4$$

$$L = 16 - 4$$

$$L = 12 \text{ cm} \qquad \text{Length} \qquad \blacktriangleleft$$

4.4 Solving Applications with Two Equations in Two Variables

Example 5

Part of $9200 was invested at 7 percent and the rest at 6 percent interest. If the total annual interest from the investments is $606, how much was invested at each rate?

Let x = amount invested at 7 percent.

Let y = amount invested at 6 percent.

Then,

$$\begin{pmatrix} \text{Amount invested} \\ \text{at 7 percent} \end{pmatrix} + \begin{pmatrix} \text{Amount invested} \\ \text{at 6 percent} \end{pmatrix} = \begin{pmatrix} \text{Total} \\ \text{investment} \end{pmatrix}$$

$$x \quad + \quad y \quad = \quad 9200 \quad (1)$$

To calculate interest we use the formula for simple interest:

$$I = P \cdot R \cdot T$$

where
I = amount of *interest* earned

P = *principal*, amount invested

R = annual *rate* of interest

T = *time*, in years

In this problem, we have *annual* interest to calculate, so $t = 1$ and

$$I = P \cdot R$$

Now, the interest earned by the 7 percent account is:

$$I = P \cdot R$$
$$I = x \cdot (0.07)$$
$$I = 0.07x$$

The interest earned by the 6 percent account is:

$$I = P \cdot R$$
$$I = y \cdot (0.06)$$
$$I = 0.06y$$

$$\begin{pmatrix} \text{Interest earned} \\ \text{at 7 percent} \end{pmatrix} + \begin{pmatrix} \text{Interest earned} \\ \text{at 6 percent} \end{pmatrix} = \begin{pmatrix} \text{Total interest} \\ \text{earned} \end{pmatrix}$$

$$0.07x \quad + \quad 0.06y \quad = \quad 606 \quad (2)$$

The system used to solve this problem is:

$$x + y = 9200 \quad (1)$$
$$0.07x + 0.06y = 606 \quad (2)$$

Multiply equation (1) by 6 and equation (2) by 100 to make the coefficients of y the same.

$$6x + 6y = 55200$$
$$7x + 6y = 60600$$
Subtract
$$-x = -5400$$
$$x = \$5400$$

Substitute $x = 5400$ into equation (1) to find y.

$$x + y = 9200$$
$$5400 + y = 9200$$
$$y = \$3800$$

The amount invested at 7 percent is $5400.
The amount invested at 6 percent is $3800. ◀

Exercise Set 4.4

Solve each of the problems below by writing and solving a system of two equations in two variables.

1. Find two numbers whose sum is 38 and whose difference is 12.

2. One number is twice another and their sum is 42. What are the numbers?

3. A trigonometry text and a calculator together cost $55. What was the cost of each if the calculator cost $21 less than the text?

4. A sport coat and a pair of slacks cost $329. The coat cost $5 more than three times the cost of the slacks. What was the cost of each?

5. Two complementary angles have a difference of 36°. What is the measure of each? (Complementary angles add up to 90°.)

6. What is the measure of two supplementary angles whose difference is 16°? (Supplementary angles add up to 180°.)

7. A basketball team scored 69 times for a total of 122 points. How many field goals (worth 2 points each) and how many free throws (worth 1 point each) were scored?

8. The home team lost a football game by 3 points. If the two teams scored a total of 59 points, then what was the final score? _for each team_

9. The length of a rectangle is 3 cm less than twice the width. If the perimeter is 36 cm, what are the dimensions?

10. The length of a certain rectangle is 1 cm longer than twice the width, and the perimeter is 50 cm. What are the dimensions?

11. Part of $10,000 was invested at 6 percent and the rest at 5 percent. If the annual interest from the 6 percent investment was $270 more than

4.4 Solving Applications with Two Equations in Two Variables

the interest of the 5 percent investment, then how much was invested at each rate?

12. Part of $6600 was invested at 5 percent and the rest at 6 percent. How much was invested at each rate if the two annual interests were the same?

13. A man is 24 years older than his son. Ten years ago he was 7 times as old. How old is each now?

14. A girl is 3 years older than her brother, and the sum of their ages is 27. How old is each?

15. Fourteen nickels and dimes have a value of $1.25. How many of each coin are there?

16. A collection of dimes and quarters has a value of $3.85. If there are 22 coins in the collection, how many of each type are there?

17. Some 25-cent and 15-cent stamps cost $5.20. How many of each were there if the number of 25-cent stamps was twice the number of 15-cent stamps?

18. How many quarters and dimes would you have if the number of quarters was four more than the number of dimes and the total value of the coins was $3.80?

19. How much of a 15 percent solution and how much of a 40 percent solution should be used to make 50 liters of a 25 percent solution?

20. How much of a 20 percent solution and how much of a 50 percent solution should be used to make 60 liters of a 40 percent solution?

21. How many pounds of 55-cent per pound candy and how many pounds of $1.00 per pound candy should be combined to make 50 pounds of a mixture to sell at 73 cents per pound?

22. How many pounds of $1.00 per pound and how many pounds of $1.80 per pound candy should be combined to make 70 pounds of a mixture to sell for $1.40 per pound?

Review

23. Solve each one-variable inequality and graph on a number line:

    a. $3x - 5 \geq x + 11$ b. $-5 < 8 - 2x \leq 14$

24. Solve each two-variable inequality and graph in an x-y plane:

    a. $2x - 5y < 20$ b. $y \geq -\frac{2}{3}x + 3$ c. $x \geq -3$

25. Solve the following equation for B:

    $$A = \frac{h}{2}(b + B)$$

4.5 Systems of Inequalities

In preparation for solving systems of inequalities, we quickly review how to graph inequalities such as $Ax + By \ \textcircled{R}\ C^*$ (see Section 3.5). Recall the graph of the equation $Ax + By = C$ is the boundary line that divides the x-y plane into two *half-planes*. (See Figure 4.5-1.) The set of all points satisfying an inequality such as $Ax + By < C$ is all the points in one of the two half-planes.

FIGURE 4.5-1
Half-Planes Determined by a Line

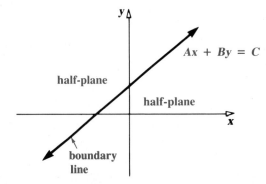

To graph the solution of $Ax + By \ \textcircled{R}\ C$:

1. Draw the graph of the boundary line $Ax + By = C$
   a. *Solid* if the "=" is included in $\textcircled{R}$.
   b. *Dashed* if the "=" is not included in $\textcircled{R}$.

2. Pick an arbitrary point in one of the half-planes determined by the boundary line. Substitute its coordinates into $Ax + By \ \textcircled{R}\ C$.
   a. If the inequality is made *true* by the selected point, shade the half-plane containing that point.
   b. If the inequality is made *false* by the selected point, shade the half-plane on the opposite side of the line from that point.

Example 1

Graph the inequality $2x - 3y \leq 6$.

Draw the graph of the corresponding equation $2x - 3y = 6$ using a solid line since $\leq$ is used. Test the coordinates of the origin (0, 0). This is a good test point since it is not on the boundary line $2x - 3y = 6$. The statement becomes

$$2x - 3y \leq 6$$
$$2(0) - 3(0) \leq 6$$

---

*Recall the symbol $\textcircled{R}$ is meant to include any of the symbols $<$, $\leq$, $>$, or $\geq$.

4.5 Systems of Inequalities

$$0 - 0 \leq 6$$
$$0 \leq 6$$

which is true. Therefore, we shade the side of the line (half-plane) that includes the origin.

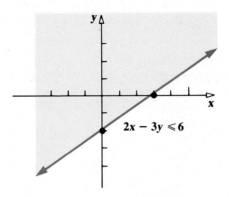

The graph of the inequality involving only one variable $x \geq 4$ is the set of all points whose x-coordinates are greater than or equal to 4; that is, all points that are to the right of or on the line $x = 4$. Similarly, the graph of the inequality $-2 < x < 5$ is the set of all points whose x-coordinates are between the lines $x = -2$ and $x = 5$. The lines themselves are included or excluded depending on whether the equality symbol is included or excluded. See Figure 4.5-2. Similarly, graphs of $y < 3$ and $-1 \leq y \leq 4$ are as shown in Figure 4.5-2.

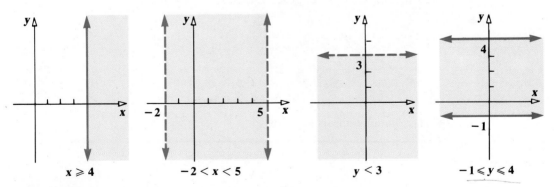

FIGURE 4.5-2
Graph of Inequalities with Only One Variable

Systems of Inequalities

A set of two (or more) inequalities in two variables is called a **system of inequalities**. To solve a system of inequalities is to find all ordered pairs of numbers that satisfy every inequality in the system. Since the solution to a single inequality is obtained by graphing, it follows that systems of inequalities are also solved graphically.

Steps for solving systems of inequalities

1. Graph all inequalities in the system in the same coordinate plane. It is recommended that a different shading pattern be used to graph each inequality.

2. The solution to the system consists of all points lying in the "overlap" of the shaded regions from Step 1. Shade this area heavily, indicating the graph of the complete solution.

Example 2

Solve the system:

$$2x - 3y \geq 6$$
$$x + y < 4$$

Graph both inequalities in the same x-y plane.

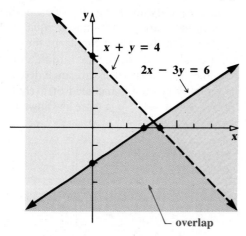

The solution appears as follows:

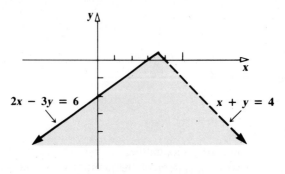

4.5 Systems of Inequalities

Example 3

Solve the system:

$$-3 \leq x < 2$$
$$2 \leq y < 5$$

Graph both inequalities on the same x-y plane.

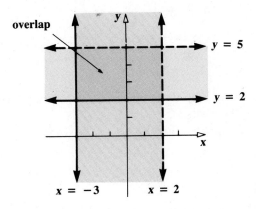

The solution appears as follows:

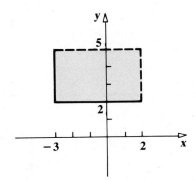

Example 4

Graph the system:

$$2x + y < 4$$
$$x \geq 0$$
$$1 < y < 3$$

Graph all inequalities in the same x-y plane.

218   FOUR   Systems of Equations and Inequalities

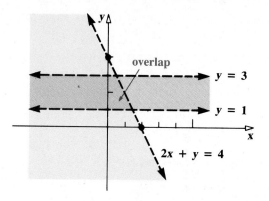

Note $x \geq 0$ implies we keep those points in the "overlap" region, which are to the right of or on the y-axis.

The solution appears as follows:

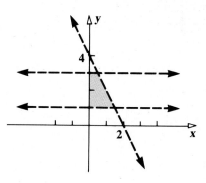

Exercise Set 4.5

Draw the graph of each of the following inequalities:

1. $2x + 3y < 12$
2. $2x - 3y \geq 12$
3. $-x + 3y \leq 6$
4. $-x + 3y > 6$
5. $y < \frac{2}{3}x + 3$
6. $y \geq \frac{2}{3}x - 5$
7. $x > 5$
8. $x \geq 7$
9. $y \leq 4$
10. $y < -6$
11. $x \geq 0$
12. $x > 0$
13. $y > 0$
14. $y \geq 0$
15. $-2 < x \leq 3$
16. $-1 \leq y < 5$
17. $-1 \leq y < 4$
18. $3 < x \leq 6$

Draw the graph of the solution of each of the following systems of inequalities:

19. $2x + y < 4$
    $x + 2y \geq 2$
20. $3x + 2y \leq 6$
    $x + 4y > 4$
21. $2x - 3y > 6$
    $3x + 2y \leq 12$

4.6 Review and Chapter Test

22. $x + 2y \geq 2$
    $3x - y < -1$

23. $3x + y < 3$
    $y \geq 3x - 1$

24. $x - 4y \geq 8$
    $y < -2x + 3$

25. $2x - y < 6$
    $x + y > 4$
    $x \geq 0$

26. $x - y \geq 3$
    $2x + 3y < 6$
    $x \geq 0$

27. $x - y > -3$
    $3x + 2y < 6$
    $y \geq 0$

28. $x + 2y < 4$
    $x - y < 2$
    $x \geq 1$

29. $x + y < 3$
    $x > 0$
    $y \geq 0$

30. $x - 2y > -4$
    $x \leq -1$
    $y \geq 0$

31. $x + y > -3$
    $x \leq 1$
    $y < -1$

32. $2x - 3y < 6$
    $x \geq 0$
    $y < 0$

Review

33. Write each of the following expressions in expanded form:

    a. b^3
    b. $6mn^3$
    c. $2x^2y$
    d. $5(x + y)^2$

34. Write each of the following expressions in exponential form:

    a. $2 \cdot 2 \cdot 2 \cdot x \cdot x$
    b. $a \cdot a \cdot a \cdot b \cdot c \cdot c \cdot c \cdot c$
    c. $2 \cdot 2 \cdot 3 \cdot 5 \cdot 5$
    d. $(x - 2)(x - 2)(x - 2)$

35. Evaluate each expression for the given value of the variable(s):

    a. $x^3 - 3x^2 + 6x + 1$ for $x = 2$
    b. $x^2 + 2xy - y^2$ for $x = 3, y = -1$
    c. $(a + 3)(a - 4)$ for $a = 6$
    d. $(2a + b)(a - 3b)$ for $a = 4, b = -2$

4.6 Review and Chapter Test

A set of two equations in two variables is called a *system of equations*. To *solve* a system of equations is to find all ordered pairs of numbers that satisfy both equations in the system.

Example 1

The following equations form a system of two linear equations in two variables, x and y.

$$x + y = -1$$
$$-x + y = 5$$

Example 2

Determine whether the ordered pair $(-3, 2)$ is a solution to the system:

$$x + y = -1$$
$$-x + y = 5$$

We substitute $(-3, 2)$ into each equation.

$$\begin{array}{cc} x + y = -1 & -x + y = 5 \\ (-3) + 2 \stackrel{?}{=} -1 & -(-3) + 2 \stackrel{?}{=} 5 \\ -1 = -1 & 5 = 5 \\ \text{Yes} & \text{Yes} \end{array}$$

Since $(-3, 2)$ is a solution to both equations in the system, then $(-3, 2)$ is a solution to the system. ◀

(4.1) A system of two linear equations in two variables, x and y, may be solved *by graphing*. Draw the graphs of both equations in an x-y plane. The points of intersection of the two lines (if any) are the solution to the system.

Example 3

Solve by graphing:

$$3x + y = 5$$
$$x - 2y = -3$$

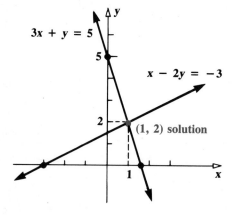

The point of intersection of the two graphs is found by inspection to be $(1, 2)$ and is the solution to the system. This is checked by substituting $x = 1$ and $y = 2$ into both equations in the system. ◀

Example 4

Solve by graphing:

$$2x + 3y = 6$$
$$4x = -3 - 6y$$

4.6 Review and Chapter Test

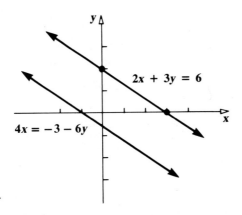

The lines are parallel and do not intersect, indicating that this system has *no solution*. ◄

Example 5

Solve by graphing:

$$x - 3y = 2$$
$$3x = 9y + 6$$

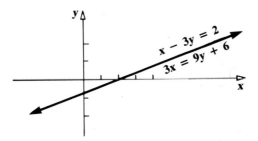

The graphs of these two equations are collinear. Every point that satisfies one equation must satisfy the other. Thus, this system has *an infinite number of solutions*. ◄

(4.2) A system of two equations in two variables, x and y, can be solved by using the *elimination method*:

1. Write each equation in the form $Ax + By = C$. Multiply both sides of the equation by the LCD to remove fractions, if present.

2. Select a variable to be eliminated and, if necessary, multiply each equation in the system by an appropriate constant to make the coefficient of that variable the same in each equation, except possibly for the sign.

3. Add or subtract the equations, as appropriate, to eliminate one variable and create a single equation in the remaining variable.

4. Solve the single equation for the remaining variable.
5. Substitute the result from Step 4 into one of the original equations to determine the value of the variable that was eliminated.
6. Check the results in both of the original equations.

Example 6

Solve by the elimination method:

$$5x + 3y = 5 \quad (1)$$
$$3x + 2y = 4 \quad (2)$$

We will eliminate y.

$5x + 3y = 5$ Multiply by 2
$3x + 2y = 4$ Multiply by 3

$$\left.\begin{array}{r} 10x + 6y = 10 \\ 9x + 6y = 12 \end{array}\right\} \text{Subtract}$$

$$x = -2$$

Substitute $x = -2$ into equation (2).

$$3x + 2y = 4$$
$$3(-2) + 2y = 4$$
$$-6 + 2y = 4$$
$$2y = 10$$
$$y = 5$$

Solution: $(-2, 5)$

Note: If the result of adding or subtracting the two equations yields an expression of the type $0 = 0$, this implies there are an infinite number of solutions. Also, if the result is an expression of the type $0 = 5$, this implies the system has no solution. ◀

(4.3) A system of two equations in two variables, x and y, can be solved by using the *substitution method*:

1. Select *one of the equations* in the system and rewrite it to isolate one of the variables.
2. Substitute the resulting expression for the isolated variable into *the other equation*. This gives us a single equation in one variable.
3. Solve the equation in Step 2 for its variable.

4.6 Review and Chapter Test

4. Substitute the result from Step 3 into the rewritten equation in Step 1 to determine the value of the second variable.
5. Check the solution by substituting into the original equations.

Example 7

Solve by substitution:

$$2x - y = 5 \quad (1)$$
$$3x + 4y = 13 \quad (2)$$

Solve equation (1) for y because its coefficient is -1.

$$2x - y = 5 \quad (1)$$
$$-y = -2x + 5$$
$$y = 2x - 5$$

Substitute this result for y into equation (2).

$$3x + 4y = 13 \quad (2)$$
$$3x + 4(2x - 5) = 13$$
$$3x + 8x - 20 = 13$$
$$11x - 20 = 13$$
$$11x = 33$$
$$x = \frac{33}{11}$$
$$x = 3$$

Substitute $x = 3$ into equation (1) to find y.

$$2x - y = 5$$
$$2(3) - y = 5$$
$$6 - y = 5$$
$$-y = -1$$
$$y = 1$$

Solution: (3, 1)   ◀

As in the elimination method, when substitution yields a result such as $0 = 0$ or $5 = 5$, the system has an infinite number of solutions. When substitution yields an impossible result such as $0 = -7$, the system has no solution.

(4.4) Many applied problems may be solved by writing and solving a system of two equations in two variables.

Steps for solving applied problems

1. Read the problem carefully, until you understand the known facts and exactly what it is that you are to find.
2. Draw a sketch (where appropriate) or jot down the known facts.
3. Identify the unknowns; then let each unknown be represented by a variable.
4. From the stated facts of the problem, write a system with two equations, using the two variables.
5. Solve the system of equations by the elimination or substitution method.
6. Check the solution by returning to the original problem.

Example 8

The home basketball team won a game by 16 points. If the two teams scored a total of 134 points, then what was the final score?

Let x = points scored by the home team.

Let y = points scored by the visiting team.

Then,

$x + y = 134$ (1)   Total points

$x - y = 16$ (2)   Since home score is higher by 16 points

Let's use the elimination method to eliminate y.

$$\left.\begin{array}{r} x + y = 134 \\ x - y = 16 \end{array}\right\} \text{Add}$$

$2x \quad = 150$

$x = 75$ Points scored by home team

Substitute into equation (1) to find y.

$x + y = 134$

$(75) + y = 134$

$y = 59$ Points scored by visiting team

The home team won the game by a score of 75 to 59.   ◀

(4.5)

To graph the solution of $Ax + By \; ℝ \; C$ (where ℝ is $<$, $>$, $\leq$, or $\geq$):

1. Draw the graph of the boundary line $Ax + By = C$
   a. *Solid* if the "$=$" is included in ℝ.

b. *Dashed* if the "=" is not included in Ⓡ.
2. Pick an arbitrary point in one of the half-planes determined by the boundary line. Substitute its coordinates into $Ax + By$ Ⓡ C.
   a. If the inequality is made *true* by the selected point, shade the half-plane containing that point.
   b. If the inequality is made *false* by the selected point, shade the half-plane on the opposite side of the boundary line from that point.

Example 9

Graph the inequality $3x + 2y > 6$.

Draw the graph of the corresponding equation $3x + 2y = 6$ using a dashed line since ">" is used. Test the coordinates of the origin $(0, 0)$. The statement becomes $0 + 0 > 6$, which is a false statement. Therefore, we shade the half-plane on the *opposite* side of the line from $(0, 0)$.

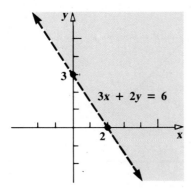

(4.6) A set of two or more inequalities in two variables is a *system of inequalities*. Systems of inequalities are also solved graphically.

Steps for solving systems of inequalities

1. Graph all inequalities in the system in the same x-y plane. It is recommended that a different shading pattern be used to graph each inequality.
2. The solution to the system consists of all points lying in the "overlap" of the shaded regions from Step 1. Shade this area heavily.

Example 10

Solve the system:

$$x + y \leq 5$$
$$2x - 3y > 6$$

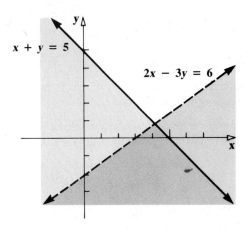

The solution appears as follows:

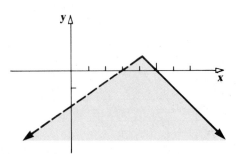

Exercise Set 4.6 Review

Solve each of the systems in Problems 1–4 by graphing:

1. $x + 2y = 7$
   $2x + y = 8$

2. $3x - 2y = 6$
   $4y = 6x - 12$

3. $x - 3y = 5$
   $-2x + 6y = 7$

4. $3x + 4y = -2$
   $-x + y = 3$

Solve each of the systems in Problems 5–14 by the elimination method:

5. $2x - y = 3$
   $x + y = 5$

6. $x - 2y = 3$
   $2x - y = 4$

7. $x + 3y = 5$
   $-x - y = -3$

8. $3x + 5y = -2$
   $-9x - 15y = 10$

9. $4x - 3y = 13$
   $3x + 5y = -12$

10. $3y = 2 - 2x$
    $6x + 9y = 6$

11. $y = \frac{2}{3}x - 1$
    $y = +\frac{3}{4}x + 2$

12. $9x - 6y = -15$
    $2y = 3x + 5$

13. $\dfrac{x}{6} + \dfrac{y}{5} = 7$
    $\dfrac{2x}{3} - \dfrac{y}{3} = 11$

14. $\dfrac{3x}{5} - \dfrac{y}{2} = 6$
    $\dfrac{5x}{2} + \dfrac{2y}{3} = 25$

Solve each of the systems in Problems 15–20 by the substitution method:

15. $7x - 5y = 3$
    $4x - y = -2$

16. $7x + 4y = 26$
    $-x + 3y = 7$

17. $y = 3x - 2$
    $y = -2x + 3$

18. $2x + 6y = 3$
    $x = -3y + 2$

19. $4x - 8y = 12$
    $x = 2y + 3$

20. $4x + y = 7$
    $-x - 3y = 6$

21. A suit and a pair of shoes cost $430. The suit cost $10 more than three times the cost of the shoes. What was the cost of each?

22. A collection of nickels and quarters has a value of $3.75. If there are 27 coins in the collection, how many of each type are there?

Draw the graph of the inequalities in Problems 23–26:

23. $2x - y \geq 5$
25. $x \leq 2$

24. $x - 3y < 6$
26. $y > 3$

Draw the solution to the systems in Problems 27–32:

27. $x + 2y > 4$
    $x - y \leq 2$

28. $2x - y \leq 6$
    $x + y > 3$

29. $-2 < x \leq 3$
    $1 \leq y \leq 4$

30. $1 < x < 4$
    $0 < y \leq 4$

31. $2x + y \geq 3$
    $-x + 2y < 2$
    $y \geq 0$

32. $3x - 2y < 6$
    $2x + 3y \geq 6$
    $x \geq 0$

Chapter 4 Review Test

1. Without solving the system, determine whether any of the given ordered pairs is a solution of the system:

$$3x - y = -14$$
$$-2x + 5y = 18$$

   a. $(-3, 5)$ b. $(1, 4)$ c. $(2, -4)$ d. $(-4, 2)$

2. Solve the following systems of equations, if possible, by graphing:

   a. $2x + y = -2$
      $y = 3x + 13$

   b. $x - 2y = -4$
      $3x + 2y = 12$

   c. $2x - y = 5$
      $y = 2x - 3$

3. Solve the following systems of equations, if possible, *by elimination*:

   a. $3x + y = -2$
   $-6x + 3y = -18$

   b. $x + y = 16$
   $0.05x + 0.1y = 1.15$

4. Solve the following systems of equations, if possible, *by substitution*:

   a. $2x - y = 14$
   $3x - y = 21$

   b. $2x + 3y = 9$
   $y = -\dfrac{2}{3}x + 3$

5. Write a system of two equations in two variables to solve the following problems. Show the equations and solve:

   a. Bruce is three years older than his wife, Joanne. The sum of their ages is 105. How old is each?

   b. Susan invested part of $7000 at 8 percent annual interest and the rest of it at 6 percent. In one year she earned the same interest in each account. How much did she invest at each rate?

6. Graph the solution of each system of inequalities:

   a. $x + y > 2$
   $2x - y \geq 11$

   b. $-3 < x < 4$
   $1 \leq y \leq 3$

   c. $2x + 3y < 12$
   $2x - y < 4$
   $x \geq 0$

CHAPTER FIVE

POLYNOMIALS

Chapter 1 presented a general discussion of algebraic expressions as statements combining constants, variables, operations of arithmetic, and symbols of grouping. To study algebraic expressions more thoroughly, it is convenient to classify them according to some property they share. This chapter introduces the type of algebraic expression called a *polynomial*.

Some definitions and procedures from earlier sections (most notably, Section 2.1) are repeated here since they apply directly to the operations on polynomials.

5.1 Definitions

A **polynomial** is an algebraic expression formed by combining real number constants and variables using the operations of addition, subtraction, and multiplication. As a result of this definition, we recognize a polynomial by noting that any variable that appears must have a whole number exponent, and a variable may not appear in a denominator. The following expressions are polynomials:

$$2, \quad x^2 + 3, \quad \sqrt{2}xy, \quad \frac{x^2}{3} + 4y, \quad xyz + 2x^2, \quad 3x^2 + 4xy - 5y^2$$

Note: $\frac{x^2}{3}$ can be interpreted as $\frac{1}{3} \cdot x^2$, so it satisfies the definition. The following expressions are *not* polynomials for the reason indicated:

$2\sqrt{x}$ May not have roots of variables. (We will see in a later chapter that this is equivalent to having fractional exponents.) Exponents of the variables must be whole numbers.

$\dfrac{3 + x}{y}$ Variables may not appear in the denominator.

$2x^{-1} + y^{-1/2}$ Exponents of the variables must be whole numbers

The parts of a polynomial that are added (or subtracted) together are *terms*. The terms of a polynomial may be written in any order, but the sign preceding each term must still precede it in the rewritten form (commutative property of addition). Consider the following polynomial and several alternative forms in which it may be written:

$$2x^3 - 5 + x - 3x^2 = -5 + x - 3x^2 + 2x^3$$
$$= 2x^3 - 3x^2 + x - 5 \quad \text{Generally preferred}$$
$$= x - 5 + 2x^3 - 3x^2$$

Each term of a polynomial generally is the product of a constant, called the *numerical coefficient*, and a letter or product of letters, called the *literal part*. A term that has no literal part is called a *constant term*. If a term does not have a numerical coefficient written, the coefficient is understood to be one (1). This is consistent with the property of real numbers (identity element) that $1 \cdot a = a$ for any real number a.

Example 1

Identify the numerical coefficient and the literal part of each term of the following polynomials:

Polynomial	Term		Numerical Coefficient	Literal Part
A. $3x^2 + 2xy + y^2$	1st	$(3x^2)$	3	x^2
	2nd	$(2xy)$	2	xy
	3rd	(y^2)	1	y^2
B. $x^2 - 5$	1st	(x^2)	1	x^2
	2nd	(-5)	-5	none
C. $3x^3 - 2xy - y^2 + x$	1st	$(3x^3)$	3	x^3
	2nd	$(-2xy)$	-2	xy
	3rd	$(-y^2)$	-1	y^2
	4th	(x)	1	x

Classification of Polynomials

For convenience, there are several ways of classifying polynomials:

1. By number of terms
2. By number of variables
3. By degree

5.1 Definitions

A polynomial of one term is a **monomial**; one of two terms is a **binomial**; and one of three terms is a **trinomial**. No special name is used for a polynomial of more than three terms.

To classify a polynomial by number of variables, only the number of different variables appearing in the polynomial is considered. It doesn't matter how many terms the variables appear in.

The **degree of a term** of a polynomial is the sum of the exponents on the variable(s). A non-zero constant term (one with no literal part) is defined as being of degree zero (0). The constant zero is said to have *no* degree. The **degree of a polynomial** is the same as the degree of the term with the highest degree.

Example 2

Classify the following polynomials by number of terms, number of variables, and degree:

Polynomial	Number of Terms, Name	Number of Variables	Degree
A. $a^2 - 3a + 4$	3, trinomial	1	2
B. $2x^3 - 3xy^2 + 2y - 4$	4, polynomial	2	3
C. 7 (monomial)	1, monomial	0	0
D. $3xy^3$	1, monomial	2	4
E. $xy + 2z^2$	2, binomial	3	2
F. 0	1, monomial	0	none

◂

Recall that an algebraic expression such as a polynomial may be evaluated if numeric values of the variables are given. That is, polynomials are used to represent various real numbers.

Example 3

Evaluate the polynomial $x^2y + 2xy - xy^2$ for:

A. $x = 2, y = 3$

$x^2y + 2xy - xy^2$
$= (2)^2(3) + 2(2)(3) - (2)(3)^2$ Substitute
$= (4)(3) + 4(3) - (2)(9)$ Evaluate
$= 12 + 12 - 18$
$= 6$

B. $x = -2, y = 4$

$x^2y + 2xy - xy^2$

$= (-2)^2(4) + 2(-2)(4) - (-2)(4)^2$ Substitute

$= (4)(4) + (-4)(4) + 2(16)$ Evaluate

$= 16 - 16 + 32$

$= 32$ ◀

Note that when we change the value of the variable(s), the value of the polynomial will usually change also.

In the same way that real numbers are combined using the operations of arithmetic, so can polynomials be added, subtracted, multiplied, and divided, under certain restrictions. The next three sections discuss the arithmetic of polynomials.

Exercise Set 5.1

Which of the following algebraic expressions are polynomials, and which are not?

1. $2x + 3$
2. $3 - 4y$
3. $3x^2 + 5x - 1$
4. $2y^2 + 3y - 8$
5. $\dfrac{3x + 4}{2x}$
6. $\dfrac{4y - 1}{2xy}$
7. $3x + \sqrt{y}$
8. $\sqrt{xy} - 2$
9. $2x^{-2} + 3x^2$
10. $x^{1/2} + y - 2xy^2$

State the numerical coefficient and the literal part of each term of each polynomial. Then classify each polynomial by number of terms, number of variables, and degree.

11. $2x^2 - 5x + 1$
12. $4y^2 + 3y + 2$
13. $x + 2y - 3$
14. $2x + y + 6$
15. $x^2 + xy - y^2$
16. $2xy - y^2$
17. $2xy$
18. $3z^2$

Evaluate each of the following polynomials for $x = 2, y = 3$:

19. $x^2 + 2x - 8$
20. $y^2 - 5y + 6$
21. $2x^2 + xy - y^2$
22. $x^2 + 2xy - 3y^2$
23. $x^3 + 3x^2 + x - 4$
24. $y^3 - 3y^2 + 2y - 1$
25. $x^2 + 2xy + 3y^2 - 2x + y - 3$
26. $x^2 - xy - 2y^2 + 3x + y + 4$

27. Differentiate between the degree of a term and the degree of a polynomial.

Review

28. Remove symbols of grouping and combine like terms.
    a. $(6x - 4y) + (x + 3y) - (2x - 5y)$
    b. $(x^2 + 3x + 5) + (2x^2 - 3x - 11)$
    c. $(m^2 - 8)(m^2 - 3m - 4) - (2m^2 + 5m - 4)$

29. Solve the following equation:
    $$3a - 2(a - 3) = 9 - 4a$$

30. Solve and graph the following inequality:
    $$3a - 2(a - 3) < 9 - 4a$$

31. Draw the graph of $3x - y = 6$.

5.2 Addition and Subtraction of Polynomials

[handwritten: Sum of coefficients variable does not change]

In a polynomial, two or more terms that have identical literal parts are called **like** (or **similar**) **terms**. Like terms may be combined into a single term by adding their numerical coefficients while the literal part remains unchanged. The distributive property of real numbers is the key to this process. Recall

$$ab + ac = a(b + c) \quad \text{or equivalently} \quad ba + ca = (b + c)a$$

Note that if the sum of the numerical coefficients is zero, the term need not be written.

Example 1

Simplify each of the following polynomials by combining like terms:

A. $8xy + 4xy - 3xy + xy$

Since all terms have the same literal part, xy, all four terms are like terms.

— Understood coefficient of 1

$$8xy + 4xy - 3xy + xy = (8 + 4 - 3 + 1)xy$$
$$= (10)xy$$
$$= 10xy$$

Note that the two middle steps are generally performed mentally as in the following examples.

B. $3x + 2y + 5x - y$

$3x + 2y + 5x - y$ Add coefficients of like terms

$= 8x + y$

C. $5a + 3b - c + 2a - 3b + 4c$

$5a + 3b - c + 2a - 3b + 4c = 7a + 3c$

Note that the terms containing literal part b combine to have a coefficient of zero.

D. $x^2 - 4x + 3 + 5x - 4 + x^2 - x + 3$

$x^2 - 4x + 3 + 5x - 4 + x^2 - x + 3 = 2x^2 + 2$

E. $x + y - 2 + 3x - 4y + x - 3y + 5$

$x + y - 2 + 3x - 4y + x - 3y + 5 = 5x - 6y + 3$

F. $x^3 + x^2 + x + 1$
Terms cannot be combined. ◀

In general, addition and subtraction of polynomials relies on combining like terms and may be done either vertically (one polynomial written under the other), or horizontally (on a single line). Both commonly used methods will be shown in this section.

Horizontal Addition of Polynomials

Recall that if a polynomial in parentheses is preceded by a "+" sign (either written or understood), the parentheses and the preceding "+" sign may be removed by keeping the sign of *each term* of the polynomial unchanged. To illustrate, the sum of $3x^2 - 5$ and $x^2 + 2x - 1$ is written

$(3x^2 - 5) + (x^2 + 2x - 1)$

$= 3x^2 - 5 + x^2 + 2x - 1$ Remove parentheses

$= 4x^2 + 2x - 6$ Combine like terms

Example 2

Add the polynomials by removing parentheses and simplify by combining like terms:

A. $(5x^3 + 2x^2 - 3), (x^3 - 2x^2 + 3x),$ and $(-2x^2 - 5x + 7)$

$(5x^3 + 2x^2 - 3) + (x^3 - 2x^2 + 3x) + (-2x^2 - 5x + 7)$

$= 5x^3 + 2x^2 - 3 + x^3 - 2x^2 + 3x - 2x^2 - 5x + 7$

$= 6x^3 - 2x^2 - 2x + 4$

5.2 Addition and Subtraction of Polynomials

B. $(3a^2 - 5ab) + (3ab - 6b^2) + (a^2 + 2ab + 4b^2)$

$(3a^2 - 5ab) + (3ab - 6b^2) + (a^2 + 2ab + 4b^2)$
$= 3a^2 - 5ab + 3ab - 6b^2 + a^2 + 2ab + 4b^2$
$= 4a^2 - 2b^2$ ◀

Vertical Addition of Polynomials

Polynomials are added vertically by writing them one under the other. In doing so, write only *like terms* in each column, which may require rearranging terms. The sum is then found by combining like terms in each column, as in the next examples.

Example 3

Add the polynomials vertically:

A. $(2x^2 + 3y), (x^2 - y), (x^2)$

$$\begin{array}{r} 2x^2 + 3y \\ x^2 - y \\ x^2 \\ \hline 4x^2 + 2y \end{array}$$

B. $(a^2 + 3ab), (2a^2 - 4ab + b^2), (-a^2 + ab - 4b^2)$

$$\begin{array}{r} a^2 + 3ab \\ 2a^2 - 4ab + b^2 \\ -a^2 + ab - 4b^2 \\ \hline 2a^2 - 3b^2 \end{array}$$

C. $(x^2 + 3x), (3 - 5x^2), (2x + x^2 - 7)$

$$\begin{array}{r} x^2 + 3x \\ -5x^2 + 3 \\ x^2 + 2x - 7 \\ \hline -3x^2 + 5x - 4 \end{array}$$

Note, rearrangement of terms is necessary to align like terms in each column ◀

Horizontal Subtraction of Polynomials

To show the subtraction of a polynomial, the polynomial is enclosed in parentheses preceded by a "−" sign. Recall that when a polynomial in parentheses is preceded by a "−" sign, the parentheses may be removed by changing the sign of each term. That is, when a polynomial is to be subtracted, the sign of each term in the polynomial is to be changed and the result is simplified by combining like terms.

Example 4

Subtract as indicated:

A. $x^2 + 5xy - 2y^2$ from $2x^2 - 3xy + 7y^2$

$(2x^2 - 3xy + 7y^2) - (x^2 + 5xy - 2y^2)$ Note the order

Sign of each term is unchanged Sign of each term is changed

$= 2x^2 - 3xy + 7y^2 - x^2 - 5xy + 2y^2$

$= x^2 - 8xy + 9y^2$ Combine like terms

B. $2a^2 - 3b^2$ from $a^2 + 2ab - 6b^2$

$(a^2 + 2ab - 6b^2) - (2a^2 - 3b^2)$

$= a^2 + 2ab - 6b^2 - 2a^2 + 3b^2$ Remove parentheses

$= -a^2 + 2ab - 3b^2$

C. $x^2 - 3x + 5$ from $-x^2 - 3x - 6$

$(-x^2 - 3x - 6) - (x^2 - 3x + 5)$

$= -x^2 - 3x - 6 - x^2 + 3x - 5$ Remove parentheses

$= -2x^2 - 11$ ◂

Vertical Subtraction of Polynomials

To subtract one polynomial from another vertically, place the polynomial being subtracted under the other polynomial. Again, write only *like terms* in a column; this may require rearranging terms. Then *mentally* change the sign of *each term* of the polynomial being subtracted, and add to the top one.

Example 5

Subtract the polynomials $2x^2 + 7x - 4$ from $-8x^2 + 7x + 3$.

Think of the subtraction: as the addition:

$$\begin{array}{r} -8x^2 + 7x + 3 \\ -(2x^2 + 7x - 4) \end{array}$$ change signs $$\begin{array}{r} -8x^2 + 7x + 3 \\ -2x^2 - 7x + 4 \\ \hline -10x^2 \qquad + 7 \end{array}$$ ◂

The rewriting of the preceding example was only done to emphasize the procedure of changing signs that should be done mentally. The ability to perform these subtractions mentally will be helpful when we study division of polynomials in Section 5.4.

5.2 Addition and Subtraction of Polynomials

Example 6

Subtract the following polynomials vertically:

A. Subtract $x^2 + 5$ from $2x^2 - 4x + 2$

$$\begin{array}{r} 2x^2 - 4x + 2 \\ \underline{x^2 + 5} \\ x^2 - 4x - 3 \end{array}$$ Subtract (change signs and add)

B. Subtract $-4x^2 + 3$ from $3x^3 - 3x + 6$

$$\begin{array}{r} 3x^3 - 3x + 6 \\ \underline{ - 4x^2 + 3} \\ 3x^3 + 4x^2 - 3x + 3 \end{array}$$ Subtract (change signs and add)

In these examples, it was necessary to rearrange terms so that like terms were aligned in the same columns. Note that "missing" terms in either the top or bottom polynomial were accounted for by leaving a space where the term would normally be written. ◀

In problems that include both addition and subtraction of polynomials, each pair of parentheses is removed as described earlier. If the parentheses are preceded by a plus sign, leave the sign of each term unchanged. If the parentheses are preceded by a minus sign, change the sign of each term.

Example 7

Remove parentheses and simplify by combining like terms:

A. $(2a^2 + 3b^2) - (a^2 - 2ab) + (-2a^2 - b^2) - (5ab - b^2)$

$(2a^2 + 3b^2) - (\underbrace{a^2 - 2ab}_{\text{change signs}}) + (-2a^2 - b^2) - (5ab - b^2)$

$= 2a^2 + 3b^2 - a^2 + 2ab - 2a^2 - b^2 - 5ab + b^2$ Remove parentheses

$= -a^2 - 3ab + 3b^2$

B. $(2x^2 - 3y) + (2y^2 - x^2) - (x^2 - y)$

$(2x^2 - 3y) + (2y^2 - x^2) - (x^2 - y)$

$= 2x^2 - 3y + 2y^2 - x^2 - x^2 + y$ Remove parentheses

$= -2y + 2y^2$ or $2y^2 - 2y$

C. $-(x^2 + xy + 3y^2) + (-x^2 + 2xy) - (6xy - 4y^2)$

$-(x^2 + xy + 3y^2) + (-x^2 + 2xy) - (6xy - 4y^2)$ Remove parentheses

$= -x^2 - xy - 3y^2 - x^2 + 2xy - 6xy + 4y^2$

$= -2x^2 - 5xy + y^2$

D. $5a^2 + 3a - (6a^2 - 2a) + 7 - a + (a^2 - 4)$

$5a^2 + 3a - (6a^2 - 2a) + 7 - a + (a^2 - 4)$
$= 5a^2 + 3a - 6a^2 + 2a + 7 - a + a^2 - 4$
$= 4a + 3$ ◀

Exercise Set 5.2

Simplify each of the polynomials in Problems 1–12 by combining like terms:

1. $2x + 3y - 5x + y$
2. $ax + 2ax + 5ax$
3. $2x^2 + 3x - 4 + x^2 - 5x + 4$
4. $3x^2 + xy - y^2 + x^2 - xy + y^2$
5. $2y - 3 + 3y + 4 - y + 2$
6. $3x + 2 - x + 4 + 4x - 6$
7. $y^2 + 2y - 3y^2 + 4 - y + 2$
8. $2x + 3 - x^2 + 4x - 6 + 3x^2$
9. $2xy + 3yz - 4xz$
10. $2m - 4n - m + n$
11. $3a + 4b - a - 3b + 2a - b$
12. $2x + 3y + 5 - x + y - 3 - x - 4y$

In Problems 13–14, add the monomials:

13. a. $2x$ \\ $\underline{7x}$
    b. $3y$ \\ $\underline{-5y}$
    c. $-4a$ \\ $\underline{3a}$
    d. 0 \\ $\underline{-2xy}$

14. a. $5m$ \\ $\underline{6m}$
    b. $8x$ \\ $\underline{-6x}$
    c. $7xy$ \\ $\underline{2xy}$
    d. 0 \\ $\underline{4u}$

In Problems 15–16, subtract the bottom monomial from the top one:

15. a. $6x^2$ \\ $\underline{-4x^2}$
    b. $3m$ \\ $\underline{0}$
    c. $-11z$ \\ $\underline{-6z}$
    d. $6ab$ \\ $\underline{6ab}$

16. a. $9x^2y$ \\ $\underline{2x^2y}$
    b. $-8r$ \\ $\underline{-8r}$
    c. $4ab$ \\ $\underline{0}$
    d. $6xy$ \\ $\underline{-9xy}$

17. Add each column of monomials:

    a. $2x$ \\ $-7x$ \\ x \\ $5x$ \\ $\underline{-3x}$
    b. $-6x^2$ \\ $3x^2$ \\ $-11x^2$ \\ $\underline{7x^2}$
    c. $3ab$ \\ $5ab$ \\ $4ab$ \\ $\underline{-2ab}$
    d. $3m$ \\ m \\ $2m$ \\ $\underline{m}$
    e. $9u$ \\ $-6u$ \\ $-8u$ \\ $\underline{-u}$

5.2 Addition and Subtraction of Polynomials

18. Add each column of monomials:

a.	b.	c.	d.	e.
$2a$	$11x$	$-4ab^2$	m	$11y$
$6a$	$4x$	$-6ab^2$	$3m$	$-4y$
$9a$	$-8x$	ab^2	m	$-2y$
$-a$	$-6x$	$-ab^2$	$-5m$	$+3y$
$-3a$				

19. Add the following polynomials horizontally: *vertically*

  a. $(x^2 + 2x - 3), (5x^2 - 7),$ and $(-2x^2 + 8x - 1)$
  b. $(8a^2 + 5), (6a^2 + 3a - 7),$ and $(-5a^2 + 7a)$
  c. $(3a^2 + 5a - 7), (2a - a^2), (3a^2 - 7),$ and $(a^2 + 5a)$
  d. $(x^3 - 2x^2y + y^3), (3x^3 + 5x^2y - 2xy^2),$ and $(x^2y - y^3)$

20. Add the following polynomials horizontally:

  a. $(2y^2 - 5y + 2), (-3y^2 + y - 1),$ and $(2y^2 - 5y - 8)$
  b. $(3y^2 - 8), (2y^2 + 6y - 4),$ and $(-7y^2 + 5y)$
  c. $(5 - m^2 + 2m), (3 + 5m^2), (-4 + 3m - 2m^2),$ and $(2 + 7m)$
  d. $(2a^3 - 3a^2b), (5a^2b - 4ab^2 + b^3),$ and $(4a^3 - 7a^2b - 9b^3)$

21. Horizontally subtract the second polynomial from the first:

  a. $(2y^2 - 8y - 11), (y^2 + 2y - 9)$
  b. $(3x^3 + 4x^2 - 7x), (6x^2 - 7x + 4)$
  c. $(6 - 5m^2 + 7m), (3m^2 + 7m + 6)$

22. Horizontally subtract the second polynomial from the first:

  a. $(x^2 - 5x + 9), (3x^2 + 4)$
  b. $(2y^3 - 3y^2 + y), (2y^2 - 8y + 9)$
  c. $(3x^2 + 5xy - 7y^2), (3x^2 + 7xy - 7y^2)$

23–26. Repeat Problems 19–22, but do the operations *vertically*.

In each of the following problems, perform the addition and subtraction by removing parentheses and combining like terms:

27. $(3a + 5) - (a - 2)$

28. $(6a - 1) - (2a - 2)$

29. $(x^2 + x) - (x + 3) + (2x - 1)$

30. $(3x^2 + 2) - (x^2 - 2x - 3) + (4 + x)$

31. $(2y - 5) - (3 - y^2) + (3y^2 - 5y + 2)$

32. $(4m + 3) - (-2m - 5) - (m + 1) + 2m$
33. $(x^2 - 2x) + (3x^2 + x - 6) - (x - 2)$
34. $(m^2 + 3) + (-2m^2 - 8) + (m - 1)$
35. $(x^2 + y^2) - (2x^2 + 3xy - y^2) - (2xy + 3y^2)$
36. $(2x^2 + 5) - (-2x^2 + 3x - 1) + (4 - 3x + x^2)$
37. $-(x^2 + 3x - 5) + (x^2 - 1) - (2 + 3x - x^2)$
38. $-(2x^2 - 3x - 1) - (2x + 4) + (x^2 - 3) - (2 + x + x^2)$
39. $2x + 3 - (5x^2 + 3x - 2) + (2 - x) - (-x^2 + 2x - 3)$
40. $3y + 4 - (2 + 3y - y^2) + (y^2 - 4) - 2y + 8$

Recall that the perimeter of a closed geometric figure is the distance around the figure. Find an expression for the perimeter of each of the following figures by adding the lengths of the sides:

41.

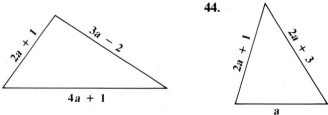

42.

43.

44.

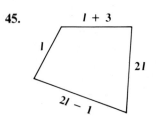

45.

46.

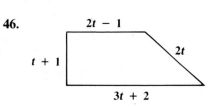

Review

47. a. Write c^3 in expanded form.
    b. Write c^5 in expanded form.
    c. Multiply $c^3 \cdot c^5$ by writing each factor in expanded form. Simplify the result.

48. Use the distributive law to multiply:
    a. $5(x - 2)$
    b. $x(x + 3)$
    c. $-3(y - x + 2)$

49. Multiply:
    a. $(-2)(6)$
    b. $(2)(-6)$
    c. $(-3)(-2)(-5)$
    d. $(-3)(5)(7)(-4)(3)(0)$

5.3 Multiplication of Polynomials

Recall that the result of a multiplication is a *product* and the expressions that are multiplied are called *factors*. For example:

$$3 \cdot 11 = 33$$

Factors of 33 ——↑ ↑—— Product of 3 and 11

In this section we multiply polynomials, and in Sections 5.6–5.11 we discuss reversing the process, or factoring.

In preparation for multiplying polynomials, we introduce a property of exponents. From the definition of exponential form, b^n indicates the product of n factors of b.

$$b^n = \underbrace{b \cdot b \cdots b}_{n \text{ factors}}$$

We can multiply two exponential expressions with the same base as in the following example.

$$x^3 \cdot x^4 = (xxx)(xxxx) \quad \text{Definition}$$
$$= \underbrace{xxx}_{3 \text{ factors}}\underbrace{xxxx}_{4 \text{ factors}} \quad \text{Parentheses may be omitted}$$
$$= x^{3+4} = x^7$$

FIVE   Polynomials

multiplication of different exponents

This intuitive example suggests a rule of exponents:

Rule

$$b^m \cdot b^n = b^{m+n}$$

like bases add exponents

In words: When multiplying exponential expressions with the same base, the product is that common base raised to the *sum* of the powers of the factors.

Example 1

Simplify each expression:

A. $x^7 \cdot x^4 = x^{7+4} = x^{11}$

B. $a \cdot a^6 = a^{1+6} = a^7$ Recall $a = a^1$

C. $m^3 \cdot m^5 \cdot m = m^{3+5+1} = m^9$

D. $(x + 3)^2 \cdot (x + 3)^4 = (x + 3)^{2+4} = (x + 3)^6$ ◀

Multiplication of polynomials is presented by considering the number of terms in each polynomial factor. We begin by learning how to find the product of monomials and then developing a general procedure for multiplying polynomials containing any number of terms.

Monomial Multiplied by a Monomial

Multiplying monomials is an extension of the property of exponents illustrated in Example 1, together with the associative and commutative laws for multiplication.

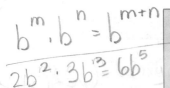

$b^m \cdot b^n = b^{m+n}$

$2b^2 \cdot 3b^3 = 6b^5$

RULE

To multiply monomials, multiply the numerical coefficients (remembering the rules for multiplying signed numbers) and then multiply the literal factors having the same bases by adding their exponents.

Example 2

Multiply the following monomials:

A. $(3x^3)(2x^4)$

$$(3x^3)(2x^4) = (3)(2)(x^3)(x^4)$$
$$= (3 \cdot 2)(x^3 \cdot x^4)$$
$$= 6x^7$$

5.3 Multiplication of Polynomials

B. $(4a^2b)(-3a^2b^3)$

$$(4a^2b)(-3a^2b^3) = (4)(-3)(a^2 \cdot a^2)(b \cdot b^3)$$
$$= -12a^4b^4$$

C. $(x^2y)(3xy^2z^2)(-2xz^3)$

$$(x^2y)(3xy^2z^2)(-2xz^3) = (1)(3)(-2)(x^2 \cdot x \cdot x)(y \cdot y^2)(z^2 \cdot z^3)$$
$$= -6x^4y^3z^5$$

In practice, the middle step in such products is not written.

D. $(-2x^2y)(9xy^3)$

$$(-2x^2y)(9xy^3) = -18x^3y^4$$

E. $(-3xy^2)(-x^2w)(4y^2w^3)$

$$(-3xy^2)(-x^2w)(4y^2w^3) = 12x^3y^4w^4$$

F. $(b^2c)(-3ab^2)(-ab)$

$$(b^2c)(-3ab^2)(-ab) = 3a^2b^5c$$ ◀

Monomial Multiplied by a Polynomial

Multiplying a monomial by a binomial is an application of the distributive property. Recall:

$$a(b + c) = ab + ac$$

Monomial   Binomial

To multiply a monomial by a binomial, use the distributive law to multiply the monomial by both terms of the binomial.

Example 3

Multiply each of the following:

A. $3x(2x^2 + 4x)$

$$3x(2x^2 + 4x) = (3x)(2x^2) + (3x)(4x)$$
$$= 6x^3 + 12x^2$$

B. $-2y(y^2 - 3y)$

$$-2y(y^2 - 3y) = (-2y)(y^2) - (-2y)(3y)$$
$$= -2y^3 + 6y^2$$

C. $(3m + 2m^2)(2m^2)$

$$(3m + 2m^2)(2m^2) = (3m)(2m^2) + (2m^2)(2m^2)$$
$$= 6m^3 + 4m^4$$

D. $(x^2y + 3xy)(3y^3)$

$$(x^2y + 3xy)(3y^3) = (x^2y)(3y^3) + (3xy)(3y^3)$$
$$= 3x^2y^4 + 9xy^4 \quad \blacktriangleleft$$

The process of multiplying a monomial by a binomial can be generalized to multiplying a monomial by a polynomial of any number of terms.

> **RULE** *every term multiplies every term*
> To multiply a monomial by a polynomial, use the distributive law to multiply the monomial by each term of the polynomial.

Example 4

Multiply each of the following:

A. $5x^2(3x^2 - 2xy - 4y^2)$

$$5x^2(3x^2 - 2xy - 4y^2) = (5x^2)(3x^2) - (5x^2)(2xy) - (5x^2)(4y^2)$$
$$= 15x^4 - 10x^3y - 20x^2y^2$$

The middle step is generally not written.

B. $3ab^2(2a^2 + 3ab - 5b^2)$

$$3ab^2(2a^2 + 3ab - 5b^2) = 6a^3b^2 + 9a^2b^3 - 15ab^4$$

C. $(3y^2 - 8y + 4x - 1)(2x)$

$$(3y^2 - 8y + 4x - 1)(2x) = 6xy^2 - 16xy + 8x^2 - 2x \quad \blacktriangleleft$$

Polynomial Multiplied by a Polynomial

To show multiplication of a polynomial by another polynomial, we will multiply

$$(x + 2)(x^2 - 3x + 6)$$

using the distributive property, $(b + c)a = b \cdot a + c \cdot a$. Each term of $x + 2$ is multiplied by $x^2 - 3x + 6$.

$$(x + 2)(x^2 - 3x + 6) = x(x^2 - 3x + 6) + 2(x^2 - 3x + 6)$$
$$= x^3 - 3x^2 + 6x + 2x^2 - 6x + 12$$
$$= x^3 - x^2 + 12$$

5.3 Multiplication of Polynomials

> **RULE**
>
> To multiply two polynomials, multiply each term of the first polynomial by every term of the second and combine like terms if possible.

Note: When forming products of individual terms of the polynomials, care must be taken to include the *signs* of the terms.

Example 5

Multiply $m^2 + m - 2$ by $2m^2 - 3m - 3$.

$(m^2 + m - 2)(2m^2 - 3m - 3)$

$= m^2(2m^2 - 3m - 3) + m(2m^2 - 3m - 3) - 2(2m^2 - 3m - 3)$

$= 2m^4 - 3m^3 - 3m^2 + 2m^3 - 3m^2 - 3m - 4m^2 + 6m + 6$

$= 2m^4 - m^3 - 10m^2 + 3m + 6$ ◀

Multiplication of polynomials can also be done using a vertical arrangement. We generally line up the left terms of the polynomials, and then multiply the top polynomial by each term of the bottom polynomial proceeding from left to right. As we write each term of the partial products, we write only like terms in any column. Study the following example.

Example 6

Multiply $3x^2 + 2x - 3$ by $2x^2 - 7$.

$3x^2 + 2x - 3$

$\underline{2x^2 - 7}$ Line up left terms

First, each term of the top polynomial will be multiplied by the term $2x^2$. Next, they will be multiplied by -7, and the terms of the second partial product are written under like terms of the first partial product.

$$\begin{array}{l} 3x^2 + 2x - 3 \\ \underline{2x^2 - 7} \\ 6x^4 + 4x^3 - 6x^2 \\ - 21x^2 - 14x + 21 \\ \hline 6x^4 + 4x^3 - 27x^2 - 14x + 21 \end{array}$$

Multiply by $2x^2$
Multiply by -7
Add   ◀

It is common practice in working with polynomials to arrange the terms from highest power to lowest power (descending powers). Notice in the preceding example that a convenient pattern of exponents developed since both factors were written with descending powers.

Example 7

Find the following products using the vertical arrangement:

A. $(x^2 - 3x + 6)(x + 2)$

$$
\begin{array}{ll}
x^2 - 3x + 6 & \\
x + 2 & \text{Line up left terms} \\
\hline
x^3 - 3x^2 + 6x & \text{Multiply by } x \\
 2x^2 - 6x + 12 & \text{Multiply by 2} \\
\hline
x^3 - x^2 + 12 & \text{Add (combine like terms)}
\end{array}
$$

B. $(m^2 + m - 2)(2m^2 - 3m - 3)$

$$
\begin{array}{ll}
m^2 + m - 2 & \\
2m^2 - 3m - 3 & \text{Line up left terms} \\
\hline
2m^4 + 2m^3 - 4m^2 & \text{Multiply by } 2m^2 \\
 - 3m^3 - 3m^2 + 6m & \text{Multiply by } -3m \\
 - 3m^2 - 3m + 6 & \text{Multiply by } -3 \\
\hline
2m^4 - m^3 - 10m^2 + 3m + 6 & \text{Add}
\end{array}
$$ ◀

The product of more than two polynomials can be found by finding the product of the first two, and then multiplying this product by the third polynomial, and so on, until all factors have been included.

Example 8

Multiply the following polynomials both vertically and horizontally:

A. $(2x + y)(3x - 2y)(-x + 3y)$

$$
\begin{array}{ll}
2x + y & \\
3x - 2y & \text{Multiply first two polynomials} \\
\hline
6x^2 + 3xy & \\
 - 4xy - 2y^2 & \\
\hline
6x^2 - xy - 2y^2 & \\
\end{array}
$$

$$
\begin{array}{ll}
6x^2 - xy - 2y^2 & \\
-x + 3y & \text{Multiply the product by the third polynomial} \\
\hline
-6x^3 + x^2y + 2xy^2 & \\
 18x^2y - 3xy^2 - 6y^3 & \\
\hline
-6x^3 + 19x^2y - xy^2 - 6y^3 & \text{Product}
\end{array}
$$

B. $(2x + y)(3x - 2y)(-x + 3y)$

$(2x + y)(3x - 2y)(-x + 3y)$

$= (6x^2 + 3xy - 4xy - 2y^2)(-x + 3y)$ Multiply first two factors

$= (6x^2 - xy - 2y^2)(-x + 3y)$ Simplify

$= -6x^3 + 18x^2y + x^2y - 3xy^2 + 2xy^2 - 6y^3$ Multiply again

$= -6x^3 + 19x^2y - xy^2 - 6y^3$ Rearrange and combine like terms ◀

5.3 Multiplication of Polynomials

Example 9

Multiply $(2x - y)(3x + 2y)^2$.

$$\begin{array}{r} 2x - y \\ 3x + 2y \\ \hline 6x^2 - 3xy \\ + 4xy - 2y^2 \\ \hline 6x^2 + xy - 2y^2 \\ 3x + 2y \\ \hline 18x^3 + 3x^2y - 6xy^2 \\ 12x^2y + 2xy^2 - 4y^3 \\ \hline 18x^3 + 15x^2y - 4xy^2 - 4y^3 \end{array}$$

Multiply first two factors

Multiply by third factor

Example 10

Multiply 25 by 31 by writing the product as $(20 + 5)(30 + 1)$.

$$(20 + 5)(30 + 1) = (20)(30) + (20)(1) + (5)(30) + (5)(1)$$
$$= 600 + 20 + 150 + 5$$
$$= 775$$

Check this result by multiplying $25 \cdot 31$ in the traditional way.

Exercise Set 5.3

Form each of the products in Problems 1–30:

1. $(3x)(2x^2)$
2. $(5y)(-2y^3)$
3. $(-3xy)(7wx^2)$
4. $(6uv)(-4uv^2)$
5. $(8a^2)(-2b^2)(ab)$
6. $(7x^3)(-x)(3x^2)$
7. $(-3u)(2u^2)^2$
8. $(-5x)^2(-3x^2)$
9. $(ab)(2ab)(-4a^2b^2)$
10. $(3ac)(-5b^2)(4abc)$
11. $2x(x + 4)$
12. $3x(2x + 5)$
13. $3x^2(x - 3)$
14. $5x^2(2x - 3)$
15. $x^2(x^2 + 2x - 3)$
16. $(2x)(3x^2 - x - 1)$
17. $-3a(a^2 - 5)$
18. $-5b(2b^2 - 4)$
19. $6x^2y(2x^2 - 7y^2)$
20. $5mn(m^2 - 3mn - 8n^2)$
21. $3xy^2(2x^2 + 3x - 11)$
22. $-6x^2y(5x^2 - 6y^2)$
23. $8(20 + 3)$
24. $5(20 + 8)$
25. $5(30 - 2)$
26. $7(4 + 3)$
27. $6x(5x^3 - 4x^2 + 3x - 7)$
28. $5y^3(3y^2 + 2xy - 4x^2)$
29. $-8a^2(a^3 - 3a^2 - 6a + 2)$
30. $-6xy(x^4 - 2x^3y - 5x^2y^2 + 3xy^3)$

Multiply horizontally:

31. $(x + 2)(x - 5)$ $x^2 - 3x - 10$
32. $(a + 3)(a + 6)$ $a^2 + 9a + 18$
33. $(x + 2)(y - 5)$
34. $(a + 3)(b + 6)$
35. $(2x - 3)(x + 4)$
36. $(3x - 1)(x + 4)$
37. $(x + 1)(x^2 - 2x + 2)$
38. $(x + 2)(x^2 + x - 3)$
39. $(2x - y)(x^2 + xy - 2y^2)$
40. $(x + 2y)(2x^2 + xy + 3y^2)$
41. $(3a - 2)(a^2 + 3a - 1)$
42. $(m + 3)(2m^2 - m - 1)$

Multiply vertically:

43. $(x + 3)(2x - 5)$
44. $(x - 5)(2x + 1)$
45. $(a^2 + a)(2a - 3)$
46. $(3a^2 + 1)(2a - 3)$
47. $(2x + y)(3x - y)$
48. $(5 - x)(2 + x + 2x^2)$
49. $(x + 2)(x^2 + 5x - 4)$
50. $(u + 3)(u^2 - 3u + 1)$
51. $(x + 4)(x + 2)(x - 1)$
52. $(x + 3)(x - 1)(x - 2)$
53. $(x^2 + x - 1)(x^2 + 3x + 2)$
54. $(x^2 - 3x + 2)(x^2 + 6x - 2)$
55. $(x^2 + 3x + 2)(x - 2)^2$
56. $(x^2 - x + 5)(x + 3)^2$
57. $(x + 2)^2(x - 1)^2$
58. $(x - 3)^2(x + 1)^2$
59. $(x^2 + 2x - 1)^3$
60. $(x^2 - 3x + 2)^3$
61. $(x + 1)(x + 2)(x + 3)(x + 4)$
62. $(x + 1)(x - 2)(x + 3)(x - 4)$

Express the area ($A = L \cdot W$) and the perimeter ($P = 2L + 2W$) of each of the rectangles shown:

63.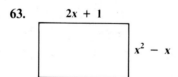
64.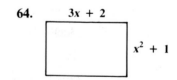

65.

66.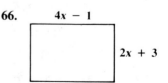

67. When multiplying polynomials, what relationship exists between the degrees of the factor polynomials and the degree of the product polynomial?

68. How many individual products are required to multiply a polynomial containing two terms with a polynomial containing three terms? In general, how many products are required to multiply a polynomial containing m terms with a polynomial containing n terms?

Review

69. Complete the multiplications in each expression and then simplify by combining like terms:

    a. $(3p)(2p) + (3p)(7) + (-5)(2p) + (-5)(7)$
    b. $(3u)(u) + (3u)(6v) + (2v)(u) + (2v)(-6v)$
    c. $(2m)(2m) + (2m)(5) + (2m)(-5) + (-5)(5)$
    d. $(3c)(3c) + (3c)(5) + (5)(3c) + (5)(5)$
    e. $(3x^3)(x^3) + (3x^3)(7) + (-2)(x^3) + (-2)(7)$

70. Solve the following system by the elimination method:

    $$-3x + 2y = -2$$
    $$6x - 5y = 4$$

71. Solve the following system by substitution:

    $$2x + 7y = -2$$
    $$x - 2y = 10$$

5.4 Special Products

We have just discussed how to multiply polynomials. Some patterns occur so frequently that it is helpful to recognize the pattern and be able to write the product directly. The products that can be completed mentally are called **special products** and are presented in this section.

FOIL

We must frequently multiply two binomials together. To do so, we use the distributive property and multiply the first term of the first binomial by each term of the second binomial, and then multiply the second term of the first binomial by each term of the second binomial as was discussed in the preceding section. This method for multiplying two binomials is called *FOIL*, which is an acronym for *F*irst, *O*utside, *I*nside, and *L*ast, indicating the order in which the terms of the binomials are multiplied.

FIVE Polynomials

Example 1

Multiply $(2x + 3)(3x - 5)$ using the FOIL method:

Step 1: Multiply the First terms.

$$(2x + 3)(3x - 5) \quad (F)$$
$$(2x)(3x) = 6x^2$$

Step 2: Multiply the Outside terms.

$$(2x + 3)(3x - 5) \quad (O)$$
$$(2x)(-5) = -10x$$

Step 3: Multiply the Inside terms.

$$(2x + 3)(3x - 5) \quad (I)$$
$$(3)(3x) = 9x$$

Step 4: Multiply the Last terms.

$$(2x + 3)(3x - 5) \quad (L)$$
$$(3)(-5) = -15$$

Step 5: Combine like terms when possible.

$$6x^2 - x - 15 \qquad \blacktriangleleft$$

Note that in this example, the outside and inside products are like terms and can be combined. This often happens, and you should become proficient with this technique and be able to perform the multiplication mentally.

Example 2

Multiply each of the following using FOIL:

A. $(2m + 3)(3m + 5)$

$$(2m + 3)(3m + 5) = \underset{\text{First}}{(2m)(3m)} + \underset{\text{Outside}}{(2m)(5)} + \underset{\text{Inside}}{(3)(3m)} + \underset{\text{Last}}{(3)(5)}$$

$$= 6m^2 + 10m + 9m + 15$$
$$= 6m^2 + 19m + 15$$

5.4 Special Products

B. $(x - 2)(y + 3)$

$$(x - 2)(y + 3) = xy + 3x - 2y - 6$$

with F, L (outer arc), I, O (inner arc) labeled on the FOIL pattern.

C. $(3a - 2b)(a + 4b)$

$$(3a - 2b)(a + 4b) = 3a^2 + 12ab - 2ab - 8b^2$$
$$= 3a^2 + 10ab - 8b^2$$

D. $(5x - 3)(4x + 1)$

$$(5x - 3)(4x + 1) = 20x^2 - 7x - 3$$

E. $(2y + 3)^2$

$$(2y + 3)(2y + 3) = 4y^2 + 12y + 9$$

F. $(x + 3)(x - 3)$

$$(x + 3)(x - 3) = x^2 - 9$$

G. $(x + a)(x + b)$ (sum/product)

$$(x + a)(x + b) = x^2 + ax + bx + ab$$
$$= x^2 + (a + b)x + ab$$

Examples **2E**, **2F**, and **2G** are examples of special products that occur frequently. We discuss each product in more detail.

From Example **2E** we have the special product called the square of a binomial.

Handwritten annotation:
① square 1st term
② twice the product of the 2 terms
③ square last term

SQUARE OF A BINOMIAL
$$(a + b)^2 = a^2 + 2ab + b^2$$
$$(a - b)^2 = a^2 - 2ab + b^2$$

It may be helpful to note that in the binomial squares, the result is always the first term squared, plus twice the product of the two terms, plus the last term squared.

We show the formula for the square of a binomial, noting that the area of the large rectangle is the sum of the areas of the smaller rectangles in Figure 5.4-1.

FIGURE 5.4-1

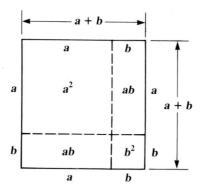

Total area   Sum of smaller areas

$(a + b)^2 = a^2 + ab + ab + b^2$
$ = a^2 + 2ab + b^2$

From Example 2F we have the special product called the product of a sum and difference.

PRODUCT OF A SUM AND DIFFERENCE

$(a + b)(a - b) = a^2 - b^2$

[handwritten: will always be a binomial F & L terms]

If we form the product of a sum and difference using FOIL, the outside and inside products will always add to zero. A sum and a difference, such as $(a + b)$ and $(a - b)$, are called *conjugates* of each other.

Finally, from Example 2G, note that when the binomial factors are of the form

(Variable + constant)(Same variable + constant)   *[handwritten: sum/product form]*

the product is of the form

$(x + a)(x + b) = x^2 + \underbrace{(a + b)}_{\text{Sum}}x + \underbrace{ab}_{\text{Product}}$

[handwritten left margin: Answer will always be trinomial / Like terms]

Example 3

Multiply as special products:

A. $(3x + 4)(3x - 4)$

$(3x + 4)(3x - 4) = 9x^2 - 16$ Product of a sum and difference (outside and inside products add to zero)

5.4 Special Products

B. $(2b - 3)^2$

$(2b - 3)^2 = (2b - 3)(2b - 3)$
$= 4b^2 - 12ab + 9$ Square of a binomial (first term squared, plus twice the product of the two terms, plus the second term squared)

C. $(x - 5)(x + 8)$

$(x - 5)(x + 8) = x^2 + 3x - 40$

          Sum   Product

$-5 + 8 = 3$ $(-5)(8) = -40$

Combined Operations

When addition, subtraction, and multiplication of polynomials appear together, multiplication must be performed before addition and subtraction, just as with real numbers.

Example 4

Complete all operations and simplify:

A. $2x + 3x(x - 2)$

$2x + 3x(x - 2) = 2x + 3x^2 - 6x$ Multiply first
$= 3x^2 - 4x$ Combine like terms

B. $-2(5 - x) + 3x$

$-2(5 - x) + 3x = -10 + 2x + 3x$ Multiply first
$= 5x - 10$ Combine like terms

C. $2(x + 4)(2x - 1) - 3(x + 1)(x - 3)$

$2(x + 4)(2x - 1) - 3(x + 1)(x - 3)$
$= 2(2x^2 + 7x - 4) - 3(x^2 - 2x - 3)$ Multiply
$= 4x^2 + 14x - 8 - 3x^2 + 6x + 9$ Multiply
$= x^2 + 20x + 1$ Combine like terms

Exercise Set 5.4

FOIL (handwritten)

Multiply directly:

1. $(a + 6)(a + 2)$
2. $(m + 3)(m + 4)$ *$m^2+7m+12$*
3. $(y + 8)(y - 1)$
4. $(u - 2)(u + 5)$ *$u^2+3u-10$*
5. $(b - 4)(b - 7)$
6. $(t - 1)(t - 6)$ *t^2-7t+6*
7. $(x - 7)(x + 2)$ *$-4x+7x$*
8. $(r - 9)(r + 3)$ *$r^2-6r-27$*
9. $(2m + 5)(m - 1)$
10. $(2x + 7)(x - 2)$ *$2x^2+3x-14$*
11. $(3y - 1)(2y - 3)$
12. $(4a - 1)(3a - 2)$
13. $(x + 3)^2$
14. $(a + 1)^2$ *a^2+2a+1*
15. $(y - 4)^2$
16. $(m - 3)^2$ *m^2-6m+9*
17. $(2x + 5y)^2$
18. $(3t + 2u)^2$ *$9t^2+12tu+4u^2$*
19. $(4b - 3)^2$
20. $(2x - 5)^2$ *$4x^2-20x+25$*
21. $(3x^2 - y)^2$
22. $(3a^2 + 2b)^2$ *$9a^4+12a^2b+4b^2$*
23. $(x - 3)(x + 3)$
24. $(r + 2)(r - 2)$ *r^2-4*
25. $(2x + 7)(2x - 7)$
26. $(3y + 1)(3y - 1)$ *$9y^2-1$*
27. $(x^2 - 1)(x^2 + 1)$
28. $(Q + 9)(Q - 9)$ *Q^2-81*
29. $(a + 2b)(a - 2b)$
30. $(x - 3y)(x + 3y)$ *x^2-9y^2*
31. $(20 + 1)(20 - 1)$
32. $(30 - 1)(30 + 1)$ *$900-1=899$*
33. $(20 + 5)^2$
34. $(30 - 2)^2$

In Problems 35–44, complete the indicated operations and simplify the result:

35. $3x + 2 + 4(1 - x)$
36. $2a + 3(a^2 - 3a - 1)$
37. $5x^2 + 2x + (x + 2)(x + 3)$
38. $3x^2 - 2(x^2 - 3x - 4)$
39. $(x + 3)(x - 2) - (x - 6)(x + 1)$
40. $x^2 - (2 + x)(3 + x) + 5x$ *$x^2-(x+2)(x+3)+5x$*   *$x^2-(x^2+5x+6)+5x$*   *$x^2-x^2-5x-6+5x$*   *-6*
41. $2u(3u^2 + 5u - 3) - 3u(u^2 - u - 2)$
42. $2B(B - 4) + 3B(B^2 + 2B - 3)$
43. $(M + 2)(3M - 1) - M(M + 2)(M - 3)$
44. $(3x + y)(x - y) - (2x + y)(x - 4y)$

Find an expression for the area of each figure in Problems 45–50:

45. rectangle: $2x + 3$ by $x + 1$ *$(2x+3)(x+1)$*

46. rectangle: $3x + 1$ by $x + 2$

47. square: $2x + 3$ by $2x + 3$

48. square: $2x - 1$ by $2x - 1$

5.4 Special Products

49.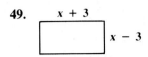

50.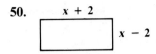

The volume of a box is given $V = LWH$ where L is the length, W is the width, and H is the height of the box.

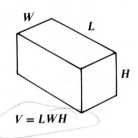

Find an expression for the volume of each figure in Problems 51–56:

51.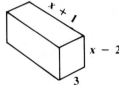

52.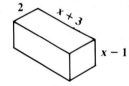

53.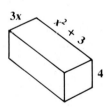

54.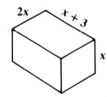

55.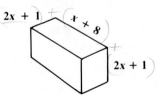

56.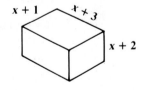

Review

57. a. Write r^9 in expanded form.

 b. Write r^4 in expanded form.

 c. Divide $\dfrac{r^9}{r^4}$ by writing r^9 and r^4 in expanded form and simplify.

58. Divide as indicated:

a. $\dfrac{-24}{3}$ b. $\dfrac{-24}{-8}$ c. $\dfrac{-24}{0}$

59. Use long division (from arithmetic) to divide: $9212 \div 28$.

5.5 Division of Polynomials

To divide polynomials, we use a property of exponents developed informally here. All the properties of exponents will be covered in more detail in Chapter 7.

$$\dfrac{b^m}{b^n} = b^{m-n} \quad \text{Where } b \neq 0,\ m > n$$

Example 1

Simplify $\dfrac{x^5}{x^2}$.

$\dfrac{x^5}{x^2} = \dfrac{xxxxx}{xx}$ Definition of exponents

$= \dfrac{\cancel{x}\cancel{x}xxx}{\cancel{x}\cancel{x}}$ Cancel common factors

$= x^3$ Definition of exponents

or $\dfrac{x^5}{x^2} = x^{5-2} = x^3$

Monomial Divided by a Monomial

Division of monomials is similar to multiplication of monomials, in that we first divide the numerical coefficients and then divide the variables using the property of exponents shown previously.

Example 2

Divide:

A. $8x^3y^2 \div 2xy$

$\dfrac{8x^3y^2}{2xy} = \left(\dfrac{8}{2}\right)\left(\dfrac{x^3}{x}\right)\left(\dfrac{y^2}{y}\right)$ Recall $x = x^1$

$= 4x^{3-1}y^{2-1}$

$= 4x^2y$

5.5 Division of Polynomials

B. $-16x^4y^3 \div (-2xy^3)$

$$\frac{-16x^4y^3}{-2xy^3} = \left(\frac{-16}{-2}\right)\left(\frac{x^4}{x}\right)\left(\frac{y^3}{y^3}\right)$$

$$= 8x^{4-1} \cdot 1 = 8x^3$$

The intermediate steps are generally done mentally.

C. $-12x^2y^5 \div 3x^2y$

$$\frac{-12x^2y^5}{3x^2y} = -4y^4$$

D. $12x^4y \div (-3x)$

$$\frac{12x^4y}{-3x} = -4x^3y$$

◀

Polynomial Divided by a Monomial

We use the next example to illustrate how to divide a polynomial by a monomial.

Example 3   Divide:

A. $46 \div 2$

$$\frac{46}{2} = 46 \cdot \left(\frac{1}{2}\right) \qquad \text{Dividing by 2 is multiplying by } \tfrac{1}{2}$$

$$= \frac{\overset{23}{\cancel{46}}}{1} \cdot \frac{1}{\cancel{2}} \qquad \text{Cancel 2's}$$

$$= \frac{23}{1} = 23$$

B. Divide $46 \div 2$ by writing 46 as the sum $40 + 6$.

$$\frac{46}{2} = \frac{40 + 6}{2}$$

$$= (40 + 6) \cdot \left(\frac{1}{2}\right) \qquad \text{Dividing by 2 is multiplying by } \tfrac{1}{2}$$

$$= (40) \cdot \left(\frac{1}{2}\right) + (6) \cdot \left(\frac{1}{2}\right) \qquad \text{Distributive property}$$

$$= 20 + 3 = 23$$

That is, $\dfrac{46}{2} = \dfrac{40 + 6}{2} = \dfrac{40}{2} + \dfrac{6}{2} = 20 + 3 = 23.$

◀

FIVE   Polynomials

The division of a polynomial by a monomial is performed in a manner similar to the division in Example **3B**. To divide a polynomial by a monomial, *each term* in the dividend (numerator) is divided by the monomial divisor.

Example 4

Divide: $(15x^4 - 3x^3 + 6x^2) \div 3x$

$$\frac{15x^4 - 3x^3 + 6x^2}{3x} = \frac{15x^4}{3x} - \frac{3x^3}{3x} + \frac{6x^2}{3x} \qquad \text{Divide each term by } 3x$$

$$= \frac{\overset{5}{\cancel{15}}x^4}{\cancel{3}x} - \frac{\cancel{3}x^3}{\cancel{3}x} + \frac{\overset{2}{\cancel{6}}x^2}{\cancel{3}x} \qquad \text{Cancel 3's}$$

$$= 5x^{4-1} - x^{3-1} + 2x^{2-1}$$

$$= 5x^3 - x^2 + 2x$$

Note: This division may be checked by multiplying the divisor $(3x)$ by the quotient $(5x^3 - x^2 + 2x)$ to obtain the dividend $15x^4 - 3x^3 + 6x^2$. ◀

Example 5

Divide: $(4y^3 - 3y^2) \div (-2y^2)$

$$\frac{4y^3 - 3y^2}{-2y^2} = \frac{4y^3}{-2y^2} - \frac{3y^2}{-2y^2} \qquad \text{Divide each term by } -2y^2$$

$$= \frac{\overset{2}{\cancel{4}}y^3}{\underset{1}{\cancel{-2}y^2}} - \frac{3\cancel{y^2}}{-2\cancel{y^2}} \qquad y^2 \text{ cancels in second fraction}$$

$$= -2y^{3-2} + \frac{3}{2}$$

$$= -2y + \frac{3}{2} \qquad \blacktriangleleft$$

Example 6

Divide: $(8y^6 - 12y^5 + 4y^4) \div 4y^4$

$$\frac{8y^6 - 12y^5 + 4y^4}{4y^4} = \frac{8y^6}{4y^4} - \frac{12y^5}{4y^4} + \frac{4y^4}{4y^4}$$

$$= 2y^2 - 3y + 1 \qquad \blacktriangleleft$$

Note in Example 6 that when one of the terms of the polynomial dividend is the same as the divisor, that term divided by an identical term yields 1.

5.5 Division of Polynomials

Example 7

Divide: $(10x^3y^2 - 5x^4y^4 + 15x^2y^3) \div 5x^2y$

$$\frac{10x^3y^2 - 5x^4y^4 + 15x^2y^3}{5x^2y} = \frac{10x^3y^2}{5x^2y} - \frac{5x^4y^4}{5x^2y} + \frac{15x^2y^3}{5x^2y}$$

$$= \frac{\overset{2}{\cancel{10}}x^3y^2}{\cancel{5}x^2y} - \frac{\cancel{5}x^4y^4}{\cancel{5}x^2y} + \frac{\overset{3}{\cancel{15}}x^2y^3}{\cancel{5}x^2y}$$

$$= 2x^{3-2}y^{2-1} - x^{4-2}y^{4-1} + 3y^{3-1}$$

$$= 2xy - x^2y^3 + 3y^2 \quad \blacktriangleleft$$

Division of a Polynomial by a Polynomial (with Two or More Terms)

The division of one polynomial by another is much like the division of one whole number by another. We begin this section by dividing one whole number by another as done in arithmetic and noting the steps.

Example 8

Divide 3995 by 17.

```
        235
    17 ) 3995
         34↓
         59
         51↓
         85
         85
          0  R
```

Comments

a. 17 divides into 39 two times. Place the 2 in the quotient, multiply 2 times 17 (divisor), and write the product under the 39; subtract; then bring down the next digit (9).

b. 17 divides into 59 three times. Place the 3 in the quotient, multiply 3 times 17, and write the product under the 59; subtract; and then bring down the next digit (5).

c. 17 divides into 85 five times. Place the 5 in the quotient, multiply 5 times 17, and write the product under the 85; subtract. The number 0 is the remainder. ◀

Polynomials are divided using similar steps. The polynomials divided in the rest of this section will have only one variable (this is the most common type of division problem).

FIVE   Polynomials

Example 9

Divide $(2x^3 + 13x^2 + 11x - 6)$ by $(2x + 3)$.

To set up the division of one polynomial by another, the terms must be arranged from highest power to lowest power of the variable (descending powers). For each missing power of the variable in the dividend, write in the term with a coefficient of zero. Note in this example, both the dividend and divisor are already written in descending powers.

In each step of the division, only the leading *term* of the dividend or the remainder is divided by the first *term* of the divisor. The quotient term is the *exact* quotient of the first dividend term or remainder term divided by the first divisor term. After writing this term in the quotient, it is multiplied by the *entire* divisor, and the terms of the product are lined up in columns of like terms under the dividend or remainder. Then, as with whole numbers, we subtract, bring down the next *term* and continue. We show the process by detailing the steps of the division.

$2x + 3 \overline{\smash{\big)}\, 2x^3 + 13x^2 + 11x - 6}$ Divisor outside; dividend inside; both written in descending powers

$ x^2$
$2x + 3 \overline{\smash{\big)}\, 2x^3 + 13x^2 + 11x - 6}$ First term of divisor into first term of dividend; record quotient term

$ x^2$
$2x + 3 \overline{\smash{\big)}\, 2x^3 + 13x^2 + 11x - 6}$
$ \underline{2x^3 + 3x^2}$ Multiply and subtract (*mentally change signs and add*)
$ 10x^2$

$ x^2$
$2x + 3 \overline{\smash{\big)}\, 2x^3 + 13x^2 + 11x - 6}$
$ \underline{2x^3 + 3x^2 \quad \downarrow}$ Bring down next term(s)
$ 10x^2 + 11x$ Remainder

$ x^2 + 5x$
$2x + 3 \overline{\smash{\big)}\, 2x^3 + 13x^2 + 11x - 6}$ First term of divisor into first term of remainder
$ \underline{2x^3 + 3x^2}$
$ 10x^2 + 11x$

$ x^2 + 5x$
$2x + 3 \overline{\smash{\big)}\, 2x^3 + 13x^2 + 11x - 6}$
$ \underline{2x^3 + 3x^2}$ Multiply and subtract
$ 10x^2 + 11x$
$ \underline{10x^2 + 15x \quad \downarrow}$ Bring down next term
$ -4x - 6$ Remainder

5.5 Division of Polynomials

$$
\begin{array}{r}
x^2 + 5x - 2 \\
2x + 3 \overline{)\, 2x^3 + 13x^2 + 11x - 6\,}\\
\underline{2x^3 + 3x^2 }\\
10x^2 + 11x \\
\underline{10x^2 + 15x }\\
-\,4x - 6
\end{array}
$$

First term of divisor into first term of remainder

$$
\begin{array}{r}
x^2 + 5x - 2 \\
2x + 3 \overline{)\, 2x^3 + 13x^2 + 11x - 6\,}\\
\underline{2x^3 + 3x^2 }\\
10x^2 + 11x \\
\underline{10x^2 + 15x }\\
-\,4x - 6\\
\underline{-\,4x - 6}\\
0
\end{array}
$$

The entire division is generally shown in this "condensed" form

The quotient is $x^2 + 5x - 2$ with remainder 0.

The division process is continued until the remainder is 0 (in which case we say the division "comes out even") or until the degree of the remainder is less than the degree of the divisor. The check is the same as for arithmetic division.

Check:

$$
\begin{aligned}
\text{Divisor} \cdot \text{Quotient} \;+\; & \text{Remainder} = \text{Dividend}\\
(2x + 3) \cdot (x^2 + 5x - 2) + 0 &= 2x^3 + 13x^2 + 11x - 6\\
2x^3 + 10x^2 - 4x + 3x^2 + 15x - 6 + 0 &= 2x^3 + 13x^2 + 11x - 6\\
2x^3 + 13x^2 + 11x - 6 &= 2x^3 + 13x^2 + 11x - 6
\end{aligned}
$$

The dividend is not always evenly divisible by the divisor. Consider the division $23 \div 4$:

$$
\begin{array}{r}
5\\
4 \overline{)\, 23\,}\\
\underline{20}\\
3 = R
\end{array}
$$

◀

The solution of the preceding example may either be written $23 \div 4 = 5$, $R = 3$; or $23 \div 4 = 5 + \frac{3}{4}$ or $5\frac{3}{4}$. Similarly, the quotient of polynomials can be written in either form as in the examples that follow.

Example 10

Divide $(3x^2 + 2x - 18)$ by $(x - 2)$.

$$
\begin{array}{r}
3x \\
x - 2 \overline{)\, 3x^2 + 2x - 18\,}
\end{array}
$$

First term of divisor into first term of dividend

$$\begin{array}{r} 3x \\ x-2\overline{\smash{)}3x^2+2x-18} \\ \underline{3x^2+6x\downarrow} \\ 8x-18 \end{array}$$ Multiply, subtract, bring down next term of dividend

Remainder

$$\begin{array}{r} 3x+8 \\ x-2\overline{\smash{)}3x^2+2x-18} \\ \underline{3x^2-6x} \\ 8x-18 \end{array}$$ First term of divisor into first term of remainder

$$\begin{array}{r} 3x+8 \\ x-2\overline{\smash{)}3x^2+2x-18} \\ \underline{3x^2-6x} \\ 8x-18 \\ \underline{8x-16} \\ -2 \end{array}$$ Multiply, subtract

Remainder

The quotient is $3x + 8$ with remainder -2.

Check:

$$\text{Divisor} \cdot \text{Quotient} + \text{Remainder} = \text{Dividend}$$

$$(x-2) \cdot (3x+8) + (-2) = 3x^2 + 2x - 18$$

$$3x^2 + 2x - 16 + (-2) = 3x^2 + 2x - 18$$

$$3x^2 + 2x - 18 = 3x^2 + 2x - 18$$

The preceding result may be written either as

$$(3x^2 + 2x - 18) \div (x-2) = 3x + 8, \text{ R} = -2$$

or $\quad (3x^2 + 2x - 18) \div (x-2) = 3x + 8 + \dfrac{-2}{(x-2)}$

Let's agree to accept this second form for the answer as the preferred form to make further algebraic manipulations simpler. ◀

Example 11

Divide $x^3 - 4x^2 + 3x - 12$ by $x - 4$.

$$\begin{array}{r} x^2+3 \\ x-4\overline{\smash{)}x^3-4x^2+3x-12} \\ \underline{x^3-4x^2\downarrow\downarrow} \\ 0+3x-12 \\ \underline{3x-12} \\ 0 \end{array}$$ Remainder

Note: The first subtraction gave a difference of 0. When this happens, write a 0 or leave a blank space in the quotient, bring down the next term(s), and

5.5 Division of Polynomials

continue as usual

$$(x^3 - 4x^2 + 3x - 12) \div (x - 4) = x^2 + 3$$

[Margin note: Remainder must be 1 less degree than divisor]

Example 12

Divide $2x^4 + 5x^3 - 3x + 1$ by $x^2 + x - 2$.

$$\begin{array}{r}
2x^2 + 3x + 1 \\
x^2 + x - 2 \overline{\smash{\big)} 2x^4 + 5x^3 + 0x^2 - 3x + 1} \\
\underline{2x^4 + 2x^3 - 4x^2 } \\
3x^3 + 4x^2 - 3x \\
\underline{3x^3 + 3x^2 + 6x } \\
x^2 + 3x + 1 \\
\underline{x^2 + x - 2} \\
2x + 3 \quad \text{Remainder}
\end{array}$$

Notice there was no x^2 term in the dividend so the $0x^2$ was written. This allows the proper alignment of like terms. Also note that the highest power of the divisor is 2, so the division is complete when the remainder equals 0 or has a degree less than 2. Here, the remainder $2x + 3$ is of degree 1.

$$(2x^4 + 5x^3 - 3x + 1) \div (x^2 + x - 2)$$
$$= 2x^2 + 3x + 1 + \frac{2x + 3}{x^2 + x - 2} \qquad \text{Preferred form}$$

Example 13

Divide $3x^2 + 5x^3 - 2 + 3x$ by $x - 2$.

The terms of the dividend must be arranged in descending powers.

$$\begin{array}{r}
5x^2 + 13x + 29 \\
x - 2 \overline{\smash{\big)} 5x^3 + 3x^2 + 3x - 2} \\
\underline{5x^3 - 10x^2 } \\
13x^2 + 3x \\
\underline{13x^2 - 26x } \\
29x - 2 \\
\underline{29x - 58} \\
56 \quad \text{Remainder}
\end{array}$$

[Margin note: when dividend by 2 or more terms — long division]

$$(3x^2 + 5x^3 - 2 + 3x) \div (x - 2) = 5x^2 + 13x + 29 + \frac{56}{x - 2}$$

When dividing polynomials, remember:

1. The degree of the divisor must be less than or equal to the degree of the dividend.

2. When subtracting, mentally change the sign of each term in the subtrahend and add.

3. Continue to divide until the remainder is 0 or the degree of the remainder is less than the degree of the divisor.

Example 14 Find an expression for the width of the rectangle pictured if the area of the rectangle is represented by the expression $6x^2 + 7x - 3$.

Area is $6x^2 + 7x - 3$

width = ?

length = $2x + 3$

Recall the area of a rectangle is the product of its length and width. That is,

$$A = L \cdot W$$

Therefore, to obtain the width we must divide the area by the length.

$$W = \frac{A}{L} = \frac{6x^2 + 7x - 3}{2x + 3}$$

$$\begin{array}{r} 3x - 1 \\ 2x+3 \overline{\smash{)}6x^2 + 7x - 3} \\ \underline{6x^2 + 9x} \\ -2x - 3 \\ \underline{-2x - 3} \\ 0 \end{array}$$

The width is given by $3x - 1$.

Exercise Set 5.5

Complete each of the following divisions:

1. $\dfrac{6x^5}{3x^2}$

2. $\dfrac{9x^4}{3x}$

3. $\dfrac{8xy^3}{-2x}$

4. $\dfrac{4ab^6}{-2a}$

5. $\dfrac{-24xy^4}{-4xy^3}$

6. $\dfrac{-20a^3b}{4ab}$

7. $(4x^3 + 8x^2) \div 2x$

8. $(6y^3 - 9y^2) \div 3y$

9. $\dfrac{4x^4 + 3x^2 - 7x}{x}$

5.5 Division of Polynomials

10. $\dfrac{2y^3 - 4y^2 + 5y}{y}$ 11. $\dfrac{8x^3 - 5x^2 + 4x}{-4x}$ 12. $\dfrac{7y^3 + 6y^2 - 3y}{-3y}$

13. $(9x^3y^2 - 18x^2y^3 + 27x^2y^2) \div (18x^2y^2)$

14. $(12m^4n^3 + 8m^3n^2 - 9m^2n^2) \div (6m^2n^2)$

15. $\dfrac{8m^7 + 4m^6 - 5m^5}{2m^3}$ 16. $\dfrac{5y^5 + 6y^7 - 9y^8}{3y^4}$

Perform each of the following divisions:

17. $(x^2 + 5x + 6) \div (x + 3)$
18. $(2x^2 + 11x + 12) \div (x + 4)$
19. $(2x^2 + 5x - 3) \div (2x - 1)$
20. $(2x^2 - 11x - 21) \div (x - 7)$
21. $(x^3 - 3x^2 - 9x - 5) \div (x - 5)$
22. $(3x^2 - 10x - 8) \div (3x + 2)$
23. $(2x^3 + 3x^2 - 8x + 3) \div (2x - 1)$
24. $(2a^3 - a^2 - 31a + 35) \div (2a - 7)$
25. $(8a^3 - 4a + 2) \div (2a - 1)$
26. $(y^3 + 5y^2 - 2y - 24) \div (y - 2)$
27. $\dfrac{a^2 - 9}{a + 3}$
28. $\dfrac{x^2 - 16}{x + 4}$
29. $\dfrac{4x^2 - 25}{2x - 5}$
30. $\dfrac{8x^2 - 50}{4x + 10}$
31. $(15x^5 - 27x^2 - 7x^4 - 7x + 6) \div (5x^2 + x - 1)$
32. $(b^4 + 3b^3 + 2b + 2) \div (b + 1)$
33. $(4y^2 - 2y + 2y^3 + y^4 - 5) \div (y^2 - 1)$
34. $(x^3 - 3x^2 + 2x + 1) \div (x - 2)$
35. $(6a^3 + a - 10 + 17a^2) \div (2a + 5)$
36. $(6a^3 + 5a^2 + 2a - 3) \div (2a + 1)$
37. $(6x^2 + 3x^3 - 4 + 5x) \div (x^2 + x - 1)$
38. $(6x^4 + 5x^3 - 8x^2 - 7x + 17) \div (2x^2 + 3x - 4)$
39. $(5x^4 - 3x^2 + 2x - 1) \div (x^2 + x - 1)$
40. $(3x^4 - 3 + 2x^3) \div (x^2 + 2x + 1)$

Find an expression for the width of a rectangle if its length and area are represented by the following polynomials:

41. Length $= y + 5$; Area $= 2y^2 + 7y - 15$
42. Length $= 3m + 2$; Area $= 6m^2 + 7m + 2$

Find an expression for the length of a rectangle if its width and its area are as indicated:

43. Width $= 4x - 2$; Area $= 8x^2 + 16x - 10$
44. Width $= 4a - 3$; Area $= 8a^2 + 14a - 15$
45. Width $= 3x$; Area $= 3x^3 + 12x^2 - 3x$
46. Width $= 4y$; Area $= 12y^4 - 8y^3 + 4y^2$

Review

47. Multiply as indicated:
    a. $3(x^2 - 2x + 4)$
    b. $3m(m^2 + 3m - 2)$
    c. $x^2(x^2 + 3x - 8)$
    d. $(b + 3)(2b - 5)$
    e. $(2a - 5b)(2a - b)$

48. Draw the graph of $3x - y = 9$

49. Solve for the variable:
    a. $3(t - 2) - 6 = 2t - 2(3t - 1)$
    b. $\dfrac{y}{4} - 1 = \dfrac{3y}{2} + 4$

50. Solve the following system *by graphing*:
    $2x + y = 7$
    $x - 3y = 4$

5.6 Factoring

We now begin our study of factoring polynomials with integer coefficients. It is important that you master this skill as it is used throughout the remainder of this course and in future mathematics courses. To *factor* a polynomial is to express it as the product of two or more other polynomials, when possible. We continue to factor until all of the polynomial factors are *prime polynomials*. A prime polynomial is a polynomial that cannot be expressed as a product of polynomials other than itself and 1. In such a form we say we have *completely factored* the polynomial, or that the polynomial is in *completely factored form*.

Factoring Out the Largest Common Monomial Factor

The first step in successful factoring is to remove the largest common monomial factor. Recall that in finding the product $3x(x + 2) = 3x^2 + 6x$, we use the

5.6 Factoring

distributive law to multiply $3x$ by both terms of the second polynomial. But imagine being presented with the polynomial $3x^2 + 6x$ and being asked to express it as the product of its prime factors. To do this, we identify the largest factor that is common to all terms in the polynomial and then remove this common factor.

$$3x^2 + 6x = (3x)(x) + (3x)(2)$$
$$= 3x(x + 2)$$

As you can see, we write the given polynomial as the product of a monomial and a polynomial when possible. This is a very important procedure since it should always be the first step in any factoring. We factor out the largest common monomial factor from a polynomial as follows:

1. Find the largest constant that is a factor of each numerical coefficient in the polynomial.

2. Find each variable that appears as a factor in every term (if any). The smallest power of each common variable factor may be factored out of each term.

3. The product of the constant (Step 1) and the variables (Step 2) is the largest common monomial factor.

[Handwritten margin note: double check that answer does not have any common factor / if it does you haven't found the LCF]

Example 1

Factor out the largest common monomial factor:

A. $6x^2 - 8x$

$$6x^2 - 8x = (2x)(3x) - (2x)(4)$$
$$= 2x(3x - 4)$$

B. $21y^3 + 14xy^2$ *[handwritten: $7y^2(3y + 2x)$]*

$$21y^3 + 14xy^2 = (7y^2)(3y) + (7y^2)(2x)$$
$$= 7y^2(3y + 2x)$$

C. $6x^3y + 15x^2y - 9xy$ *[handwritten: $3xy(2x^2 + 5x - 3)$]*

$$6x^3y + 15x^2y - 9xy = (3xy)(2x^2) + (3xy)(5x) - (3xy)(3)$$
$$= 3xy(2x^2 + 5x - 3)$$

Generally, the middle step is done mentally, as in the following examples.

D. $12xy - 8x^2y + 4x^3y = 4xy(3 - 2x + x^2)$

E. $6x^2 - 6x - 12 = 6(x^2 - x - 2)$

F. $3x^4 - 9x^3 - 21x^2 = 3x^2(x^2 - 3x - 7)$ ◂

Example 2

Factor out the largest common monomial factor:

A. $8a^3 - 4a^2 + 2a$

Note the largest common monomial factor, $2a$, is one of the terms of the polynomial itself. Care must be taken when factoring $2a$ from the third term, particularly.

$$8a^3 - 4a^2 + 2a = (2a)(4a^2) - (2a)(2a) + (2a)(1) \quad \text{Third term must be written as a product}$$

$$= 2a(4a^2 - 2a + 1)$$

B. $3x^2 + 9x^3 - 3x^4 = 3x^2(1 + 3x - x^2)$

C. $4xy - 8x^2y + 4x^3y = 4xy(1 - 2x + x^2)$

D. $2x^3y - 6xy^2 - 2xy = 2xy(x^2 - 3y - 1)$ ◀

Example 3

Factor out the largest common monomial factor from the polynomial $-3x^3 - 6x^2 + 9x$.

Based upon the previous examples, we may factor $3x$ from the polynomial, obtaining

$$-3x^3 - 6x^2 + 9x = 3x(-x^2 - 2x + 3)$$

Note, however, that since a factor of -1 can always be factored from any term, we may also factor $-3x$ from the polynomial. We obtain

$$-3x^3 - 6x^2 + 9x = -3x(x^2 + 2x - 3)$$

Both factorizations of $-3x^3 - 6x^2 + 9x$ are correct. Whether we remove $3x$ or $-3x$ as a common factor depends on how we want the product to appear, although generally if the leading coefficient of the polynomial is negative, we remove a negative numerical coefficient. ◀

Exercise Set 5.6

Find each of the following products:

1. $4x(2x - 3)$
2. $5x(3x + 5)$
3. $2b(3b^2 - 2b + 1)$
4. $4y(y^2 - 3y + 2)$
5. $2xy(x + 2y - 3)$
6. $3ab(2a - b + 1)$
7. $-3m(-2m^2 - 3m - 4)$
8. $-2m(-2m^2 - 4m - 6)$
9. $3x^2y(2x - 3y + 4)$
10. $3ab^2(3a + 2b - 5)$

Factor out the largest common monomial factor in each of the following polynomials:

11. $6x^2 - 8x$
12. $5x^2y + 15xy^2$
    $5xy(x + 3y)$
13. $6x^3 + 15x^2 + 6x$

5.6 Factoring

14. $15x^2 - 5x - 5$
15. $10x^4y^2 - 15x^2y^3 - 5xy^4$
16. $30x^3y^2 + 12x^2y^3 - 18xy^4$
17. $7a^6 - 7a^5 + 21a^4 - 14a^2$
18. $18m^5 - 30m^4 + 12m^3 - 24m^2$
19. $-6y^3 - 9y^2 + 24y$
20. $-20x^3 + 12x^2 + 8x$
21. $15x^4y^3 + 9x^3y^3 - 3x^2y^3$
22. $4x^4y^4 + 12x^3y^4 - 8x^2y^4$
23. $3a^2 + 9a^3 - 12a^4$
24. $5x^4 + 10x^3 - 20x^2$
25. $9x^4y - 12x^3y^2 + 15x^2y^3$
26. $16x^6 - 12x^5 + 4x^4 - 20x^3$
27. $14x^5y - 21x^4y^2 + 35x^3y^3 + 7x^2y^4$
28. $12a^7 + 30a^6b - 42a^5b^2 + 18a^4b^3$
29. $16x^4y^2 - 72x^3y^3 + 8x^2y^4$
30. $33a^3b - 11a^2b^2 + 77ab^3$
31. $12x^8 - 8x^7 + 20x^6 - 16x^5 + 4x^4$
32. $18y^5 + 27y^4 - 81y^3 + 9y^2$
33. $30x^5y - 18x^4y + 12x^3y - 6x^2y$
34. $3x^{10} - 7x^9 + 11x^7 - 3x^6$

Find an expression for the width of a rectangle if its length and area are as indicated. Recall: $A = LW$.

35. Length = $5x$; Area = $10x^3 - 15x^2 + 5x$
36. Length = $3y$; Area = $6y^3 + 15y^2 + 9y$
37. Length = $4x^2y$; Area = $4x^3y^3 - 12x^3y + 8x^2y^3$
38. Length = $2x^2y^2$; Area = $2x^4y^3 - 8x^3y^4 + 2x^3y^5$

Review

39. Multiply as indicated:
    a. $(x + 3)(x + 5)$
    b. $(x + 3)(x - 5)$
    c. $(x - 3)(x + 5)$
    d. $(x - 3)(x - 5)$

40. Multiply as indicated:
    a. $(y + 7)(y + 2)$
    b. $(y + 7)(y - 2)$
    c. $(y - 7)(y + 2)$
    d. $(y - 7)(y - 2)$

41. List all pairs of integers whose product is
    a. 12
    b. -15

FIVE **Polynomials**

42. What is the degree of each polynomial?

    a. $x - 3$ b. $x^2 + 2x - 1$

43. a. Multiply $(x - 3)(x^2 + 2x - 1)$

    b. What is the degree of the product?

5.7 Factoring Trinomials: $x^2 + bx + c$

This section discusses procedures for factoring trinomials of the form $x^2 + bx + c$ (the coefficient of x^2 is 1). Trinomials of the form $ax^2 + bx + c$ (the coefficient of x^2 is different from 1) will be factored in the next section.

Recall from Section 5.3 the product

$$(x + a)(x + b) = x^2 + \underbrace{(a + b)}_{\text{Sum}}x + \underbrace{ab}_{\text{Product}}$$

So, to factor a trinomial such as $x^2 + 5x + 6$ into a product of binomials, we attempt to find two *factors* of 6 whose *sum* is 5. First, we write parentheses to contain the binomial factors and enter the variable as follows:

$$x^2 + 5x + 6 = (x \quad)(x \quad)$$

1. The *factors* of 6:

    $1 \cdot 6$

    $2 \cdot 3$

    $-2 \cdot -3$

    $-1 \cdot -6$

2. We must find a pair of these factors whose sum is 5. So the constants must be 2 and 3 since $2 + 3 = 5$.

Therefore, $x^2 + 5x + 6 = (x + 2)(x + 3)$. Check by FOIL.

The Use of Signs

To help in the task of factoring (which is the reverse of multiplication), let's study the signs in the following four simple products:

a. $(x + 2)(x + 4) = x^2 + 6x + 8$

b. $(x + 2)(x - 4) = x^2 - 2x - 8$

c. $(x - 2)(x + 4) = x^2 + 2x - 8$

d. $(x - 2)(x - 4) = x^2 - 6x + 8$

5.7 Factoring Trinomials: $x^2 + bx + c$

Note that when the sign of the third term of the trinomial is *positive*, the two signs in the binomials must be *the same*.

a. $x^2 + 6x + 8 = (x + 2)(x + 4)$
   Positive   Like signs

d. $x^2 - 6x + 8 = (x - 2)(x - 4)$
   Positive   Like signs

Similarly, when the sign of the third term of the trinomial is *negative*, the two signs in the binomials must be *opposites*:

b. $x^2 - 2x - 8 = (x + 2)(x - 4)$
   Negative   Opposite signs

c. $x^2 + 2x - 8 = (x - 2)(x + 4)$
   Negative   Opposite signs

Consider the product $(x + a)(x + b) = x^2 + (a + b)x + ab$. We are again reminded that the sum of the constant terms from the binomials must be the coefficient of the middle term of the trinomial.

$$(x + a)(x + b) = x^2 + (a + b)x + ab$$
 Sum Product

Recall:

$x^2 + 5x + 6 = (x + 2)(x + 3)$

sum $= 2 + 3 \qquad$ product $= 2 \cdot 3$

Example 1

Factor $x^2 + 2x - 15$.

First we write the parentheses

$x^2 + 2x - 15 = (x \quad)(x \quad)$

The constants required to complete the binomials must be factors of -15.

Choices:

$1 \cdot (-15)$
$3 \cdot (-5)$
$(-1) \cdot 15$
$(-3) \cdot 5$

The constants must have a sum of 2, the coefficient of the middle term. The choice

must be $(-3), 5$ since $-3 + 5 = 2$. We write
$$x^2 + 2x - 15 = (x - 3)(x + 5)$$
Check by FOIL.

Example 2

Factor $x^2 - x - 12$.

Write the parentheses
$$x^2 - x - 12 = (x \quad)(x \quad)$$

The constants required to complete the binomials must have a product of -12 (opposite signs).

Choices:

$1 \cdot (-12)$

$(-1) \cdot 12$

$2 \cdot (-6)$

$(-2) \cdot 6$

$\boxed{3 \cdot (-4)}$

$(-3) \cdot 4$

The constants must háve a sum of -1 since the middle term is $-1x$. The choice must be $3, (-4)$ since $3 + (-4) = -1$. We write
$$x^2 - x - 12 = (x + 3)(x - 4)$$
Check by FOIL.

Example 3

Factor $y^2 + 6y + 10$.

Write the parentheses
$$y^2 + 6y + 10 = (y \quad)(y \quad)$$

The constants required to complete the binomials must have a product of 10 (like signs).

Choices:

$1 \cdot 10$

$2 \cdot 5$

$(-1) \cdot (-10)$

$(-2) \cdot (-5)$

The constants must have a sum of 6 since the middle term is $6y$. However, there

5.7 Factoring Trinomials: $x^2 + bx + c$

are no constants from the preceding choice list whose sum is 6. Therefore, $y^2 + 6y + 10$ is not factorable. It is a prime polynomial. ◀

You should become so proficient at factoring trinomials of the form $x^2 + bx + c$ that the constants can be determined mentally. The resulting factorization can be checked quickly using **FOIL**. Remember two things:

1. The constants selected must have a *product* equal to the third term of the trinomial.

2. The constants selected must have a *sum* equal to the coefficient of the middle term of the trinomial.

Example 4

Factor $3a^2 + 27a^2 + 60a$.

Remove the largest common monomial factor (always our first step when possible).

$$3a^3 + 27a^2 + 60a = 3a(a^2 + 9a + 20)$$

Write the parentheses

$$= 3a(a \quad)(a \quad)$$

The constants required to complete the binomials must have a product of 20 (like signs) and a sum of 9.

The correct choice is 4 and 5.

$$= 3a(a + 4)(a + 5)$$

Check by FOIL. ◀

Trinomials in two variables of the form $x^2 + bxy + cy^2$ are factored in a similar fashion. When we write the parentheses, we insert both variables. That is, we write

$$x^2 + bxy + cy^2 = (x \quad y)(x \quad y)$$

We must now determine the coefficients of y in the same way we determined the constants in the earlier examples.

Example 5

Factor $x^2 - 4xy - 21y^2$.

Write the parentheses

$$x^2 - 4xy - 21y^2 = (x \quad y)(x \quad y)$$

The coefficients on y required to complete the binomials must have a product of -21 (opposite signs) and a sum of -4. The correct choice is 3 and -7.

$$= (x + 3y)(x - 7y)$$

Check by FOIL. ◀

Exercise Set 5.7

Write each of the following trinomials in completely factored form. If the trinomial is not factorable, indicate it to be a prime polynomial.

1. $y^2 + 2y - 15$
2. $x^2 + x - 6$
3. $m^2 - 7m + 10$
4. $a^2 - 9a + 20$
5. $3x^2 + 9x - 15$
6. $4y^2 - 8y + 28$
7. $a^3 + 4a^2 - 21a$
8. $m^3 + 3m^2 - 40m$
9. $2x^3 + 24x^2 + 70x$
10. $3x^3 + 21x^2 + 36x$
11. $a^2 + ab - 6b^2$
12. $a^2 - 2ab - 8b^2$
13. $x^2 - 7xy + 12y^2$
14. $m^2 - 8mn + 12n^2$
15. $m^2 + 5mn + 4n^2$
16. $x^2 + 12xy + 35y^2$
17. $k^2 - 11km + 30m^2$
18. $x^2 + 3xy - 28y^2$
19. $4x^4 - 4x^3 - 24x^2$
20. $5y^3 + 15y^2 - 50y$
21. $2x^2y^4 + 22x^2y^3 + 56x^2y^2$
22. $3m^4n^2 - 15m^3n^2 + 18m^2n^2$
23. $3x^3y - 3x^2y^2 - 36xy^3$
24. $4a^3b + 28a^2b^2 + 48ab^3$
25. $m^4n^3 + 11m^3n^4 + 28m^2n^5$
26. $x^5y^2 - 11x^4y^3 + 3x^4y^4$
27. $2x^3y - 6x^2y^2 + 10xy$
28. $3m^3n^2 - 3m^2n^3 + 12mn^2$
29. $3x^3y + 6x^2y^2 + 15x^2y$
30. $4x^2y^2 - 2xy^3 - 6xy^2$
31. $x^2 - 13x + 40$
32. $y^2 - 13y + 42$
33. $x^2 + xy - 6y^2$
34. $m^2 - 5mn - 14n^2$
35. $2x^3y - 8x^2y + 6xy$
36. $2xy^3 - 12xy^2 + 10xy$
37. $-2a^4 - 18a^3b - 36a^2b^2$
38. $-3k^4 - 18k^3m - 15k^2m^2$

Review

39. Multiply:
    a. $(2a + 3)(3a + 4)$
    b. $(2a + 3)(3a - 4)$
    c. $(2a - 3)(3a + 4)$
    d. $(2a - 3)(3a - 4)$

40. Multiply:
    a. $(3r + t)(4r + 3t)$
    b. $(3r + t)(4r - 3t)$
    c. $(3r - t)(4r + 3t)$
    d. $(3r - t)(4r - 3t)$

41. Find a pair of integers whose product and whose sum are as listed:

	Product	Sum
a.	12	7
b.	12	−4
c.	−12	1
d.	−12	−11

42. Find the slope of the line through the given pair of points:

a. $(-3, 0), (2, 3)$ b. $(-2, -1), (3, -4)$
c. $(4, 1), (4, -3)$ d. $(3, 5), (-1, 5)$

5.8 Factoring Trinomials: $ax^2 + bx + c$, $a \neq 1$

We continue studying the factorization of trinomials by considering trinomials of the form $ax^2 + bx + c$ where the coefficient of x^2 is not 1.

Example 1

Recall the multiplication $(2x + 3)(3x + 1)$.

$$(2x + 3)(3x + 1) = 6x^2 + 11x + 3$$

Note: Product of two firsts = first term of trinomial.
Product of two lasts = last term of trinomial.
Sum of inner and outer products = middle term of trinomial.

We stress a technique for factoring these trinomials that we arbitrarily label as the *guess-check-guess again* method. That is, we

1. Make an educated *guess* as to the factorization.
2. *Check* our guess by a FOIL expansion.
3. *Guess again* at the factorization if our first guess is wrong.

We will review those points that allow our first guess to be an "educated" one. But remember, if our first attempt is not correct, we merely guess again.

To factor a trinomial such as $6x^2 + 7x - 3$, we write parentheses to contain the binomial factors and enter the variable as follows:

$$6x^2 + 7x - 3 = (\underline{\quad} x \quad)(\underline{\quad} x \quad).$$

Constants are now required to complete each parentheses as a binomial. We need constants to form the second term of each binomial as in the previous section, but we also require constants (perhaps other than 1) to be coefficients of x in each binomial. As we consider $6x^2 + 7x - 3$, keep in mind the technique of finding the product from Example 1.

$$6x^2 + 7x - 3 = (\underline{\quad}x \quad)(\underline{\quad}x - \underline{\quad})$$

Product = 6

Choices:

$6 \cdot 1$

$2 \cdot 3$

and $\quad 6x^2 + 7x - 3 = (\underline{\quad}x \quad)(\underline{\quad}x \quad)$

Product = -3

Choices:

$\left. \begin{array}{r} -1 \cdot 3 \\ 1 \cdot -3 \end{array} \right\}$ (Opposite signs needed)

First guess:

$$6x^2 + 7x - 3 = (\underline{6}x \underline{\,-\,3})(\underline{1}x \underline{\,+\,1})$$

Check:

$$(6x - 3)(x + 1) = 6x^2 + \underset{\uparrow}{3x} - 3$$
$$\text{Wrong}$$

Note, the middle term is wrong. Try another factorization of 6. Let's use $2 \cdot 3$.
Second guess:

$$6x^2 + 7x - 3 = (2x - 3)(3x + 1) = 6x^2 \underset{\uparrow}{- 7x} - 3$$
$$\text{Wrong sign}$$

The middle term is again incorrect, but is off only by the sign. It has the correct "size." Let's try reversing the signs of 3 and 1.
Third guess:

$$6x^2 + 7x - 3 = (2x + 3)(3x - 1)$$

5.8 Factoring Trinomials: $ax^2 + bx + c, a \neq 1$

Check:

$$(2x + 3)(3x - 1) = 6x^2 + 7x - 3$$

Our third try is the correct factorization of $6x^2 + 7x - 3$.

We must not get discouraged if our guessing produces an error. As in our discussion on factoring $6x^2 + 7x - 3$, our error may be nothing more than an incorrect position of the constants or signs, or an inappropriate factorization on one of the constants in the trinomial. In the following examples, only the correct factorization is shown. The results can be checked using FOIL.

Example 2

Factor each trinomial:

A. $3x^2 - 13x - 10$

Write the parentheses

Product = 3

Choice: $1 \cdot 3$

$$3x^2 - 13x - 10 = (\underline{}x)(\underline{}x)$$

Product = 3

Product = -10

Choices:

$(-1) \cdot 10; \quad (-2) \cdot 5$

$1 \cdot (-10); \quad 2 \cdot (-5)$

Factor:

$$3x^2 - 13x - 10 = (3x + 2)(x - 5)$$

Check by FOIL.

B. $8x^2 - 2x - 15$

Product = 8

$$8x^2 - 2x - 15 = (2x - 3)(4x + 5)$$

Product = -15

Check by FOIL.

FIVE   Polynomials

$(2x+y)(3x+2y)$

C. $6x^2 + 7xy + 2y^2$

This trinomial in two variables is factored in a manner similar to trinomials in one variable (just as in Section 5.6).

$$6x^2 + 7xy + 2y^2 = (2x + y)(3x + 2y)$$

Product = 6

Product = 2

Check by FOIL.

D. $20^3b^2 + 22a^2b^3 - 12ab^4$

Note in this example, there is a common monomial factor that must be removed first.

$$20a^3b^2 + 22a^2b^3 - 12ab^4 = 2ab^2(10a^2 + 11ab - 6b^2)$$

Now, factor the trinomial

$$= 2ab^2(2a + 3b)(5a - 2b)$$

Product = 10

Product = −6

Check by FOIL.

E. $4x^2 - 12xy + 9y^2$

$$4x^2 - 12xy + 9y^2 = (2x - 3y)(2x - 3y)$$
$$= (2x - 3y)^2$$ ◀

It is critical to your future success in algebra that you become proficient at factoring trinomials. You are encouraged to work all problems at the end of this section as a means of becoming comfortable with factoring. Even those problems for which no answer is provided in the back of the text can be checked *quickly* using FOIL.

Exercise Set 5.8

Completely factor each of the following trinomials. If a trinomial is not factorable, identify it as a prime polynomial.

1. $x^2 + 6x + 8$
2. $x^2 + 2x - 15$
3. $2x^2 - xy - 3y^2$
4. $6x^2 - 17xy + 21y^2$
5. $a^2 + 6a - 16$
6. $6m^2 - 7m + 2$

5.8 Factoring Trinomials: $ax^2 + bx + c$, $a \neq 1$

7. $6m^2 - 5mn + n^2$
8. $a^2 + 2ab + b^2$
9. $x^2 + 8x + 12$
10. $x^2 + 13x + 12$
11. $x^2 - 8x + 15$
12. $y^2 - 9y + 8$
13. $x^2 - 6x + 9$
14. $x^2 + 16x + 64$
15. $x^2 + 4x - 12$
16. $x^2 - x - 12$
17. $2x^2 + 11x - 6$
18. $2x^2 + x - 6$
19. $2x^2 + 10x + 12$
20. $2x^2 + 8x + 16$
21. $6x^2 - 7x - 5$
22. $4y^2 + 11y - 15$
23. $6x^2 + 7x - 10$
24. $4x^2 + 47x - 12$
25. $10x^3 + 45x^2 + 45x$
26. $6x^4 + 10x^3 - 4x^2$
27. $3x^2 - 24x + 48$
28. $2x^3 - 20x^2 + 50x$
29. $6x^2 + 25x - 9$
30. $5x^3 - 32x^2 + 12x$
31. $3x^3 + 6x^2 + 6x$
32. $x^3 + 4x^2 + 5x$
33. $2x^2 - 4x - 5$
34. $3x^2 + 2x - 3$
35. $9a^2 - 17a + 8$
36. $9a^2 - 18a + 8$
37. $2m^2 - 8m + 8$
38. $m^3 - 2m^2 + m$
39. $7a^2 - 19a - 6$
40. $2a^2 - 7a - 22$
41. $2x^2 + 19x + 35$
42. $3x^3 - 24x^2 - 27x$
43. $4x^4 + 4x^3 - 24x^2$
44. $36x^3 - 6x^2 - 6x$

Write each of the following polynomials in completely factored form:

45. $6x^2y^2 + 4xy^3 - 6xy^2$
46. $-3x^3y - 9x^2y^2 + 12x^2y$
47. $9m^2 + 12mn + 4n^2$
48. $16x^2 - 24xy + 9y^2$
49. $30x^3y^2 + 2x^2y^3 - 4xy^4$
50. $28m^4n + 26m^3n^2 - 24m^2n^3$
51. $4x^2 - 11xy + 6y^2$
52. $5x^2 - 22xy + 8y^2$
53. $6ma^2n + 23mna - 4mn$
54. $15a^3b + 24a^2b^2 - 12ab^3$

Review

55. Multiply:
    a. $(3x - 2)(3x + 2)$
    b. $(x^2 - 1)(x^2 + 1)$
    c. $(x + 3y)(x - 3y)$
    d. $(2x + 7)(2x - 7)$
    e. $(x + 2)^2$
    f. $(a - 4)^2$
    g. $(2ab + 3)^2$
    h. $(3m - 2n)^2$

56. Find the area of each of the following figures:

a. 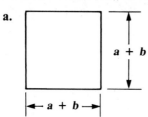   b.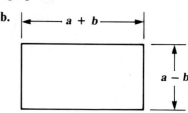

57. The length of a rectangle is 1 inch less than twice the width. Its perimeter is 40 inches. Find its dimensions.

58. Divide: $(x^3 - 3x^2 - 7x + 6) \div (x + 2)$.

5.9 Factoring Polynomials (Two Special Cases)

In general, once we have factored out the largest common monomial factor from the original polynomial, we attempt to factor the remaining polynomial further. In this section, we concentrate on two special forms of polynomials called:

a. The difference of two squares
b. Perfect square trinomials

The Difference of Two Squares

One of the special products in Section 5.4 was the product of a sum and difference:

$$(a + b)(a - b) = a^2 - b^2$$

The result of such a product is the difference of two perfect squares, a^2 and b^2. If we reverse the way we look at this special product, we observe that the difference of two perfect squares is always factorable as the product of a sum and difference.

> **DIFFERENCE OF TWO SQUARES**
> $a^2 - b^2 = (a + b)(a - b)$

We may show the factorization $a^2 - b^2 = (a + b)(a - b)$ geometrically, as follows. Draw a square whose sides are a units in length, then remove a small square b units in length, as shown in Figure 5.9-1A. We may then remove the

5.9 Factoring Polynomials (Two Special Cases)

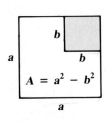

FIGURE 5.9-1A

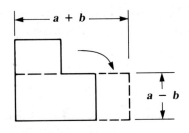

FIGURE 5.9-1B

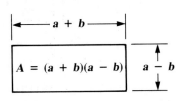

FIGURE 5.9-1C

indicated rectangle and reposition it as shown in Figure 5.9-1B. The areas of these two figures are the same (Figure 5.9-1C).

$$a^2 - b^2 = (a + b)(a - b)$$

Example 1

Factor each of the following expressions. First, remove the largest common monomial factor if one exists.

A. $x^2 - 4$

$x^2 - 4 = x^2 - 2^2$ Difference of two squares

$ = (x + 2)(x - 2)$ Sum times difference

B. $4m^2 - 16n^2$

$4m^2 - 16n^2 = 4(m^2 - 4n^2)$ Remove common monomial factor

$ = 4[m^2 - (2n)^2]$ Difference of two squares

$ = 4(m + 2n)(m - 2n)$

C. $4x^2 - 9y^2$

$4x^2 - 9y^2 = (2x)^2 - (3y)^2$

$ = (2x + 3y)(2x - 3y)$

D. $25 - 16y^2$

$25 - 16y^2 = (5)^2 - (4y)^2$

$ = (5 + 4y)(5 - 4y)$

E. $(x + y)^2 - 9$

$(x + y)^2 - 9 = (x + y)^2 - (3)^2$

$ = [(x + y) + 3][(x + y) - 3]$

$ = (x + y + 3)(x + y - 3)$

Observe that even though one of the perfect squares was a binomial, the method still applies.

FIVE   Polynomials

Note the *sum* of two perfect squares is a prime polynomial and *cannot* be factored. For example, $x^2 + y^2$, $4x^2 + 9$, and $25x^2 + 16m^2$ are all prime polynomials.

Perfect Square Trinomials

A **perfect square trinomial** is the result of squaring a binomial. For example, $(x + y)^2 = x^2 + 2xy + y^2$. That is, $x^2 + 2xy + y^2$ is a trinomial that is a perfect square and factors as a binomial multiplied by itself.

Whenever we encounter a trinomial whose first and third terms have the same sign and are perfect squares, we determine if it is a perfect square trinomial by checking whether the middle term is twice the product of the quantities whose squares are the first and third terms. If the trinomial is not a perfect square, we must try to factor it using the techniques that were covered in Sections 5.7 and 5.8.

Example 2

Factor each of the following polynomials. Remove the largest common factor if one exists.

A. $x^2 - 10x + 25$

The first term is the square of x, the third term is the square of 5, and the middle term is twice their product. We factor $x^2 - 10x + 25$ as the square of a binomial.

$$x^2 - 10x + 25 = (x - 5)(x - 5)$$
$$= (x - 5)^2$$

Checking with FOIL shows this factorization is correct. Thus, $x^2 - 10x + 25$ is a perfect square trinomial.

B. $9a^2 + 24ab + 16b^2$

The first term is the square of $3a$, the third term is the square of $4b$, and the middle term is twice their product: $2(3a)(4b) = 24ab$.

$$9a^2 + 24ab + 16b^2 = (3a + 4b)^2$$

Check with FOIL.

C. $2x^3y - 12x^2y^2 + 18xy^3$

Our first step is to remove the largest common monomial factor, $2xy$.

$$2x^3y - 12x^2y^2 + 18xy^3 = 2xy(x^2 - 6xy + 9y^2)$$

Note the first and third terms of the trinomial are the squares of x and $3y$, respectively. The middle term is twice their product.

$$= 2xy(x - 3y)^2$$

Check with FOIL.

5.9 Factoring Polynomials (Two Special Cases)

Note that the sign between the terms in the binomial is the same as the sign of the middle term in the perfect square trinomial. (See each of the factorizations in Examples **2A**, **2B**, and **2C**.)

D. $16m^2 + 20mn + 25n^2$

This appears to be a perfect square trinomial when we observe that the first term is the square of $4m$, the third term is the square of $5n$, but the middle term *is not* twice their product. Therefore,

$$16m^2 + 20mn + 25n^2$$

is not a perfect square trinomial. Attempting to factor by the methods of Section 5.7, we find the trinomial is prime.

E. $x^4 - 2x^2y^2 + y^4$

The first and third terms are the perfect squares of x^2 and y^2, respectively, and the middle term is twice their product. We write

$$x^4 - 2x^2y^2 + y^4 = (x^2 - y^2)^2$$

Although this factorization is correct, we note that the binomial factors further as the difference of two squares. That is,

$$\begin{aligned} x^4 - 2x^2y^2 + y^4 &= (x^2 - y^2)^2 \\ &= (x^2 - y^2)(x^2 - y^2) \\ &= (x + y)(x - y)(x + y)(x - y) \\ &= (x + y)^2(x - y)^2 \end{aligned}$$

This is the complete factorization of the original polynomial. ◀

Exercise Set 5.9

Completely factor each polynomial:

1. $x^2 + 9x + 18$
2. $x^2 - 8x + 12$
3. $y^2 - 25$
4. $m^2 - 16$
5. $x^2 - 100y^2$
6. $y^2 - 36z^2$
7. $2a^2 + 7a - 4$
8. $3x^2 + 9x - 12$
9. $8x^2 + 38x + 45$
10. $6x^2 - 13x - 28$
11. $8x^2 - 8$
12. $3x^3 - 12x$
13. $12x^3 - 8x^2 - 15x$
14. $16x^3 - 18x^2 - 9x$

15. $R^4 - 4$
16. $x^2y^2 - 9$
17. $4x^3 + 16x^2 - 20x$
18. $12x^2 + 44x - 16$
19. $2x^2 - 162$
20. $3x^2 - 192$
21. $2x^2 - 5x + 8$
22. $3x^2 + 7x - 4$
23. $9A^2 + 52A - 12$
24. $27M^2 + 30M - 8$
25. $10x^3 + 15x^2 - 10x$
26. $12x^4 - 6x^3 - 90x^2$
27. $2y^2 - 22y - 24$
28. $2A^2 - 2A - 144$
29. $x^4 - 81$
30. $x^4 - 16$
31. $12x^4 - 52x^3 + 35x^2$
32. $32x^3 - 4x^2 - 6x$
33. $8x^3 - 29x^2 - 12x$
34. $6x^2 + 29x + 30$
35. $5x^5 - 80x$
36. $3x^5 - 243x$

In each of Problems 37–42, determine the value(s) of k that make the expression a perfect square trinomial:

37. $4x^2 - 12x + k$
38. $9x^2 - 12x + k$
39. $9x^2 + kx + 16$
40. $16x^2 - kx + 9$
41. $kx^2 - 20x + 4$
42. $kx^2 - 30x + 25$

Review

43. Multiply:
    a. $(x + y)(x^2 - xy + y^2)$
    b. $(x + 2)(x^2 - 2x + 4)$
    c. $(x + 3)(x^2 - 3x + 9)$

44. Multiply:
    a. $(m - n)(m^2 + mn + n^2)$
    b. $(m - 2)(m^2 + 2m + 4)$
    c. $(m - 4)(m^2 + 4m + 16)$

45. Multiply. Combine like terms when possible:
    a. $(x + 3)(x + 5)$
    b. $(x^2 + 3)(x^2 - 5)$
    c. $(x^2 + 3)(x + 5)$
    d. $(x + 3)(y + 5)$

46. Sketch the solution of the following systems of inequalities:
    a. $2x + y > -1$
       $x + y < 5$
       $x \geq 0$
    b. $-1 < x \leq 3$
       $0 \leq y < 2$

5.10 Factoring the Sum and Difference of Cubes and Grouping

The Sum and Difference of Two Cubes

The following are the formulas for factorization of a sum of two cubes and the difference of two cubes. These formulas can be verified by multiplying out the right sides of the equations.

Know these formulas

$$a^3 + b^3 = (a + b)(a^2 - ab + b^2) \quad \text{Sum of two cubes}$$
$$a^3 - b^3 = (a - b)(a^2 + ab + b^2) \quad \text{Difference of two cubes}$$

Recall that to cube an expression means to use it as a factor three times. For example:

$$(3x^2y)^3 = (3x^2y)(3x^2y)(3x^2y)$$
$$= 3 \cdot 3 \cdot 3 \cdot x^2 \cdot x^2 \cdot x^2 \cdot y \cdot y \cdot y \quad \text{Multiplication of monomials}$$
$$= 3^3 x^{2+2+2} y^{1+1+1} \quad \text{Multiplication property of exponents}$$
$$= 27x^6y^3$$

The expression $27x^6y^3$ is called a *perfect cube*. Note that its numerical coefficient is the cube of an integer (see Table 5.10-1) and that the exponents of the variables are multiples of three. To reverse this process and write an expression such as $27x^6y^3$ as some base cubed, ( )^3, we find the base by writing the numerical coefficient as the cube of an integer and one-third of each of the exponents on the variables:

$$\frac{1}{3} \cdot 6 = 2 \qquad \frac{1}{3} \cdot 3 = 1$$

$$27x^6y^3 = (3x^2y^1)^3 \quad \text{Note: } 27 = 3^3;\ x^6 = (x^2)^3;\ y^3 = (y^1)^3$$

$$3^3 = 27$$

TABLE 5.10-1
The Cubes of 1–10

n	n³	n	n³
1	1	6	216
2	8	7	343
3	27	8	512
4	64	9	729
5	125	10	1000

11
12
13
14
15

FIVE Polynomials

To use either of the preceding factoring formulas, first recognize each of the two terms as a perfect cube and rewrite in the form ( )^3 by finding the number whose cube gives the numerical coefficient of the term and taking one-third of each exponent on the variable factors.

After the expression has been written as the sum of two cubes, $a^3 + b^3$, or the difference of two cubes, $a^3 - b^3$, identify the expressions as a and b and substitute using the appropriate formula.

[Margin note: $a^3 + b^3 = (a+b)(a^2 \pm ab + b^2)$; the trinomial will always be prime]

Example 1

Factor each expression:

A. $x^3 + 27$

$$x^3 + 27 = (x)^3 + (3)^3 \qquad a = x;\ b = 3$$
$$= (x + 3)(x^2 - 3x + 3^2)$$
$$= (x + 3)(x^2 - 3x + 9)$$

B. $y^3 - 512$

$$y^3 - 512 = (y)^3 - (8^3) \qquad a = y;\ b = 8$$
$$= (y - 8)(y^2 + 8y + 64)$$

C. $8x^6 + 125$

$$8x^6 + 125 = (2x^2)^3 + (5)^3 \qquad a = 2x^2;\ b = 5$$
$$= (2x^2 + 5)[(2x^2)^2 - 5 \cdot (2x^2) + (5)^2]$$
$$= (2x^2 + 5)(4x^4 - 10x^2 + 25)$$

Factoring by Grouping

Factoring by grouping is a technique by which we attempt to factor expressions with four terms. We group the first two terms in parentheses and the last two terms in parentheses and then remove the largest common monomial factor from each "group." Next, we observe whether the resulting expressions have a common binomial factor. If they do, this common factor is also factored out using the distributive property.

[Margin note: should produce common Binomial factor]

Example 2

Factor each polynomial by grouping:

A. $x^2 + 3x + 2x + 6$

$$x^2 + 3x + 2x + 6 = (x^2 + 3x) + (2x + 6) \qquad \text{Use parentheses to "group" the first two and last two terms}$$

$$= (x + 3) \cdot x + (x + 3) \cdot 2 \qquad \text{Remove common factors}$$

5.10 Factoring the Sum and Difference of Cubes and Grouping

$\qquad = (x + 3)(x + 2)$ Distributive property

Common binomial factor of $(x + 3)$

B. $x^2 + 3x - 2x - 6$

If the third term is preceded by a "−", we write this "−" outside the second parentheses, requiring that the sign of each term within the parentheses be changed.

$x^2 + 3x - 2x - 6 = (x^2 + 3x) - (2x + 6)$ Signs changed; group

$\qquad = (x + 3)x - (x + 3) \cdot 2$ Remove common factors

$\qquad = (x + 3)(x - 2)$ Distributive property

C. $y^2 - 4y - 3y + 12$

$y^2 - 4y - 3y + 12 = (y^2 - 4y) - (3y - 12)$ Group

$\qquad = (y - 4)y - (y - 4)3$ Remove common factors

$\qquad = (y - 4)(y - 3)$ Remove common binomial factor

D. $a^3 + a^2 - 4a - 4$

$a^3 + a^2 - 4a - 4 = (a^3 + a^2) - (4a + 4)$ Group

$\qquad = (a + 1)a^2 - (a + 1)4$ Remove common factors

$\qquad = (a + 1)(a^2 - 4)$ Remove common binomial factor

Observe that the second binomial can be factored further. It is the difference of two squares.

$\qquad = (a + 1)(a + 2)(a - 2)$ Complete factorization

E. $x^2 + 5x + x + 5$

$x^2 + 5x + x + 5 = (x^2 + 5x) + (x + 5)$ Group

$\qquad = (x + 5)x + (x + 5) \cdot 1$

Note that we wrote the second "group" as $(x + 5) \cdot 1$ to emphasize the 1 as a factor.

$\qquad = (x + 5)(x + 1)$ Remove common binomial factor ◀

Exercise Set 5.10

Multiply each of the following expressions:

1. $(2x)^3$
2. $(2a)^3$
3. $(xy)^3$
4. $(ab)^3$
5. $(3x)^3$
6. $(4b)^3$
7. $(5xy)^3$
8. $(4ab)^3$
9. $(7a^2)^3$
10. $(10x^2)^3$
11. $(8xy^2)^3$
12. $(6a^3b)^3$

Write each of the following monomials as a cube:

13. $8x^3$
14. $512y^3$
15. $1000x^6y^3$
16. $729a^3b^6$
17. $64m^9n^6$
18. $27p^{12}$

Factor each of the following polynomials:

19. $a^3 + 64$
20. $y^3 + 27$
21. $8x^3 - 27$
22. $8m^3 - 1$ $(2m-1)(4m^2+2m+1)$
23. $y^3 - 125$
24. $x^3 + 512$
25. $27b^3 - 8$
26. $8y^3 - 27$
27. $x^3 - y^3$
28. $a^3 - 27b^3$
29. $1000m^3 + 27n^3$
30. $729x^3y^6 + 1$

Complete the following factorizations:

31. $(x + 5)x + (x + 5)3$
32. $(y + 4)y + (y + 4)7$
33. $(a - 2)a - (a - 2)5$
34. $(m + 3)m^2 - (m + 3)7$
35. $(a + 4)x + (a + 4)y$
36. $(y + 2)x + (y + 2)6$

Factor each of the following polynomials by grouping:

37. $n^2 + 2n + 3n + 6$
38. $p^2 + 3p + 4p + 12$
39. $y^2 - 2y + 5y - 10$
40. $x^2 - 4x + 2x - 8$
41. $x^2 + 6x - x - 6$ *(MUST FACTOR out a (-1))*
42. $y^2 + 4y - y - 4$
43. $b^3 + 2b^2 + 3b + 6$
44. $u^3 + 2u^2 + 5u + 10$
45. $x^3 - 2x^2 + 4x - 8$
46. $x^3 - 3x^2 + 9x - 27$
47. $a^3 + 2a^2 - 9a - 18$
48. $y^3 + 6y^2 - 4y - 24$
49. $3x^3 - 3x^2 + 6x - 6$
50. $4m^3 - 8m^2 + 12m - 24$

Review

51. Multiply:
    a. $(y + 3)(y - 7)$
    b. $(u + 6)(u - 6)$
    c. $(2a + 5b)(a + 2b)$
    d. $(2v + 7)^2$

52. Factor:
   a. $n^2 - 11n + 24$
   b. $4a^2 - 49$
   c. $6t^2 - 5t - 4$
   d. $4y^2 + 20y + 25$

53. What is the slope of the graph of $3x - 5y = 8$?

54. Evaluate each expression:
   a. $(-3)(-2)(-5) - (-2)^2$
   b. $14 - \{11 - [6 - (2 - 8)] + 3\} - 5$
   c. $\dfrac{-6 + 9}{4(-2) + 8}$

5.11 Factoring Summary

The process of factoring has traditionally been a stumbling block for students. Therefore, we summarize the steps in factoring and give an exercise set for more practice.

Factoring summary

Given a polynomial to factor:

1. Remove the largest common monomial factor (if any).
2. Determine the number of terms there are in the remaining polynomial.

   a. If there are *two* terms, look for:
      The difference of two perfect squares, which is always factorable as the product of a sum and difference.
      $$a^2 - b^2 = (a + b)(a - b)$$
      The sum or difference of two cubes, which can be factored using the formulas
      $$a^3 + b^3 = (a + b)(a^2 - ab + b^2)$$
      $$a^3 - b^3 = (a - b)(a^2 + ab + b^2)$$

   b. If there are *three* terms, look for:
      (1) A perfect square trinomial
      $$a^2 + 2ab + b^2 = (a + b)^2$$
      $$a^2 - 2ab + b^2 = (a - b)^2$$

(2) A trinomial that requires us to use general factoring techniques (Sections 5.7 and 5.8).

c. If there are *four* terms, attempt to factor by grouping.

3. Check to see if any further factorization is possible.

The methods presented in this chapter give us the ability to factor a wide variety of polynomials. Although not all polynomials can be factored, more techniques will be introduced in later courses for factoring polynomials with higher degrees or having more than three terms. Before beginning the exercise set, you may wish to scan the preceding sections to review each of the formulas and methods. In addition, study each of the following factorizations.

Example 1

$2x^3y + 18xy^3$

$2x^3y + 18xy^3 = 2xy(x^2 + 9y^2)$ Remove common factor $2xy$; sum of two squares is prime, so no further factoring is possible. ◂

Example 2

$12a^3b - 75ab^3$

$12a^3b - 75ab^3 = 3ab(4a^2 - 25b^2)$ Remove common factor $3ab$

$= 3ab(2a + 5b)(2a - 5b)$ Factor difference of two squares ◂

Example 3

$2x^3 + 2x^2 - 24x$

$2x^3 + 2x^2 - 24x = 2x(x^2 + x - 12)$ Remove $2x$

$= 2x(x - 3)(x + 4)$ General trinomial factoring ◂

Example 4

$12c^4 - 36c^3 + 27c^2$

$12c^4 - 36c^3 + 27c^2 = 3c^2(4c^2 - 12c + 9)$ Remove $3c^2$
Perfect square trinomial

$= 3c^2(2c - 3)(2c - 3)$ Trinomial factoring

$= 3c^2(2c - 3)^2$ ◂

Example 5

$4x^4 + 12x^3 + 20x^2$

$4x^4 + 12x^3 + 20x^2 = 4x^2(x^2 + 3x + 5)$ Remove $4x^2$
No further factoring possible ◂

5.11 Factoring Summary

Example 6

$6a^3b^2 - 20a^2b^3 - 16ab^4$

$$6a^3b^2 - 20a^2b^3 - 16ab^4 = 2ab^2(3a^2 - 10ab - 8b^2) \quad \text{Remove } 2ab^2$$
$$= 2ab^2(3a + 2b)(a - 4b) \quad \text{Trinomial factoring}$$

Example 7

$3m^3 - 12m^2 + 9m - 36$

$$3m^3 - 12m^2 + 9m - 36 = 3(m^3 - 4m^2 + 3m - 12) \quad \text{Factor out 3}$$
$$= 3[(m^3 - 4m^2) + (3m - 12)] \quad \text{Group}$$
$$= 3[(m - 4)m^2 + (m - 4)3] \quad \text{Common monomial factors}$$
$$= 3(m - 4)(m^2 + 3) \quad \text{Common binomial factor}$$

Example 8

$16u^4 - 54u$

$$16u^4 - 54u = 2u(8u^3 - 27) \quad \text{Factor out } 2u$$
$$= 2u[(2u)^3 - (3)^3] \quad \text{Difference of two cubes}$$
$$= 2u(2u - 3)[(2u)^2 + (2u)(3) + (3)^2]$$
$$= 2u(2u - 3)(4u^2 + 6u + 9)$$

Exercise Set 5.11

Completely factor each polynomial:

1. $x^2 + 9x + 18$
2. $x^2 - 8x + 12$
3. $y^2 - 25$
4. $m^2 - 16$
5. $2a^2 + 7a - 4$
6. $3x^2 + 9x - 12$
7. $8x^2 + 38x + 45$
8. $6x^2 - 13x - 28$
9. $8x^2 - 8$
10. $3x^3 - 12x$
11. $12x^3 - 8x^2 - 15x$
12. $16x^3 - 18x^2 - 9x$
13. $R^4 - 4$
14. $x^2y^2 - 9$
15. $x^3 + 125$
16. $8y^3 - 1$
17. $4x^3 + 16x^2 - 20x$
18. $12x^2 + 44x - 16$
19. $2x^2 - 162$
20. $3x^2 - 192$

21. $8a^3 + 125$
22. $27x^3 + y^3$
23. $2x^2 - 5x + 8$
24. $3x^2 + 7x - 4$
25. $9A^2 + 52A - 12$
26. $27M^2 + 30M - 8$
27. $10x^3 + 15x^2 - 10x$
28. $12x^4 - 6x^3 - 90x^2$
29. $x^2 + 2x + 5x + 10$
30. $x^2 - 3x - 4x + 12$
31. $2y^2 - 22y - 24$
32. $2A^2 - 2A - 144$
33. $x^4 - 81$
34. $x^4 - 16$
35. $y^2 - 4y + y - 4$
36. $m^2 + 3m - 2m - 6$
37. $12x^4 - 52x^3 + 35x^2$
38. $32x^3 - 4x^2 - 6x$
39. $8x^3 - 29x^2 - 12x$
40. $6x^2 + 29x + 30$
41. $3y^4 + 24y$
42. $54y^4 + 2y$
43. $5x^5 - 80x$
44. $3x^5 - 243x$
45. $3x^4 - 18x^3 + 6x^2 - 36x$
46. $2x^3 + 4x^2 - 8x - 16$

Review

47. Solve the equation: $3a - 2(3a + 1) = 2(a - 5) - 2$
48. Solve and graph the inequality: $3a - 2(3a + 1) < 2(a - 5) - 2$
49. Graph the two variable inequalities:

  a. $y > \frac{1}{2}x + 3$
  b. $2x - 5y \geq 20$
  c. $-3 \leq x < 4$

50. Solve the system:

$$3x + 4y = 4$$
$$x - 6y = -17$$

5.12 Review and Chapter Test

(5.1) A *polynomial* is an algebraic expression formed by combining real number constants and variables using the operations of addition, subtraction, and multiplication. Any variable that appears must have a whole number exponent, and a variable may not appear in a denominator.

Example 1

A. $\sqrt{3}xy^2$; $2x^3 - x^2 + 2$; $\frac{1}{3}x^2 + y$ are all polynomials.

B. $3\sqrt{xy}$; $\frac{4}{x} - 3$; $3x^{-2} - y^{1/2}$ are not polynomials.

5.12 Review and Chapter Test

The parts of a polynomial that are added (or subtracted) together are the *terms* of the polynomial. Each term is generally the product of a constant called the *numerical coefficient* and one or more variables called the *literal part*.

Example 2

$4x^2 + 2xy^2 - 3y - 4$

Term	Numerical Coefficient	Literal Part
$4x^2$	4	x^2
$2xy^2$	2	xy^2
$-3y$	-3	y
-4	-4	none

◀

A polynomial of one term is a *monomial*; a polynomial of two terms is a *binomial*; and a polynomial of three terms is *trinomial*. The *degree of a term* of a polynomial is the sum of the exponents on the variables in the literal part of the term. The *degree of a polynomial* is equal to the degree of the term with the highest degree.

Example 3

$4x^3y^2 - 2x^2y^4 + 5$

This is a trinomial in the two variables, x and y. The degree of the first term is 5; the degree of the second term is 6; the degree of the third term is 0. The degree of the polynomial is 6. ◀

(5.2) Two or more terms are *like terms* if they have identical literal parts. Like terms are combined into a single term by adding their numerical coefficients.

Simplify the following polynomial by combining like terms:

Example 4

$x^3 - x^2y + x - 3x^3 + 2x^2 - 3 + 4x$

$x^3 - x^2y + x - 3x^3 + 2x^2 - 3 + 4x$

$= -2x^3 - x^2y + 2x^2 + 5x - 3$ ◀

To add or subtract polynomials horizontally, remove parentheses and combine like terms.

Example 5

A. Add:

$(x^2 - 3x - 2) + (-2x^2 + 5x) + (7x - 3)$
$= x^2 - 3x - 2 - 2x^2 + 5x + 7x - 3$ Remove parentheses
$= -x^2 + 9x - 5$ Combine like terms

B. Subtract:

$$(y^2 + 2y - 3) - (2y^2 - 4y) - (-3y - 5)$$
$$= y^2 + 2y - 3 - 2y^2 + 4y + 3y + 5 \qquad \text{Note the sign changes}$$
$$= -y^2 + 9y + 2$$ ◀

Polynomials are added vertically by writing them one under the other where only like terms appear in each column. The sum is found by combining like terms in each column.

Example 6

Add: $(x^2 - 3xy) + (x^2 + 5xy - 2) + (2xy + 3)$

$$\left.\begin{array}{r} x^2 - 3xy \\ x^2 + 5xy - 2 \\ 2xy + 3 \end{array}\right\} \qquad \text{Align like terms}$$
$$\overline{2x^2 + 4xy + 1} \qquad \text{Add}$$ ◀

To subtract polynomials vertically, place the polynomial being subtracted under the other polynomial. Write only like terms in each column. Then, mentally change signs on each term of the polynomial being subtracted and add to the top polynomial.

Example 7

Subtract: $3y^2 - 4y - 2$ from $5y^2 - 6y + 7$

$$\begin{array}{r} 5y^2 - 6y + 7 \\ 3y^2 - 4y - 2 \end{array} \qquad \text{Mentally change signs and add}$$
$$\overline{2y^2 - 2y + 9}$$ ◀

(5.3) To multiply powers of the same base:

$$b^m \cdot b^n = b^{m+n} \text{ where } m \text{ and } n \text{ are natural numbers}$$

Example 8

Multiply:

A. $x^5 \cdot x^3 = x^8$

B. $y \cdot y^3 \cdot y^4 = y^8$

C. $x^2 \cdot y \cdot x^3 \cdot y^2 = x^5 \cdot y^3$ ◀

To multiply monomials, multiply the coefficients and then multiply the literal factors with the same bases by adding their exponents.

Example 9

Multiply: $(-3x^2yz^3)(4xy^2z^2)$

$$(-3x^2yz^3)(4xy^2z^2) = -12x^3y^3z^5$$ ◀

5.12 Review and Chapter Test

To multiply a monomial by a polynomial, multiply the monomial by each term in the polynomial.

Example 10

Multiply: $2x^2y(3x^2 - 2xy + 4y^2)$

$2x^2y(3x^2 - 2xy + 4y^2) = (2x^2y)(3x^2) - (2x^2y)(2xy) + (2x^2y)(4y^2)$

Note: The preceding step is generally done mentally.

$= 6x^4y - 4x^3y^2 + 8x^2y^3$ ◀

To multiply two polynomials either horizontally or vertically, multiply each term of the first polynomial times every term of the second polynomial.

Example 11

Multiply: $(2x^2 - 3x + 2)(x^2 + x - 5)$

$$\begin{array}{r} 2x^2 - 3x + 2 \\ x^2 + x - 5 \\ \hline \end{array}$$

$2x^4 - 3x^3 + 2x^2$	Multiply by x^2
$\quad\quad 2x^3 - 3x^2 + 2x$	Multiply by x
$\quad\quad\quad\quad -10x^2 + 15x - 10$	Multiply by -5
$2x^4 - x^3 - 11x^2 + 17x - 10$	Add ◀

(5.4) When dealing with special products, use the FOIL method to multiply a binomial by a binomial.

Example 12

Multiply: $(2x - 3y)(3x + 4y)$

$$\overset{F\quad\quad L}{(2x - 3y)(3x + 4y)}$$
$$\underset{O}{\quad\quad I}$$

$= 6x^2 + 8xy - 9xy - 12y^2$

$= 6x^2 - xy - 12y^2$ ◀

Other special products

1. $(a + b)(a - b) = a^2 - b^2$ Product of a sum and difference
2. $(a + b)^2 = a^2 + 2ab + b^2$ ⎫
3. $(a - b)^2 = a^2 - 2ab + b^2$ ⎭   Square of a binomial

When an expression contains combinations of addition, subtraction, and multiplication, the multiplication is done first.

Example 13

Complete all operations and simplify:

$2(x - 3)^2 + (x - 4)(x + 4)$

$= 2(x^2 - 6x + 9) + (x^2 - 16)$ Multiply

$= 2x^2 - 12x + 18 + x^2 - 16$ Clear parentheses

$= 3x^2 - 12x + 2$ Simplify   ◀

(5.5) To divide a polynomial by a monomial, *each term* in the dividend (numerator) is divided by the monomial divisor (denominator). To divide powers of the same base,

$$\frac{b^m}{b^n} = b^{m-n}, b \neq 0, m > n$$

Example 14

A. $\dfrac{y^8}{y} = y^{8-1} = y^7$

B. $\dfrac{16x^9 y}{-2x^5 y} = -8x^4$ ◀

Example 15

Divide: $(4x^4 y^3 - 8x^3 y^2 + 10x^2 y) \div 2x^2 y$

$\dfrac{4x^4 y^3 - 8x^3 y^2 + 10x^2 y}{2x^2 y}$

$= \dfrac{4x^4 y^3}{2x^2 y} - \dfrac{8x^3 y^2}{2x^2 y} + \dfrac{10x^2 y}{2x^2 y}$ Divide each term by $2x^2 y$

$= \dfrac{\overset{2}{\cancel{4}} x^4 y^3}{\cancel{2x^2 y}} - \dfrac{\overset{4}{\cancel{8}} x^3 y^2}{\cancel{2x^2 y}} + \dfrac{\overset{5}{\cancel{10}} x^2 y}{\cancel{2x^2 y}}$ Cancel common factors

$= 2x^{4-2} y^{3-1} - 4x^{3-2} y^{2-1} + 5$ Division property of exponents

$= 2x^2 y^2 - 4xy + 5$ ◀

To divide two polynomials:

1. Arrange the divisor and dividend in descending powers, writing any "missing" powers with a coefficient of 0.

2. Divide the first term of the dividend (remainder) by the first term of the divisor.

3. Multiply and subtract. Bring down the next term.

4. Repeat Steps 2 and 3 until the remainder equals 0 or the degree of the remainder is less than the degree of the divisor.

5.12 Review and Chapter Test

Example 16

Divide: $3x^4 - x^2 + x - 1$ by $x + 2$

$$
\begin{array}{r}
3x^3 - 6x^2 + 11x - 21 \\
x+2 \overline{\smash{\big)}\, 3x^4 + 0x^3 - x^2 + x - 1} \\
\underline{3x^4 + 6x^3} \\
-6x^3 - x^2 \\
\underline{-6x^3 - 12x^2} \\
11x^2 + x \\
\underline{11x^2 + 22x} \\
-21x - 1 \\
\underline{-21x - 42} \\
41 \quad \text{Remainder}
\end{array}
$$

(5.6) To completely factor a polynomial is to express it as a product of factors that can be factored no further, called *prime factors*. The first step in factoring is to remove the largest common monomial factor.

Example 17

Remove the largest common monomial factor:

$$4m^3 - 8m^2 + 2m = (2m)(2m^2) - (2m)(4m) + (2m)(1)$$
$$= 2m(2m^2 - 4m + 1)$$

(5.7) A trinomial of the form $x^2 + bx + c$, when factorable, will factor as the product of two binomials.

$$x^2 + bx + c = (x \quad)(x \quad)$$

Two constants are required to complete the binomials.

1. The constants selected must have a product equal to the third term of the trinomial (c).

2. The constants selected must have a sum equal to the coefficient of the middle term of the trinomial (b).

Example 18

Factor: $x^2 - x - 20$

Write the parentheses: $x^2 - x - 20 = (x \quad)(x \quad)$.
The constants required to complete the binomials must have a product of -20 and a sum of -1. Use -5 and 4 since $(-5)(4) = -20$ and $-5 + 4 = -1$.

$$x^2 - x - 20 = (x - 5)(x + 4)$$

(5.8) To factor a trinomial of the form $ax^2 + bx + c$, $a \neq 1$, into a product of binomials, we:

1. Make an educated *guess* as to the factorization. In factored form, the coefficients of x are factors of a, and the constant terms are factors of c.

2. *Check* our guess by a FOIL expansion.
3. *Guess again* if our first guess is wrong.

Example 19

Factor each trimonial:

A. $4x^2 + 9x + 2$

Product = 4

$4x^2 + 9x + 2 = (x + 2)(4x + 1)$

Product = 2

B. $12x^2 + x - 6$

Product = 12

$12x^2 + x - 6 = (3x - 2)(4x + 3)$

Product = -6   ◀

(5.9) The *difference of two squares* factors as the product of a sum and difference.
$$a^2 - b^2 = (a + b)(a - b)$$

Example 20

Factor: $9a^2 - 25b^2$

$$9a^2 - 25b^2 = (3a)^2 - (5b)^2$$
$$= (3a + 5b)(3a - 5b)$$ ◀

A *perfect square trinomial* factors as the square of a binomial.
$$a^2 - 2ab + b^2 = (a - b)^2$$
$$a^2 + 2ab + b^2 = (a + b)^2$$

Example 21

Factor: $x^2 - 6xy + 9y^2$

The first term is the square of x, the third term is the square of $3y$, and the middle term is twice their product.

$$x^2 - 6xy + 9y^2 = (x - 3y)(x - 3y)$$
(Perfect squares)
$$= (x - 3y)^2$$

Check with FOIL.   ◀

5.12 Review and Chapter Test

(5.10) The sum or difference of two cubes are factored using the following formulas:
$$a^3 + b^3 = (a + b)(a^2 - ab + b^2)$$
$$a^3 - b^3 = (a - b)(a^2 + ab + b^2)$$

Example 22

Factor: $125x^3 + 8y^3$

$$125x^3 + 8y^3 = (5x)^3 + (2y)^3 \qquad a = 5x, b = 2y$$
$$= (5x + 2y)[(5x)^2 - (5x)(2y) + (2y)^2]$$
$$= (5x + 2y)(25x^2 - 10xy + 4y^2) \qquad \blacktriangleleft$$

A polynomial with four terms may be factored by grouping.

Example 23

Factor: $a^2 + 2a - 3a - 6$

$$\begin{aligned} a^2 + 2a - 3a - 6 &= (a^2 + 2a) - (3a + 6) && \text{Group} \\ &= (a+2)a - (a+2)3 && \text{Monomial factors} \\ &= (a+2)(a-3) && \text{Distributive property} \quad \blacktriangleleft \end{aligned}$$

(5.11) Factoring summary

Given a polynomial to factor:

1. Remove the largest common monomial factor (if any).
2. Consider the number of terms there are in the remaining polynomial.

   a. If *two* terms, look for:

   (1) The difference of two perfect squares, which is always factorable as the product of a sum and difference.
   $$a^2 - b^2 = (a + b)(a - b)$$

   (2) The sum or difference of two cubes, which are factored by the formulas
   $$a^3 + b^3 = (a + b)(a^2 - ab + b^2)$$
   $$a^3 - b^3 = (a - b)(a^2 + ab + b^2)$$

   b. If *three* terms, look for:

   (1) A perfect square trinomial
   $$a^2 + 2ab + b^2 = (a + b)^2$$
   $$a^2 - 2ab + b^2 = (a - b)^2$$

(2) A trinomial that requires us to use general factoring techniques (see Sections 5.6 and 5.7).

c. If *four* terms, try to factor by grouping.

3. Check to see if any further factorization is possible.

Exercise Set 5.12 Review

Which of the following are polynomials, and which are not?

1. $\sqrt{5} - 2y + y^2$
2. $x^2 + 3x^{1/2} - 2x$
3. $2x - \sqrt{y}$
4. $\dfrac{x^2 - 4}{x^2 + 2}$

Classify each of the following polynomials by: (A) number of terms; (B) number of variables; and (C) degree: (sum of exponents) then take highest # of exp, that is the degree

5. $4x^2 - 9$
6. $2x^2y - 3xy + y^2$
7. $3xy$
8. -3
9. $x - 5$
10. $x^2y - 3xy^3$

Evaluate the polynomials in Problems 11–14 for $x = -3$ and $y = 2$:

11. $x^2 - 3x + 5$
12. $y^3 - 3y^2 + 4y - 7$
13. $x^2y - 3xy - x$
14. $x^2 - y^2$

Perform the following operations:

15. Add: $3x^2 + 4x$; $6 + x^2 - 5x$; $3 + x - 4x^2$
16. Add: $y^2 - 6$; $3 + 2y$; $3y^2 + 4y$
17. Subtract: $m^2 + 3 + 6m$ from $m^2 + 2m + 3$
18. Subtract: $x^2 - 3xy + y^2$ from $3x^2 - y^2$
19. Simplify: $p^2 - (3 + 2p^2) + (3p^2 - 6p)$
20. Simplify: $u^2 + (5 - 3u) - (-5 - 3u + u^2)$

Form each of the products in Problems 21–28:

21. $(5xy)(-2x^2y)$
22. $3x^2y(x^2 - 3xy + y^2)$
23. $(x - 4)(x + 5)$
24. $(x - 2)(x + 2)(x - 3)$
25. $(x + 2)(x^2 - 3x + 5)$
26. $(y^2 + y - 1)(3y^2 + 4y - 8)$
27. $(2x - y)^2$
28. $(3a^2 - 2)(a^2 + 4)$

Divide:

29. $(4x^3 - 8x^2) \div 2x$
30. $(5x^2y^3 + 7x^3y^2 - xy^2) \div xy^2$

31. $(x^2 + 5x + 6) \div (x + 2)$
32. $(x^3 - 8) \div (x - 2)$
33. $(2m - 3m^2 + 2m^4) \div (m - 1)$
34. $(2c^5 - c^3 + 4) \div (c^2 - 3c + 1)$

Factor each of the following polynomials by removing the largest common monomial factor and/or identifying it to be a perfect square trinomial or difference of two perfect squares:

35. $27x^3 - 3x$
36. $7m^5 + 21m^4 + 35m^3$
37. $4x^2 - 12xy + 9y^2$
38. $25m^2 - 49y^2$
39. $8x^2 + 18y^2$
40. $6x^3y - 8xy^2 + 10x^2y^2 - 4y$
41. $2m^4n + 20m^3n^2 + 50m^2n^3$
42. $x^4 - 16$

Write each of the following polynomials in completely factored form. If the polynomial is not factorable, indicate it to be a prime polynomial.

43. $x^2 + 8x + 15$
44. $y^2 + 2y - 35$
45. $6x^2 + xy - 2y^2$
46. $m^2 + 5m + 2$
47. $2x^4 - 14x^3 + 24x^2$
48. $5x^2 + 13xy - 6y^2$
49. $18x^2 - 3x - 10$
50. $36x^4 - 48x^3 + 15x^2$
51. $18x^2 - 17x + 4$
52. $12x^2 + 28x + 15$
53. $a^3 - 125$
54. $x^3 - 5x^2 + 3x - 15$

Chapter 5 Review Test

1. Consider the polynomial: $3xy^2 + x^2y^2 - x^2y + 3$.
    a. How many terms does this polynomial have?
    b. How many variables?
    c. What is the degree of the third term?
    d. What is the degree of the polynomial?
    e. Evaluate the polynomial for $x = 2$, $y = -1$.
2. Add: $2m^2 - 3m + 4$; $3 - m^2 + m$; $2 - 4m - 3m^2$.
3. Simplify: $(x^2 - 3x) + (6 - 2x) - (-2 + 3x - 2x^2)$.
4. Form each of the following products:
    a. $(3a^2b)(-2ab)$
    b. $(3xy^2)(x^2y)(-2x^2)$
    c. $2ab^2(a^2 - 3ab + 4b^2)$
    d. $(y^2 + 3)(y^2 - 2)$
    e. $(3x + 2y)(2x - 3y)$
    f. $(2m - 3)(m^2 - 2m + 4)$

g. $(m + 3q)^2$

h. $(x^2 - 3x + 2)(2x^2 + x + 3)$

i. $(2x + y)(2x - y)$

5. Divide:

   a. $\dfrac{10x^2y^3 - 5xy^2 + 5x^3y^3}{-5xy^2}$

   b. $(2x^2 - 3x + 2) \div (x - 2)$

   c. $(y^3 - 3y^2 + 4) \div (y + 1)$

   d. $(2x^3 - 4x^2 + 5x - 1) \div (x^2 + 2x + 1)$

6. Write each of the following polynomials in completely factored form:

   a. $18x^2 - 9x$

   b. $8x^3 - 2x$

   c. $9a^2 - 16b^2$

   d. $x^2 - 2x - 15$

   e. $3x^2 - 10xy - 8y^2$

   f. $6y^4 - 15y^3 - 9y^2$

   g. $x^4 - 81$

   h. $4x^2 + 12x + 9$

   i. $6x^2 + 4x - 3x - 2$

   j. $16x^4 + 54x$

CHAPTER SIX

RATIONAL EXPRESSIONS

Recall that in arithmetic we first learned to add, subtract, multiply, and divide whole numbers. Once these concepts were mastered, we moved ahead to consider the arithmetic of rational numbers or fractions. In much the same way, we now move ahead from the algebra of polynomials (Chapter 5) to the algebra of rational expressions. A **rational expression** is an algebraic expression that may be written in fractional form with the numerator and denominator as polynomials. It is defined for all values of the variables that *do not* make the denominator zero. Throughout the remainder of this chapter we will use *fraction* and *rational expression* interchangeably.

Example 1

The following are examples of rational expressions:

A. $\dfrac{x^2 - 3x + 1}{x - 2}$, $x \neq 2$ Since $x = 2$ would make the denominator zero

B. $\dfrac{2x - 5}{3xy}$, $x \neq 0, y \neq 0$

C. $\dfrac{4}{x^2 + 6x + 9} = \dfrac{4}{(x + 3)^2}$, $x \neq -3$

Note that a rational expression can contain a variable in the denominator, whereas a polynomial cannot.

Recall that a constant is a polynomial of degree 0 and that any expression can be written as itself over 1. The following are also rational expressions:

D. $x^2 + 5x - 7 = \dfrac{x^2 + 5x - 7}{1}$

E. $\dfrac{2y + 5}{8}$

F. $\dfrac{2}{3}$

G. $3 = \dfrac{3}{1}$

H. $0 = \dfrac{0}{1}$ ◀

Essentially, much of the material of the first five sections of this chapter is a review of arithmetic on fractions. It has been found to be an area of difficulty to many students, so you are encouraged to study the examples carefully and to be diligent in your homework.

6.1 Fundamental Principle of Fractions

A rational expression must frequently be written in an alternative form so it can be compared to or combined with other rational expressions. To rewrite rational expressions we use the Fundamental Principle of Fractions, which states that the numerator and denominator of a fraction can be multiplied or divided by the same non-zero expression without changing the value of the fraction.

FUNDAMENTAL PRINCIPLE OF FRACTIONS

Let $\dfrac{A}{B}$ represent a fraction where $B \neq 0$. Let C be an expression (polynomial or fractional) where $C \neq 0$. Then,

a. $\dfrac{AC}{BC} = \dfrac{A\cancel{C}}{B\cancel{C}} = \dfrac{A}{B}$ A fraction is reduced to "lower terms" by dividing the numerator and denominator by C. We also call this operation *canceling*.

b. $\dfrac{A}{B} = \dfrac{AC}{BC}$ A fraction is raised to "higher terms" by multiplying the numerator and denominator by the same non-zero factor.

A fraction is in *lowest* (or *simplest*) *terms* when all *factors* common to the numerator and the denominator have been removed by use of the Fundamental Principle of Fractions:

$$\dfrac{AC}{BC} = \dfrac{A}{B}$$

in which C represents the largest common factor of the numerator and denominator. The key word here is *factor*. We may *only* remove (or "cancel") common factors. Note, the phrase "highest terms" is meaningless because we can always multiply the numerator and denominator by another factor.

6.1 Fundamental Principle of Fractions

Example 1

The following fractions are rewritten using the Fundamental Principle of Fractions:

A. $\dfrac{12}{18} = \dfrac{6 \cdot 2}{9 \cdot 2} = \dfrac{6}{9}$ (Lower terms)

B. $\dfrac{12}{18} = \dfrac{4 \cdot 3}{6 \cdot 3} = \dfrac{4}{6}$ (Lower terms)

C. $\dfrac{12}{18} = \dfrac{2 \cdot 6}{3 \cdot 6} = \dfrac{2}{3}$ (Lowest terms)

D. $\dfrac{2x^2}{6x} = \dfrac{(2x) \cdot x}{(2x) \cdot 3} = \dfrac{x}{3}$ (Lowest terms)

E. $\dfrac{8xy^3}{12y^2} = \dfrac{(4y^2) \cdot 2xy}{(4y^2) \cdot 3} = \dfrac{2xy}{3}$ (Lowest terms)

F. $\dfrac{3}{7} = \dfrac{3 \cdot 2}{7 \cdot 2} = \dfrac{6}{14}$ (Higher terms)

G. $\dfrac{3}{7} = \dfrac{3 \cdot 5}{7 \cdot 5} = \dfrac{15}{35}$ (Higher terms)

H. $\dfrac{3}{7} = \dfrac{3 \cdot 2xy^2}{7 \cdot 2xy^2} = \dfrac{6xy^2}{14xy^2}$ (Higher terms) ◀

When rewriting fractions it is helpful to use the following properties of signs.

$$\dfrac{-A}{-B} = -\dfrac{A}{-B} = -\dfrac{-A}{B} = \dfrac{A}{B}$$

$$\dfrac{-A}{B} = \dfrac{A}{-B} = -\dfrac{A}{B}$$

We think of a fraction as having three signs: the sign of the numerator, the sign of the denominator, and the sign of the fraction itself. We can change any *two* of these signs without changing the value of the fraction. Generally, we avoid writing fractions with a negative denominator.

When reducing fractions, the removal of common factors is generally shown as in arithmetic.

Example 2

Each of the following fractions is reduced to lowest terms:

A. $\dfrac{-8}{12} = \dfrac{-2 \cdot \cancel{4}}{3 \cdot \cancel{4}} = \dfrac{-2}{3} = -\dfrac{2}{3}$

B. $\dfrac{-(-8)}{-10} = \dfrac{8}{-10} = -\dfrac{8}{10} = -\dfrac{\cancel{2} \cdot 4}{\cancel{2} \cdot 5} = -\dfrac{4}{5}$

SIX Rational Expressions

C. $\dfrac{12}{18} = \dfrac{\cancel{12}^{2}}{\cancel{18}_{3}} = \dfrac{2}{3}$

D. $\dfrac{-8}{12} = \dfrac{\cancel{-8}^{-2}}{\cancel{12}_{3}} = -\dfrac{2}{3}$

E. $\dfrac{2x^2}{6x} = \dfrac{\cancel{2x^2}^{x}}{\cancel{6x}_{3}} = \dfrac{x}{3}$

F. $\dfrac{8xy^3}{12y^2} = \dfrac{\cancel{8xy^3}^{2xy}}{\cancel{12y^2}_{3}} = \dfrac{2xy}{3}$ ◀

If the numerator or denominator of a fraction contains a polynomial of more than one term, we must *first* factor where possible, and *then* remove any common factors. Remember, only *factors* can be canceled, *not terms!*

Example 3

Reduce each of the following rational expressions to lowest terms:

A. $\dfrac{2x^4 - 4x^3}{6x^2} = \dfrac{\cancel{2x^3}(x-2)}{\cancel{6x^2}_{3}} = \dfrac{x(x-2)}{3}$

B. $\dfrac{2x^3y + 6x^2y^2}{4x^2y^2} = \dfrac{\cancel{2x^2y}(x+3y)}{\cancel{4x^2y^2}_{2}} = \dfrac{x+3y}{2y}$

C. $\dfrac{x^2 - 2x}{x^2 - 4} = \dfrac{x\cancel{(x-2)}}{(x+2)\cancel{(x-2)}}$ Factor numerator and denominator

$= \dfrac{x}{x+2}$ Remove common binomial factor

D. $\dfrac{2x^2 + 5x - 3}{x^3 + 2x^2 - 3x} = \dfrac{(2x-1)(x+3)}{x(x^2 + 2x - 3)}$ Factor

$= \dfrac{(2x-1)\cancel{(x+3)}}{x\cancel{(x+3)}(x-1)}$ Factor and reduce

$= \dfrac{2x-1}{x(x-1)}$ Lowest terms

Note: We will generally leave the numerator and denominator of a reduced fraction in factored form.

6.1 Fundamental Principle of Fractions

E. $\dfrac{x^2 + x - 2}{x^2 + 4x + 3} = \dfrac{(x+2)(x-1)}{(x+3)(x+1)}$ ◀ No common factors. Already in lowest terms.

Recall that a fraction is considered to have three signs, the sign of the numerator, the sign of the denominator, and the sign of the fraction itself. We may change the sign of any two of these without changing the value of the fraction. A useful variation of this principle is expressed

$$\dfrac{a-b}{b-a} = -1$$

This is proved quite simply. See if you can follow each step.

$\dfrac{a-b}{b-a} = \dfrac{(a-b)(-1)}{(b-a)(-1)}$ Multiply numerator and denominator by -1 using the Fundamental Principle of Fractions

$= \dfrac{(a-b)(-1)}{-b+a}$ Distributive property in denominator

$= \dfrac{(a-b)(-1)}{(a-b)}$ Commutative property of addition in denominator

$= \dfrac{\cancel{(a-b)}(-1)}{\cancel{(a-b)}}$ Reduce

$= -1$ Lowest terms

For example,

$$\dfrac{7-3}{3-7} = \dfrac{4}{-4} = -1$$

An expression divided by its negative equals -1. Therefore, both the factors $a - b$ and $b - a$ are canceled, provided one of them is replaced by -1.

Example 4

Reduce each of the following rational expressions to lowest terms:

A. $\dfrac{3(x-5)}{4(5-x)} = \dfrac{3\overset{-1}{\cancel{(x-5)}}}{4\cancel{(5-x)}} = \dfrac{-3}{4}$

B. $\dfrac{x-2}{2x-x^2} = \dfrac{x-2}{x(2-x)}$ Factor

$= \dfrac{\overset{-1}{\cancel{x-2}}}{x\cancel{(2-x)}}$ Reduce

$= \dfrac{-1}{x}$ Lowest terms

C. $\dfrac{(x-2)(x+3)}{(2-x)(2+x)} = \dfrac{\overset{-1}{\cancel{(x-2)}}(x+3)}{\cancel{(2-x)}(2+x)}$

$= \dfrac{-(x+3)}{2+x}$

D. $\dfrac{x^2 - x - 6}{6x^2 - 2x^3} = \dfrac{(x-3)(x+2)}{2x^2(3-x)}$ Factor

$= \dfrac{\overset{-1}{\cancel{(x-3)}}(x+2)}{2x^2\cancel{(3-x)}}$ Reduce

$= \dfrac{-(x+2)}{2x^2}$ Lowest terms ◀

Example 5

Raise each of the following rational expressions to higher terms using the Fundamental Principle of Fractions to make the fraction on the left equal to the fraction on the right by replacing the ? with the appropriate expression:

A. $\dfrac{3}{5} = \dfrac{?}{35}$

$\searrow \dfrac{?}{5 \cdot 7}$ Factor

therefore

$\dfrac{3}{5} = \dfrac{3 \cdot 7}{5 \cdot 7}$ Multiply numerator and denominator by 7

$= \dfrac{21}{35}$

B. $\dfrac{3x}{4y} = \dfrac{?}{8xy^2}$

$\searrow \dfrac{?}{4y(2xy)}$ Factor

therefore

$\dfrac{3x}{4y} = \dfrac{3x(2xy)}{4y(2xy)}$ Multiply numerator and denominator by $2xy$

$= \dfrac{6x^2 y}{8xy^2}$

6.1 Fundamental Principle of Fractions

C. $\dfrac{3mn}{m-2n} = \dfrac{?}{2m^2 - 4mn}$

$\quad\quad\quad\quad\quad\quad \hookrightarrow \dfrac{?}{2m(m-2n)} \quad$ Factor

therefore

$\dfrac{3mn}{m-2n} = \dfrac{3mn(2m)}{(m-2n)(2m)} \quad$ Multiply numerator and denominator by $2m$

$\quad\quad\quad = \dfrac{6m^2n}{2m^2 - 4mn}$

D. $\dfrac{2x+y}{x-y} = \dfrac{?}{2x^2 - 3xy + y^2}$

$\quad\quad\quad\quad\quad\quad \hookrightarrow \dfrac{?}{(x-y)(2x-y)} \quad$ Factor

therefore

$\dfrac{2x+y}{x-y} = \dfrac{(2x+y)(2x-y)}{(x-y)(2x-y)} \quad$ Multiply numerator and denominator by $(2x-y)$

$\quad\quad\quad = \dfrac{4x^2 - y^2}{2x^2 - 3xy + y^2}$ ◀

Example 6

Use the Fundamental Principle of Fractions to make the following fractions equal by replacing the ? with an appropriate expression:

$$\dfrac{x^2 - 3x}{2x - 6} = \dfrac{x}{?}$$

Note that we must reduce the fraction on the left to make it equal to the one on the right.

$$\dfrac{x^2 - 3x}{2x - 6} = \dfrac{x(x-3)}{2(x-3)} = \dfrac{x}{2}$$

The common factor, $x - 3$, has been canceled. Therefore, the ? is replaced with 2. ◀

When we reflect upon our experience with fractions, we generally recall the process of reducing fractions to lowest terms as being more important than raising fractions to higher terms. But in fact, they are of equal importance since we usually cannot add or subtract fractions without being able to raise fractions to higher terms (see Section 6.4).

Exercise Set 6.1

Reduce each of the following fractions to lowest terms:

1. $\dfrac{28}{36}$
2. $\dfrac{35}{30}$
3. $\dfrac{3x^3}{2x}$
4. $\dfrac{5x^2}{10}$

5. $\dfrac{3y^2}{9y}$
6. $\dfrac{8x^4}{2x}$
7. $\dfrac{12x^3}{6x^2}$
8. $\dfrac{4xy}{12xy^3}$

9. $\dfrac{6xy}{9xy}$
10. $\dfrac{8x^3y}{8xy}$
11. $\dfrac{14x^2}{21y}$
12. $\dfrac{15y^3}{25x}$

13. $\dfrac{6x^2y}{21xy^2}$
14. $\dfrac{3x^2 - 3x}{6x^2 + 6x}$
15. $\dfrac{10x^2 + 15x}{4x^3 + 6x^2}$

16. $\dfrac{x^2 + 3x + 2}{3x + 6}$
17. $\dfrac{2x - 1}{2 - 4x}$
18. $\dfrac{x - 3}{3x^2 - x^3}$

19. $\dfrac{2x^2 + x - 3}{8x^2 + 12x}$
20. $\dfrac{x^2 - 5x + 4}{5x - 5}$
21. $\dfrac{2x + 4}{x^2 + 7x + 10}$

22. $\dfrac{3x^2 - 15x}{x^2 - 2x - 15}$
23. $\dfrac{4 - x^2}{x^2 - 5x + 6}$
24. $\dfrac{9 - x^2}{x^2 - 9x + 18}$

25. $\dfrac{x^2 - 16}{x^2 + 8x + 16}$
26. $\dfrac{y^2 - 49}{y^2 + 14y + 49}$
27. $\dfrac{x^2 - x - 12}{x^2 + 6x + 9}$

28. $\dfrac{x^2 + 4x + 4}{x^2 - x - 6}$
29. $\dfrac{16 - x^2}{x^2 - 8x + 16}$
30. $\dfrac{25 - a^2}{a^2 - 10a + 25}$

31. $\dfrac{x^2 - 2x + xy - 2y}{x^2 + x - 6}$
32. $\dfrac{y^2 + y - 6}{xy - y^2 + 3x - 3y}$

33. $\dfrac{a^2 - ba + 2a - 2b}{ab + a - b^2 - b}$
34. $\dfrac{ab - 3a + b^2 - 3b}{a^2 - a + ab - b}$

35. $\dfrac{4a^2 + 4ab + b^2}{4a^2 - b^2}$
36. $\dfrac{m^2 - 9n^2}{m^2 + 6mn + 9n^2}$

37. $\dfrac{2xy - 4y^2}{3x^2 - 6xy}$
38. $\dfrac{10ab + 5b^2}{8a^2 + 4ab}$

39. $\dfrac{3x^2 - 4xy + y^2}{6x^2 + xy - y^2}$
40. $\dfrac{m^2 - 2mn - 3n^2}{m^2 - mn - 6n^2}$

In Problems 41–50 replace the ? with an appropriate expression to make the equation true. Some of the problems involve raising rational expressions to higher terms while other problems will require you to reduce fractions.

41. $\dfrac{3}{2y} = \dfrac{?}{10xy}$

42. $\dfrac{2}{5m^2} = \dfrac{?}{10m^2n}$

43. $\dfrac{9x^3y^2}{15x^2y^3} = \dfrac{?}{5y}$

44. $\dfrac{16a^3b}{28a^2b^2} = \dfrac{?}{7b}$

45. $\dfrac{3a^2b}{2x^2y} = \dfrac{6a^2b^2xy}{?}$

46. $\dfrac{6ab^3}{5mn^2} = \dfrac{18a^3b^4m}{?}$

47. $\dfrac{2a^2b}{5mn} = \dfrac{?}{10amn - 5m^2n}$

48. $\dfrac{3xy^2}{2b} = \dfrac{?}{2ab - 4bx}$

49. $\dfrac{2x - 3y}{3y} = \dfrac{2x^2 + xy - 6y^2}{?}$

50. $\dfrac{3a - b}{2a} = \dfrac{6a^2 + 7ab - 3b^2}{?}$

51. In a short paragraph, describe the difference between a rational expression and a polynomial.

52. What does it mean for a fraction to be reduced to lowest terms?

Review

53. Multiply:

   a. $3x(2x - 1)$

   b. $(3x + 4)(2x - 1)$

   c. $(x + 2)(3x + 4)(2x - 1)$

   d. $(-1)(a - b)$

54. Solve each of the following equations for x:

   a. $2x - 5 = 7 - 4x$

   b. $2x - 5 = 7y - 4x$

   c. $ax - b = cy - dx$

6.2 Multiplication and Division of Rational Expressions

Multiplication of Rational Expressions

Rational expressions are multiplied using the same technique used for multiplying rational numbers. The product of rational expressions is formed by multiplying the numerators and denominators, respectively.

SIX   Rational Expressions

MULTIPLICATION OF RATIONAL EXPRESSIONS

Let $\dfrac{A}{B}$ and $\dfrac{C}{D}$ be rational expressions where $B \neq 0$ and $D \neq 0$.

$$\frac{A}{B} \cdot \frac{C}{D} = \frac{AC}{BD}$$

Since products of rational expressions are to be reduced to lowest terms, it is important to follow these steps:

1. Factor all numerators and denominators.

2. Cancel common factors. Remember, in a product of fractions, a factor in any numerator may cancel with a common factor in any denominator (Fundamental Principle of Fractions).

3. Multiply. (If Steps 1 and 2 have been done completely, the product will be in lowest terms.)

Example 1

Multiply each of the following rational expressions and reduce to lowest terms:

A. $\dfrac{3x}{-5y} \cdot \dfrac{-15y}{x^2}$

$$\frac{3x}{-5y} \cdot \frac{-15y}{x^2} = \frac{3\cancel{x}}{-\cancel{5}\cancel{y}} \cdot \frac{-\overset{3}{\cancel{15}}\cancel{y}}{\cancel{x^2}}$$

$$= \frac{9}{x} \qquad \text{Notice the two negatives cancel}$$

B. $\dfrac{4a^2b}{9xy^3} \cdot \dfrac{6x^2y^2}{10ab}$

$$\frac{4a^2b}{9xy^3} \cdot \frac{6x^2y^2}{10ab} = \frac{\overset{2}{\cancel{4}}a^{\cancel{2}}\cancel{b}}{\underset{3}{\cancel{9}}\cancel{x}y^{\cancel{3}}} \cdot \frac{\overset{2}{\cancel{6}}x^{\cancel{2}}y^{\cancel{2}}}{\underset{5}{\cancel{10}}\cancel{a}\cancel{b}}$$

$$= \frac{4ax}{15y}$$

C. $\dfrac{-3y}{8x^3} \cdot 2x$

$$\frac{-3y}{8x^3} \cdot 2x = \frac{-3y}{8x^3} \cdot \frac{2x}{1}$$

6.2 Multiplication and Division of Rational Expressions

Recall that any expression that is not in fractional form may be written as itself over 1, as $a = \frac{a}{1}$.

$$= \frac{-3y}{\underset{4}{\cancel{8x^3}}} \cdot \frac{\cancel{2x}}{1} = \frac{-3y}{4x^2}$$

Multiply STEPS
(1) FACTOR
(2) REDUCE/CANCEL
(3) Multiply

D. $\dfrac{x+2}{3x^2} \cdot \dfrac{2x}{2x+4}$

$$\frac{x+2}{3x^2} \cdot \frac{2x}{2x+4} = \frac{x+2}{3x^2} \cdot \frac{2x}{2(x+2)} \qquad \text{Factor}$$

$$= \frac{\cancel{x+2}}{3x^{\cancel{2}}} \cdot \frac{\cancel{2x}}{2(\cancel{x+2})} \qquad \text{Cancel}$$

$$= \frac{1}{3x}$$

E. $\dfrac{x^2 - 4}{x+3} \cdot \dfrac{x^2 + 4x + 3}{x - 2}$

$$\frac{x^2 - 4}{x+3} \cdot \frac{x^2 + 4x + 3}{x - 2} = \frac{(x+2)(x-2)}{x+3} \cdot \frac{(x+3)(x+1)}{x-2}$$

$$= \frac{(x+2)\cancel{(x-2)}}{\cancel{x+3}} \cdot \frac{\cancel{(x+3)}(x+1)}{\cancel{x-2}}$$

$$= \frac{(x+2)(x+1)}{1}$$

$$= (x+2)(x+1)$$

F. $\dfrac{x^2 + 2x}{3 - x} \cdot \dfrac{2x}{3y} \cdot \dfrac{xy - 3y}{x^2 + 3x + 2}$

$$\frac{x^2 + 2x}{3-x} \cdot \frac{2x}{3y} \cdot \frac{xy-3y}{x^2+3x+2} = \frac{x(x+2)}{3-x} \cdot \frac{2x}{3y} \cdot \frac{y(x-3)}{(x+2)(x+1)}$$

$$= \frac{x\cancel{(x+2)}}{\cancel{3-x}} \cdot \frac{2x}{3\cancel{y}} \cdot \frac{\overset{-1}{\cancel{y(x-3)}}}{\cancel{(x+2)}(x+1)}$$

$$= \frac{-2x^2}{3(x+1)} \qquad \blacktriangleleft$$

Division of Rational Expressions

Recall that the rule for division of fractions calls for us to "invert the divisor and multiply." The rule for division or rational expressions is the same and is justified by expressing the division in fractional form and then using the Fundamental

314   **SIX**   **Rational Expressions**

Principle of Fractions to simplify. Consider

$$\frac{A}{B} \div \frac{C}{D} = \frac{\dfrac{A}{B}}{\dfrac{C}{D}} \qquad \text{Rewrite}$$

$$= \frac{\dfrac{A}{B} \cdot \dfrac{D}{C}}{\dfrac{C}{D} \cdot \dfrac{D}{C}} \qquad \text{Multiply numerator and denominator by } \frac{D}{C}$$
$$\text{(Fundamental Principle)}$$

$$= \frac{\dfrac{A}{B} \cdot \dfrac{D}{C}}{1} \qquad \text{Denominator multiplies to 1}$$

$$\frac{A}{B} \div \frac{C}{D} = \frac{A}{B} \cdot \frac{D}{C}$$

 That is, to divide one rational expression by another, we invert (take the reciprocal of) the divisor and multiply by the dividend.

DIVISION OF RATIONAL EXPRESSIONS

Let $\dfrac{A}{B}$ and $\dfrac{C}{D}$ be rational expressions where $B \neq 0$, $D \neq 0$, and $C \neq 0$. Then,

$$\frac{A}{B} \div \frac{C}{D} = \frac{A}{B} \cdot \frac{D}{C} = \frac{AD}{BC}$$

As in all rational expressions, once we invert and multiply, we should then look for common factors and reduce.

Example 2   Perform each of the following divisions and reduce to lowest terms:

A. $\dfrac{6x^2}{-5y} \div \dfrac{-2x}{y^3}$

$$\frac{6x^2}{-5y} \div \frac{-2x}{y^3} = \frac{\overset{3}{\cancel{6x^2}}}{-5\cancel{y}} \cdot \frac{y^{\cancel{3}\,2}}{-\cancel{2x}} \qquad \text{Invert and multiply, then reduce}$$

$$= \frac{3xy^2}{5}$$

6.2 Multiplication and Division of Rational Expressions

[Handwritten margin notes: Division Steps 1) inverse expression following sign 2) factor 3) reduce 4) multiply]

B. $\dfrac{5x^2}{4y} \div 2x$

$\dfrac{5x^2}{4y} \div 2x = \dfrac{5x^2}{4y} \div \dfrac{2x}{1}$ Write in fractional form $a = \dfrac{a}{1}$

$= \dfrac{5x^2}{4y} \cdot \dfrac{1}{2x}$ Invert and multiply, then reduce

$= \dfrac{5x}{8y}$

C. $\dfrac{x^2 - 4}{x} \div \dfrac{x - 2}{5x}$

$\dfrac{x^2 - 4}{x} \div \dfrac{x - 2}{5x} = \dfrac{(x-2)(x+2)}{x} \cdot \dfrac{5x}{x-2}$ Factor and invert divisor, then multiply

$= \dfrac{(x-2)(x+2)}{x} \cdot \dfrac{5x}{x-2}$ Reduce

$= \dfrac{5(x+2)}{1}$

$= 5(x+2)$

D. $\dfrac{3x - 6}{x^2 + x - 6} \div (x + 3)$

$\dfrac{3x-6}{x^2+x-6} \div (x+3) = \dfrac{3(x-2)}{(x+3)(x-2)} \cdot \dfrac{1}{x+3}$ Factor and invert divisor

$= \dfrac{3(x-2)}{(x+3)(x-2)} \cdot \dfrac{1}{x+3}$ Reduce

$= \dfrac{3}{(x+3)^2}$

E. $(4 - x) \div \dfrac{x^2 - 3x - 4}{3x^2 + 3x}$

$(4-x) \div \dfrac{x^2-3x-4}{3x^2+3x} = \dfrac{(4-x)}{1} \cdot \dfrac{3x(x+1)}{(x-4)(x+1)}$ Factor and invert divisor

$= \dfrac{\overset{-1}{(4-x)}}{1} \cdot \dfrac{3x(x+1)}{(x-4)(x+1)}$ Reduce

$= \dfrac{-3x}{1}$

$= -3x$

SIX Rational Expressions

In practice, the first and second step of the division process are shown as one. Example 2E would appear as follows:

$$(4-x) \div \frac{x^2 - 3x - 4}{3x^2 + 3x} = \frac{\overset{-1}{(4-x)}}{1} \cdot \frac{3x(x+1)}{(x-4)(x+1)} = -3x$$

If a problem includes a combination of multiplication and division of more than two fractions, <u>we must perform all multiplications and divisions as they are encountered from left to right</u>, *unless* symbols of grouping are used to indicate a different sequence of operations.

Example 3 Divide each of the following rational expressions and reduce to lowest terms:

A. $\dfrac{x^2 + x - 6}{x^2 + 3x + 2} \div \dfrac{x^2 - 9}{x^2 + 2x} \div \dfrac{x+4}{2x^2 + 2x}$ *4x*

$\dfrac{x^2 + x - 6}{x^2 + 3x + 2} \div \dfrac{x^2 - 9}{x^2 + 2x} \div \dfrac{x+4}{2x^2 + 2x}$ *4x*

$= \dfrac{(x+3)(x-2)}{(x+2)(x+1)} \cdot \dfrac{x(x+2)}{(x+3)(x-3)} \div \dfrac{x+4}{2x^2 + 2x}$ *4x*   Factor and invert first divisor (work left to right)

$= \dfrac{x(x-2)}{(x+1)(x-3)} \div \dfrac{x+4}{2x(x+2)}$ *2x² + 4x*

$= \dfrac{x(x-2)}{(x+1)(x-3)} \cdot \dfrac{2x(x+2)}{x+4}$ Invert last divisor

$= \dfrac{2x^2(x-2)(x+2)}{(x+1)(x-3)(x+4)}$ Leave answer in factored form

B. $\dfrac{x^2 + x - 6}{x^2 + 3x + 2} \div \left(\dfrac{x^2 - 9}{x^2 + 2x} \div \dfrac{x+4}{2x^2 + 2x} \right)$

This problem changes the order of operations for Example 3A.

$\dfrac{x^2 + x - 6}{x^2 + 3x + 2} \div \left(\dfrac{x^2 - 9}{x^2 + 2x} \div \dfrac{x+4}{2x^2 + 2x} \right)$

$= \dfrac{x^2 + x - 6}{x^2 + 3x + 2} \div \left(\dfrac{(x-3)(x+3)}{x(x+2)} \cdot \dfrac{2x(x+2)}{x+4} \right)$ Factor and invert the divisor in the parentheses

$= \dfrac{x^2 + x - 6}{x^2 + 3x + 2} \div \dfrac{2(x-3)(x+3)}{x+4}$

$= \dfrac{(x+3)(x-2)}{(x+2)(x+1)} \cdot \dfrac{x+4}{2(x-3)(x+3)}$ Invert and multiply

$= \dfrac{(x-2)(x+4)}{2(x+2)(x+1)(x-3)}$

Notice that even though the fractions and operations are the same in Example 3A and 3B, the answers are *not* the same. This is because the sequence of operations is different for each problem, as determined by the parentheses. Division is not an associative operation.

Exercise Set 6.2

Multiply and divide as indicated. Write all answers in lowest terms.

1. $\dfrac{9}{15} \cdot \dfrac{18}{12}$

2. $\dfrac{8}{16} \cdot \dfrac{9}{12}$

3. $\dfrac{21}{40} \div \dfrac{7}{8}$

4. $\dfrac{12}{32} \div \dfrac{9}{16}$

5. $\dfrac{4}{5} \cdot \dfrac{10}{9} \cdot \dfrac{12}{16}$

6. $\dfrac{3}{7} \cdot \dfrac{4}{3} \cdot \dfrac{21}{12}$

7. $\dfrac{8}{9} \cdot \dfrac{3}{5} \div \dfrac{2}{5}$

8. $\dfrac{9}{7} \cdot \dfrac{3}{4} \cdot \dfrac{14}{8}$

9. $\dfrac{3}{7} \div \dfrac{9}{10} \cdot \dfrac{21}{5}$

10. $\dfrac{7}{8} \cdot \dfrac{2}{12} \div \dfrac{14}{6}$

11. $\dfrac{5x^2}{3} \cdot \dfrac{9}{2x}$

12. $\dfrac{4x^2y}{3} \cdot \dfrac{9}{16x^2}$

13. $\dfrac{6x}{5y} \cdot \dfrac{15y^2}{2x}$

14. $\dfrac{9x^2}{2y} \cdot \dfrac{6y^2}{x^3}$

15. $\dfrac{16x^2}{3y} \cdot \dfrac{3}{2x} \cdot \dfrac{5x}{4}$

16. $\dfrac{5x^2}{3y} \cdot \dfrac{4y}{5x} \cdot \dfrac{6xy}{4}$

17. $\dfrac{x}{15y} \cdot \dfrac{3xy}{6x^3} \cdot \dfrac{10x}{y^2}$

18. $\dfrac{3x}{15y^3} \cdot \dfrac{5y}{6x^2} \cdot \dfrac{1}{8x^2y}$

19. $\dfrac{9x^4}{-2y} \div \dfrac{-3x}{8y^2}$

20. $\dfrac{11x^3}{-5} \div \dfrac{22x}{15}$

21. $\dfrac{6x^3}{5} \div \dfrac{18x^4}{7}$

22. $\dfrac{5xy}{3} \div \dfrac{7xy}{9}$

23. $\dfrac{x+2}{3} \cdot \dfrac{6x}{x+2}$

24. $\dfrac{x-5}{6y} \cdot \dfrac{2y^2}{x-5}$

25. $\dfrac{x^2 + 5x + 6}{x - 1} \cdot \dfrac{x^2 - 1}{x + 2}$

26. $\dfrac{x^2 - 9}{x + 2} \cdot \dfrac{x^2 - x - 6}{x + 3}$

27. $\dfrac{2x + 6}{(x + 1)^2} \div \dfrac{(x + 3)(x - 1)}{(x + 1)}$

28. $\dfrac{x^3(x + 2)}{x - 5} \div \dfrac{2x^2(x + 2)}{x + 1}$

29. $(2x + 4) \div \dfrac{x^2 + 6x + 8}{x + 4}$

30. $3(x + 3) \div \dfrac{x^2 - 9}{x - 3}$

31. $\dfrac{(2x + 1)(x - 1)}{x + 3} \div \dfrac{x + 4}{(x + 3)(2x + 1)}$

32. $\dfrac{x + 2}{x - 1} \cdot \dfrac{x^2 + 2x - 3}{6x - 2} \div \dfrac{x + 2}{3x - 1}$

33. $\dfrac{x-2}{5} \div \dfrac{x+3}{15} \div \dfrac{x+3}{2}$

34. $\dfrac{x+1}{6} \div \dfrac{x+1}{2x-1} \div \dfrac{2x-1}{x-1}$

35. $\dfrac{x^2+3x+2}{2x} \div \left(\dfrac{2x-1}{x} \div \dfrac{x+2}{2}\right)$

36. $\dfrac{x^2-9}{x+2} \div \left(\dfrac{x-3}{x-1} \div \dfrac{x+2}{x+3}\right)$

37. $\dfrac{x^2-9}{x^2+6x+5} \cdot \dfrac{x+5}{2x+8} \div \dfrac{3-x}{x^2+4x+3}$

38. $\dfrac{2x^2+11x-6}{3x+2} \cdot \dfrac{x-3}{2x^2+7x-4} \div \dfrac{x^2+3x-18}{3x+2}$

39. $\dfrac{x-2}{x+5} \div \dfrac{6-3x}{x+7} \cdot \dfrac{x^2+6x-7}{6x-6}$

40. $\dfrac{x^2-7x+10}{4x+2} \div \dfrac{4x+8}{2x+1} \cdot \dfrac{x+2}{x^2-7x+10}$

41. $\dfrac{ab+a-b-1}{ab+b+a+1} \cdot \dfrac{a^2+4a+3}{a^2+a-2}$

42. $\dfrac{xy+2x-2y-4}{x^2+4x+3} \cdot \dfrac{x^2+2x-3}{xy+3x-2y-6}$

The area of a rectangle having length L and width W is given by the formula $A = L \cdot W$.

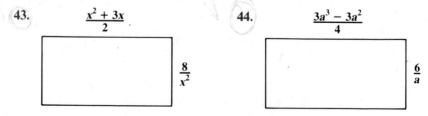

Use this formula to write an expression for the area of the following rectangles. Write all answers in lowest terms.

43. $\dfrac{x^2+3x}{2}$, $\dfrac{8}{x^2}$

44. $\dfrac{3a^3-3a^2}{4}$, $\dfrac{6}{a}$

45.

$\dfrac{x^2 + 5x + 6}{x + 2}$

$\dfrac{x - 1}{x^2 - 1}$

46.

$\dfrac{y^2 - y - 2}{y + 3}$

$\dfrac{y^2 + 4y + 3}{y - 2}$

Review

47. Find the least common multiple (LCM) of the following groups of numbers or expressions:

   a. 6, 8, 3
   b. 4, 12, 8   24
   c. 8, 12, 16
   d. $2x, 3x^2$ $6x^2$

48. Write each of the following polynomials in completely factored form:

   a. $7x^3 - 21x^2$
   b. $a^2 - 2a - 8$
   c. $8m^2 + 22m + 9$
   d. $p^3 - 36p$
   e. $y^3 - 27$
   f. $xy - 5x + 2y - 10$

6.3 Prime and Composite Numbers: Prime Factorization and Least Common Multiples

To add and subtract rational expressions in Section 6.4, it will be necessary to rewrite fractions by raising them to higher terms (as in Section 6.1). Therefore, in this section we study some basic relationships in the natural numbers that will help us to add and subtract fractions in the most efficient manner.

Multiples and Factors

If we multiply the natural numbers 3 and 4, the result is 12. That is, $3 \cdot 4 = 4 \cdot 3 = 12$. The numbers 3 and 4 are called *factors* of 12 while 12 is a *multiple* of both 3 and 4. A complete list of the natural number factors of 12 includes 1, 2, 3, 4, 6, and 12. Further, 12 is a multiple of all these factors.

In general; let $a, b, x \in N$ such that $a \cdot b = b \cdot a = x$. Then, a and b are factors of x, and x is a multiple of a and b.

Example 1

A. List all factors of the number 18:

$$18 = 1 \cdot 18 \qquad 18 = 2 \cdot 9 \qquad 18 = 3 \cdot 6$$

The factors of 18 are: 1, 2, 3, 6, 9, and 18.

B. Find the first six multiples of 4:

$$1 \cdot 4 = 4 \qquad 4 \cdot 4 = 16$$
$$2 \cdot 4 = 8 \qquad 5 \cdot 4 = 20$$
$$3 \cdot 4 = 12 \qquad 6 \cdot 4 = 24$$

The first six multiples of 4 are: 4, 8, 12, 16, 20, and 24. ◀

Prime and Composite Numbers

It is useful to separate the natural numbers into two types: **prime numbers** and **composite numbers**.

DEFINITION   *[handwritten: any 2 factors]*

1. A *prime number* is any natural number, except 1, whose only factors are 1 and the number itself.

2. A *composite number* is any natural number, except 1, which is not prime. *[handwritten: product of only prime #'s]*

3. The number 1 is neither prime nor composite.

The list of prime numbers includes: 2, 3, 5, 7, 11, 13, 17, ... The list of composite numbers includes: 4, 6, 8, 9, 10, 12, 14, 15, 16, ... A number such as 14 is composite (not prime) since it has factors 2 and 7 as well as the factors 1 and itself.

FUNDAMENTAL THEOREM OF ARITHMETIC

Every natural number, except 1, is either prime or can be expressed uniquely as a product of primes.

Expressing a number as a product of primes is called writing the *prime factorization* of that number or writing the number in *prime factored form*.

6.3 Prime Factorization and Least Common Multiples

Example 2 Find the prime factorization of each of the following numbers. Use exponential notation on repeated prime factors.

A. 24

$24 = 6 \cdot 4$ Factor 24

$ = 3 \cdot 2 \cdot 2 \cdot 2$ Factor 6 and 4

$ = 2^3 \cdot 3$ All primes

Note that there is nothing unique about the first factorization of 24. Any factorization may be used.

$24 = 8 \cdot 3$ Factor 24

$ = 4 \cdot 2 \cdot 3$ Factor 8

$ = 2 \cdot 2 \cdot 2 \cdot 3$ Factor 4

$ = 2^3 \cdot 3$ Same prime factorization as before

B. 75

$75 = 5 \cdot 15$ Factor 75

$ = 5 \cdot 3 \cdot 5$ Factor 15

$ = 3 \cdot 5^2$ All primes

or $75 = 25 \cdot 3$ Factor 75

$ = 5 \cdot 5 \cdot 3$ Factor 25

$ = 5^2 \cdot 3$

$ = 3 \cdot 5^2$ Same prime factorization as before

C. 35

$35 = 5 \cdot 7$ Factor 35, all primes

Note that no further factorization of 35 is possible since 5 and 7 are prime.

D. 17

17 is prime

E. 300

$300 = 30 \cdot 10$ Factor 300

$ = 3 \cdot 10 \cdot 2 \cdot 5$ Factor 30 and 10

$ = 3 \cdot 2 \cdot 5 \cdot 2 \cdot 5$ Factor 10

$ = 2^2 \cdot 3 \cdot 5^2$ All primes

or $300 = 15 \cdot 20$ Factor 300

$ = 3 \cdot 5 \cdot 4 \cdot 5$ Factor 15 and 20

$ = 3 \cdot 5 \cdot 2 \cdot 2 \cdot 5$ Factor 4

$ = 2^2 \cdot 3 \cdot 5^2$ Same prime factorization as before ◀

Least Common Multiple

The **least common multiple**, **LCM**, of a collection of natural numbers is the smallest number that contains each of the numbers as a factor. Or equivalently, the LCM of a collection of natural numbers is the smallest number exactly divisible by all of the numbers in the collection.

Example 3

Find the least common multiple of each of the following sets of numbers by inspection:

A. 2, 3

The LCM of 2 and 3 is the number 6. We write:

LCM (2, 3) = 6

There is no number smaller than 6 which is a multiple of 2 and 3.

B. 3, 6

LCM (3, 6) = 6

C. 2, 4, 5

LCM (2, 4, 5) = 20

D. 3, 6, 8

LCM (3, 6, 8) = 24 ◀

If the LCMs in Example 3 are not easily found by inspection, or if the numbers involved are large, we require a systematic technique for finding the LCM.

TECHNIQUE FOR FINDING THE LEAST COMMON MULTIPLE (LCM) OF A GROUP OF NUMBERS

1. Write each number in prime factored form.
2. The LCM is the number composed of factors that are all the different primes from Step 1, each raised to the highest power appearing on that prime in any one factorization.

6.3 Prime Factorization and Least Common Multiples

Example 4

Find the LCM for each of the following groups of numbers:

A. 10, 8

Step 1: $10 = 2 \cdot 5$ — Prime factors
$8 = 2 \cdot 4$
$ = 2 \cdot 2 \cdot 2$
$8 = 2^3$ ← Highest exponent

Step 2: LCM (10, 8) = $2^3 \cdot 5$ — Different primes

LCM (10, 8) = $2^3 \cdot 5 = 40$

B. 24, 28

Step 1: $24 = 6 \cdot 4$
$ = 2 \cdot 3 \cdot 2 \cdot 2$
$24 = 2^3 \cdot 3$ — Prime factors

$28 = 4 \cdot 7$
$ = 2 \cdot 2 \cdot 7$
$28 = 2^2 \cdot 7$

— Highest exponent

Step 2: LCM (24, 28) = $2^3 \cdot 3 \cdot 7$ — Different primes

LCM (24, 28) = $2^3 \cdot 3 \cdot 7 = 168$

C. 12, 18, 35

Step 1: $12 = 2 \cdot 6$
$ = 2 \cdot 2 \cdot 3$
$12 = 2^2 \cdot 3$

$18 = 2 \cdot 9$
$ = 2 \cdot 3 \cdot 3$
$18 = 2 \cdot 3^2$

$35 = 5 \cdot 7$

— Prime factors

SIX Rational Expressions

$$\text{Step 2:} \quad \text{LCM }(12, 18, 35) = 2^2 \cdot 3^2 \cdot 5 \cdot 7 \quad \text{— Highest exponents}$$

$$\text{— Different primes}$$

$$\text{LCM }(12, 18, 35) = 2^2 \cdot 3^2 \cdot 5 \cdot 7 = 1260 \quad \blacktriangleleft$$

The same technique is used to find the LCM of a collection of expressions containing variable factors as well as constant factors.

TECHNIQUE FOR FINDING THE LEAST COMMON MULTIPLE (LCM) OF TWO OR MORE EXPRESSIONS CONTAINING VARIABLE FACTORS

1. Write each expression in completely factored form and each coefficient in prime factored form.
2. The LCM is the expression composed of factors that are all the different factors from Step 1, each raised to the highest power appearing on that factor in any one factorization.

Example 5

Find the LCM for each of the following expressions:

A. $3x, 2x^3, 6x^2$

Step 1:
$$\left.\begin{array}{l} 3x = 3x \\ 2x^3 = 2x^3 \\ 6x^2 = 2 \cdot 3x^2 \end{array}\right\} \quad \text{Prime factors}$$

Step 2: $\text{LCM }(3x, 2x^3, 6x^2) = 2 \cdot 3 \cdot x^3$ — Highest exponent

— Different factors

$$\text{LCM }(3x, 2x^3, 6x^2) = 6x^3$$

B. $9xy, 5xy^2, 15x^2$

Step 1:
$$\left.\begin{array}{l} 9xy = 3 \cdot 3xy = 3^2 xy \\ 5xy^2 = 5xy^2 \\ 15x^2 = 3 \cdot 5x^2 \end{array}\right\} \quad \text{Prime factors}$$

Step 2:   LCM $(9xy, 5xy^2, 15x^2) = 3^2 \cdot 5 \cdot x^2 \cdot y^2$ ← Highest exponents
                                                   ↑ ↑ ↑ ↑
                                                   └─┴─┴─┴── Different factors

LCM $(9xy, 5xy^2, 15x^2) = 3^2 \cdot 5 \cdot x^2 \cdot y^2 = 45x^2y^2$

C. $a^3bc^2, 4ab^2c^2, 6a^2b$

We determine the LCM of the numerical coefficients to be 12, by inspection.

LCM $(a^3bc^2, 4ab^2c^2, 6a^2b) = 12a^3 \cdot b^2 \cdot c^2$ ← Highest exponents
                                        ↑   ↑   ↑
                                        └───┴───┴── Different factors

$= 12a^3b^2c^2$

D. $x^2 - 3x, 2x - 6, x^2 - 6x + 9$

$$x^2 - 3x = x(x - 3)$$
$$2x - 6 = 2(x - 3)$$
$$x^2 - 6x + 9 = (x - 3)(x - 3)$$
$$= (x - 3)^2$$

Complete factorization

LCM $(x^2 - 3x, 2x - 6, x^2 - 6x + 9) = 2x(x - 3)^2$ ← Highest exponent
                                          ↑↑   ↑
                                          └┴────┴── Different factors

Note that, generally, the factored form for the LCM is preferred (when adding and subtracting fractions). That is, we write

LCM $(x^2 - 3x, 2x - 6, x^2 - 6x + 9) = 2x(x - 3)^2$

E. $8x^2, 2x^2 - 18, x^2 + x - 12$

$$8x^2 = 2^3x^2$$
$$2x^2 - 18 = 2(x + 3)(x - 3)$$
$$x^2 + x - 12 = (x + 4)(x - 3)$$

Complete factorization

LCM $= 2^3x^2(x + 3)(x - 3)(x + 4)$ ◀

The LCM is useful for determining the least common denominator for a collection of fractions. This skill, which is necessary for adding and subtracting fractions, is discussed in the next section.

Exercise Set 6.3

List all factors of each of the following numbers:

1. 10
2. 14
3. 24
4. 20
5. 11
6. 23
7. 1
8. 7
9. 42
10. 36

Find the first five multiples of each of the following numbers:

11. 3
12. 5
13. 7
14. 8
15. 2
16. 1
17. 11
18. 10
19. 15
20. 12

Find the prime factorization of each of the following numbers:

21. 15
22. 14
23. 25
24. 9
25. 11
26. 19
27. 45
28. 54
29. 72
30. 80
31. 100
32. 200
33. 750
34. 275

Find the least common multiple (LCM) of each of the following sets of numbers:

35. LCM(3, 5)
36. LCM(7, 9)
37. LCM(4, 12)
38. LCM(3, 15)
39. LCM(6, 8)
40. LCM(6, 9)
41. LCM(3, 5, 6)
42. LCM(4, 5, 8)
43. LCM(21, 35)
44. LCM(24, 15)
45. LCM(30, 18)
46. LCM(20, 28)
47. LCM(10, 15, 12)
48. LCM(9, 12, 15)
49. LCM(150, 225)
50. LCM(95, 25)

Find the least common multiple (LCM) of each of the following sets of expressions:

51. $3a^2$, $12a$, $8a^3$
52. $10y^3$, $15y^2$, $5y$
53. $5xy^2$, $10y^3$, x^2
54. $3a^2b$, $6b^3$, $9b^2$
55. $6m$, $m - n$
56. $2x$, $x - 3y$
57. $8x$, $4x^3 - 6x^2$
58. $6m$, $2m^2 - 4m$
59. $3(a + 4)$, $9(a + 4)^2$
60. $10(x - 2)$, $4(x - 2)^2$
61. $4a^3bc^2$, $12ab^3$, $8ab^2c^3$
62. $3x^3yz^2$, $8xy^3z$, $12x^2yz^2$
63. $y^2 - 9$, $6y$, $4(y + 3)$
64. $x^2 - 4$, $3x$, $6(x - 2)$
65. $9y^2 - 16$, $3y^2 - 2y - 8$
66. $4x^2 - 25$, $2x^2 - 11x + 15$

67. What is the difference between a prime number and a composite number?
68. Distinguish between the words *factor* and *multiple*.

Review

69. Multiply:

a. $\dfrac{1}{c} \cdot (a + b)$ b. $\dfrac{1}{c} \cdot (a - b)$

70. Write the original problems and the result of Problem 69 in the form of a division, recalling $\dfrac{1}{a} \cdot b = \dfrac{b}{a}$.

71. Multiply:

a. $x(x + 4)$ b. $(x - 1)(x + 4)$

c. $x(x - 1)(x + 4)$

72. Remove parentheses and combine like terms:

a. $2(x - 1) - 3(x - 4)$ b. $3(x + 2) - (2x - 5)$

c. $(x + 1)x - 2x(x - 4)$

6.4 Addition and Subtraction of Rational Expressions

We define the addition and subtraction of rational expressions having the same denominator as follows.

Let A, B, and C be polynomials where $C \neq 0$. Then,

$$\frac{A}{C} + \frac{B}{C} = \frac{A + B}{C}$$

$$\frac{A}{C} - \frac{B}{C} = \frac{A - B}{C}$$

This definition applies whether we are combining numbers of arithmetic (such as $\tfrac{2}{3}$) or rational expressions that contain variables (such as $\dfrac{2x}{x - 3}$). When the denominator contains a variable, we assume the fraction is only defined for values of the variable that do *not* make the denominator zero.

SIX Rational Expressions

Example 1

Add or subtract as indicated:

A. $\dfrac{2}{7} + \dfrac{3}{7}$ Note: Denominators are the same

$$\dfrac{2}{7} + \dfrac{3}{7} = \dfrac{2+3}{7} \qquad \text{Add numerators, denominator unchanged}$$

$$= \dfrac{5}{7}$$

B. $\dfrac{7}{9} - \dfrac{1}{9}$ Same denominators

$$\dfrac{7}{9} - \dfrac{1}{9} = \dfrac{7-1}{9} \qquad \text{Subtract numerators, denominator unchanged}$$

$$= \dfrac{6}{9}$$

$$= \dfrac{\cancel{6}^{2}}{\cancel{9}_{3}} \qquad \text{Remove common factor}$$

$$= \dfrac{2}{3}$$

C. $\dfrac{5}{11} - \dfrac{2}{11} + \dfrac{6}{11} - \dfrac{4}{11}$

$$\dfrac{5}{11} - \dfrac{2}{11} + \dfrac{6}{11} - \dfrac{4}{11} = \dfrac{5-2+6-4}{11} \qquad \text{Combine numerators}$$

$$= \dfrac{5}{11}$$

D. $\dfrac{2y+3}{3y-2} + \dfrac{y-1}{3y-2}$ Same denominators

$$\dfrac{2y+3}{3y-2} + \dfrac{y-1}{3y-2} = \dfrac{(2y+3)+(y-1)}{3y-2} \qquad \text{Add numerators}$$

$$= \dfrac{2y+3+y-1}{3y-2} \qquad \text{Simplify numerator}$$

$$= \dfrac{3y+2}{3y-2}$$

E. $\dfrac{3x-5}{x(x-3)} - \dfrac{3-2x}{x(x-3)}$ Same denominators

6.4 Addition and Subtraction of Rational Expressions

$$\frac{3x-5}{x(x-3)} - \frac{3-2x}{x(x-3)} = \frac{(3x-5)-(3-2x)}{x(x-3)} \quad \text{Subtract numerators}$$

$$= \frac{3x-5-3+2x}{x(x-3)} \quad \text{Simplify numerator}$$

$$= \frac{5x-8}{x(x-3)} \qquad \blacktriangleleft$$

Note the use of parentheses in Example **1E** to indicate that the entire numerator of the second fraction, $3 - 2x$, is to be subtracted from the numerator of the first fraction. Be very careful when combining rational expressions whose numerators have more than one term. It is recommended that parentheses be inserted around all such numerators before additions and (particularly) subtractions are performed.

Addition and Subtraction with Different Denominators

To add or subtract rational expressions that do not have like denominators, we use the Fundamental Principle of Fractions (Section 6.1) to *rewrite* the fractions so that they have a common (same) denominator. This denominator is the LCM of all denominators involved and is called the **least common denominator (LCD)** of the fractions. This process is illustrated in the following examples.

Example 2

Combine the following rational expressions and simplify if possible:

A. $\dfrac{1}{6} + \dfrac{5}{9}$

$\dfrac{1}{6} + \dfrac{5}{9}$
$\begin{aligned} 6 &= 2 \cdot 3 \\ 9 &= 3 \cdot 3 = 3^2 \\ \text{LCD} &= 2 \cdot 3^2 = 18 \end{aligned}$

$= \dfrac{1}{6} \cdot \dfrac{3}{3} + \dfrac{5}{9} \cdot \dfrac{2}{2}$ Fundamental Principle is used to rewrite each fraction to have a denominator of 18

$= \dfrac{3}{18} + \dfrac{10}{18}$ Common denominator

$= \dfrac{3 + 10}{18}$ Add numerators

$= \dfrac{13}{18}$

B. $\dfrac{3x}{5y} - \dfrac{2y}{15x}$ LCD = $15xy$

$\dfrac{3x}{5y} - \dfrac{2y}{15x} = \dfrac{3x}{5y} \cdot \dfrac{3x}{3x} - \dfrac{2y}{15x} \cdot \dfrac{y}{y}$ Use the Fundamental Principle to make denominators $15xy$

$= \dfrac{9x^2}{15xy} - \dfrac{2y^2}{15xy}$ Common denominators

$= \dfrac{9x^2 - 2y^2}{15xy}$ Combine numerators

C. $\dfrac{3}{x-2} - \dfrac{5}{2-x}$ Denominators have opposite signs

$\dfrac{3}{x-2} - \dfrac{5}{2-x} = \dfrac{3}{x-2} + \dfrac{5}{x-2}$ Rewrite denominator of second fraction and change the sign of the fraction

$= \dfrac{3+5}{x-2}$

$= \dfrac{8}{x-2}$

Note: In the previous example, we used the properties of signs to rewrite the second fraction by changing the sign of the denominator and the sign of the fraction, itself.

D. $\dfrac{x+1}{4} - \dfrac{x-3}{5} - \dfrac{x-2}{20}$

$4 = 2 \cdot 2 = 2^2$

$5 = 5$

$20 = 4 \cdot 5$

$= 2 \cdot 2 \cdot 5 = 2^2 \cdot 5$

LCD $= 2^2 \cdot 5 = 20$

$\dfrac{x+1}{4} - \dfrac{x-3}{5} - \dfrac{x-2}{20}$

$= \dfrac{(x+1)}{4} \cdot \dfrac{5}{5} - \dfrac{(x-3)}{5} \cdot \dfrac{4}{4} - \dfrac{(x-2)}{20}$ Fundamental Principle. Make all denominators 20

6.4 Addition and Subtraction of Rational Expressions

Note that parentheses were inserted in the numerators to help us keep track of signs.

$$= \frac{5(x+1)}{20} - \frac{4(x-3)}{20} - \frac{(x-2)}{20} \qquad \text{Common denominators}$$

$$= \frac{5(x+1) - 4(x-3) - (x-2)}{20} \qquad \text{Combine numerators}$$

$$= \frac{5x + 5 - 4x + 12 - x + 2}{20} \qquad \text{Remove parentheses using the distributive property}$$

$$= \frac{19}{20} \qquad \text{Simplify}$$

Note the subtraction of the *entire* second and third numerators and the careful removal of parentheses.

E. $\dfrac{4}{x-3} - \dfrac{x+2}{x^2-9}$

$$\frac{4}{x-3} - \frac{x+2}{x^2-9} = \frac{4}{x-3} - \frac{x+2}{(x+3)(x-3)} \qquad \text{Variable denominator factors}$$

LCD $= (x-3)(x+3)$

$$\frac{4}{x-3} \cdot \frac{(x+3)}{(x+3)} - \frac{x+2}{(x+3)(x-3)} \qquad \text{Fundamental Principle to make the denominators } (x+3)(x-3)$$

$$= \frac{4(x+3)}{(x-3)(x+3)} - \frac{(x+2)}{(x-3)(x+3)} \qquad (1) \text{ Common denominators; insert parentheses in the numerators}$$

By LCM & MULT.

$$= \frac{4(x+3) - (x+2)}{(x-3)(x+3)} \qquad (2) \text{ Combine numerators}$$

$$= \frac{4x + 12 - x - 2}{(x-3)(x+3)} \qquad (3) \text{ Distributive property in the numerator}$$

$$= \frac{3x + 10}{(x-3)(x+3)} \qquad (4) \text{ Simplify}$$

F. $\dfrac{9}{x+2} + \dfrac{3}{x} - 2 \qquad$ LCD $= x(x+2)$

$$\frac{9}{x+2} + \frac{3}{x} - \frac{2}{1} = \frac{9}{x+2} \cdot \frac{x}{x} + \frac{3}{x} \cdot \frac{x+2}{x+2} - \frac{2}{1} \cdot \frac{x(x+2)}{x(x+2)} \qquad \text{Fundamental Principle}$$

$$= \frac{9x}{x(x+2)} + \frac{3(x+2)}{x(x+2)} - \frac{2x(x+2)}{x(x+2)} \quad \text{Common denominators}$$

$$= \frac{9x + 3(x+2) - 2x(x+2)}{x(x+2)} \quad \text{Combine numerators}$$

$$= \frac{9x + 3x + 6 - 2x^2 - 4x}{x(x+2)} \quad \text{Distributive property}$$

$$= \frac{-2x^2 + 8x + 6}{x(x+2)} \quad \text{Simplify}$$

G. $\dfrac{3x-2}{x^2+4x+3} - \dfrac{x+2}{x^2+2x-3}$

$$\frac{3x-2}{x^2+4x+3} - \frac{x+2}{x^2+2x-3}$$

$$= \frac{(3x-2)}{(x+3)(x+1)} - \frac{(x+2)}{(x+3)(x-1)} \quad \begin{array}{l}\text{Factor denominators}\\ \text{Insert parentheses}\end{array}$$

LCD $= (x+3)(x+1)(x-1)$

$$= \frac{(3x-2)}{(x+3)(x+1)} \cdot \frac{(x-1)}{(x-1)} - \frac{(x+2)}{(x+3)(x-1)} \cdot \frac{(x+1)}{(x+1)}$$

$$= \frac{(3x^2-5x+2)}{(x+3)(x+1)(x-1)} - \frac{(x^2+3x+2)}{(x+3)(x+1)(x-1)}$$

$$= \frac{(3x^2-5x+2) - (x^2+3x+2)}{(x+3)(x+1)(x-1)}$$

$$= \frac{3x^2-5x+2-x^2-3x-2}{(x+3)(x+1)(x-1)} \quad \text{Remove parentheses}$$

$$= \frac{2x^2-8x}{(x+3)(x+1)(x-1)} \quad \text{Simplify}$$

$$= \frac{2x(x-4)}{(x+3)(x+1)(x-1)} \quad \begin{array}{l}\text{Factor. No reduction}\\ \text{possible so leave}\\ \text{in this form.}\end{array}$$

Note the last step. We factored the numerator to see if it was possible to reduce the answer to lower terms. Since the numerator and denominator share *no* common factors, the answer is already in lowest terms. Results of adding and subtracting rational expressions should *always* be expressed in lowest terms. ◀

6.4 Addition and Subtraction of Rational Expressions

In summary, to add or subtract rational expressions, we follow these steps:

1. Factor all denominators and determine the least common denominator (LCD) of all fractions involved.
2. Use the Fundamental Principle of Fractions to rewrite each fraction so that each of them has the LCD as its denominator.
3. Add or subtract the fractions by combining their numerators, leaving the denominator unchanged as the LCD.
4. Reduce the resulting rational expression to lowest terms if necessary.

Exercise Set 6.4

Add or subtract the following rational expressions. Write answers in lowest terms.

1. $\dfrac{3x}{2y} + \dfrac{5}{2y}$
2. $\dfrac{5}{3x} + \dfrac{2}{3x}$
3. $\dfrac{x}{3y} - \dfrac{2x}{3y}$

4. $\dfrac{2x}{5} - \dfrac{8x}{5}$
5. $\dfrac{2x}{y} + \dfrac{1}{3}$
6. $\dfrac{3a}{2b} + \dfrac{2}{3}$

7. $\dfrac{3}{x+2} - \dfrac{2}{x+2}$
8. $\dfrac{5}{x+1} - \dfrac{2}{x+1}$
9. $\dfrac{2x}{x-2} - \dfrac{4}{x-2}$

10. $\dfrac{3x}{x+2} + \dfrac{6}{x+2}$
11. $\dfrac{4a}{a-1} + \dfrac{4}{1-a}$
12. $\dfrac{2x}{x-3} + \dfrac{6}{3-x}$

13. $\dfrac{y}{y-1} - \dfrac{2}{y^2-1}$
14. $\dfrac{3y-1}{2y^2+y-3} - \dfrac{2}{y-1}$

15. $\dfrac{y+1}{y-1} - 1$
16. $\dfrac{x+1}{x-3} + 2$

17. $\dfrac{7}{x-3} - \dfrac{2x}{3-x}$
18. $\dfrac{2x-3}{x-2} - \dfrac{x}{4}$

19. $\dfrac{x-1}{x-2} - x + 3$
20. $\dfrac{x^2-2x}{x+2} + x - 3$

21. $\dfrac{4a}{a-3} + \dfrac{2a}{a+4}$
22. $\dfrac{4x}{x^2-9} - \dfrac{3}{x-3}$

23. $\dfrac{x}{x+1} + 5 - \dfrac{x}{x-1}$
24. $\dfrac{3}{x+1} + \dfrac{5}{x-1} - \dfrac{2}{x^2-1} + 1$

SIX   Rational Expressions

FACTOR DENOM.
SELECT LCD
CHANGE DE TO LCD
REWRITE

25. $\dfrac{2}{x+1} + \dfrac{5}{x^2+4x+3}$

26. $\dfrac{2}{x^2+3x+2} - \dfrac{5}{x^2-x-2}$

27. $\dfrac{3}{x^2+2x-3} - \dfrac{1}{x^2+3x-4}$

28. $\dfrac{3x}{x^2+7x+10} - \dfrac{3x}{x^2+3x-10}$

29. $\dfrac{2}{x^2-9} + \dfrac{2}{x+3} - \dfrac{1}{3-x}$

30. $\dfrac{2}{x^2-4} + \dfrac{1}{x+2} - \dfrac{3}{2-x}$

The perimeter of a rectangle having length L and width W is given by the formula $P = 2L + 2W$. Use this formula to write an expression for the perimeter of the following rectangles. Write all answers in lowest terms.

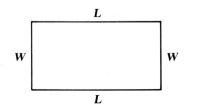

31.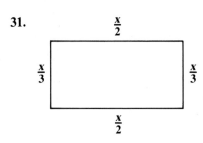

32.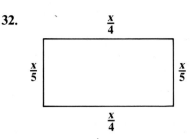

33.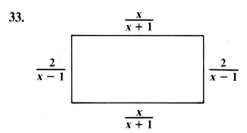

34.

Review

35. Divide:

a. $\dfrac{3}{x} \div \dfrac{x+2}{2x}$

b. $\dfrac{5}{b-1} \div \dfrac{15}{b^2-1}$

c. $\dfrac{5}{c^2+c-2} \div \dfrac{1}{c^2-4}$

36. Find the LCD of the expressions:

a. $\dfrac{1}{x}, \dfrac{3}{2x}, \dfrac{1}{x^2}$

b. $\dfrac{3}{4}, \dfrac{5}{x-1}, \dfrac{2}{2x-2}$

c. $1 - \dfrac{1}{x}, \; 3 + \dfrac{1}{x^2}$

37. Name the following properties of real numbers:

a. $a + b = b + a$

b. $a(bc) = (ab)c$

c. $1 \cdot a = a$

d. $a\left(\dfrac{1}{a}\right) = 1$

6.5 Complex Fractions

A **complex fraction** is a fractional expression in which a fraction appears in the numerator or the denominator. The following expressions are complex fractions:

$$\dfrac{\frac{2}{3}}{\frac{3}{5}}, \quad \dfrac{\frac{3}{4} - \frac{5}{6} \cdot \frac{1}{4}}{\frac{3}{5} + \frac{1}{3}}, \quad \dfrac{1 - \frac{3}{x}}{2 + \frac{1}{x} - \frac{1}{x^2}}, \quad \dfrac{1 - \frac{1}{y}}{y + \frac{1}{y}}$$

Complex fractions must be converted to simple fractions before we can compare them to other fractional forms or perform the operations of arithmetic on them. A **simple fraction** is a fraction whose numerator and denominator are polynomials with integer numerical coefficients. Complex fractions are reduced to simple fractions by one of two methods. These methods will now be explained briefly and illustrated.

Method I

A complex fraction may be reduced to a simple fraction by first writing the numerator as a single fraction and then writing the denominator as a single fraction. Then the numerator is divided by the denominator using the rules for dividing a fraction by a fraction (invert and multiply).

Example 1 Write each of the following as a simple fraction and reduce to lowest terms:

A. $\dfrac{3\frac{3}{4}}{1\frac{1}{2}}$

$\dfrac{3\frac{3}{4}}{1\frac{1}{2}} = \dfrac{3 + \frac{3}{4}}{1 + \frac{1}{2}}$ Definition of mixed numbers

$= \dfrac{\frac{12+3}{4}}{\frac{2+1}{2}}$

$= \dfrac{\frac{15}{4}}{\frac{3}{2}}$ Express numerator and denominator as a single fraction

$= \dfrac{15}{4} \div \dfrac{3}{2}$

$= \dfrac{\overset{5}{\cancel{15}}}{\underset{2}{\cancel{4}}} \cdot \dfrac{\overset{1}{\cancel{2}}}{\underset{1}{\cancel{3}}}$ Invert and multiply

$= \dfrac{5}{2}$

B. $\dfrac{y - \frac{1}{x}}{1 - \frac{1}{y}}$

$\dfrac{y - \frac{1}{x}}{1 - \frac{1}{y}} = \dfrac{\frac{y}{1} \cdot \frac{x}{x} - \frac{1}{x}}{1 \cdot \frac{y}{y} - \frac{1}{y}}$

$= \dfrac{\frac{xy}{x} - \frac{1}{x}}{\frac{y}{y} - \frac{1}{y}}$

6.5 Complex Fractions

$$= \frac{\dfrac{xy-1}{x}}{\dfrac{y-1}{y}} \qquad \text{Express numerator and denominator as single fractions}$$

$$= \frac{xy-1}{x} \cdot \frac{y}{y-1} \qquad \text{Invert and multiply}$$

$$= \frac{y(xy-1)}{x(y-1)}$$

C. $\dfrac{3y-1}{2 + \dfrac{y}{y+2}}$

$$\frac{3y-1}{2 + \dfrac{y}{y+2}} = \frac{3y-1}{\dfrac{2}{1} \cdot \dfrac{y+2}{y+2} + \dfrac{y}{y+2}}$$

$$= \frac{3y-1}{\dfrac{2(y+2)}{y+2} + \dfrac{y}{y+2}}$$

$$= \frac{3y-1}{\dfrac{2y+4+y}{y+2}}$$

$$= \frac{3y-1}{\dfrac{3y+4}{y+2}} \qquad \text{Express numerator and denominator as single fractions}$$

$$= \frac{3y-1}{1} \cdot \frac{y+2}{3y+4} \qquad \text{Invert and multiply}$$

$$= \frac{(3y-1)(y+2)}{3y+4} \qquad \blacktriangleleft$$

Method II

A complex fraction may be reduced to a simple fraction by multiplying its numerator and its denominator by the LCM of *all denominators* appearing in the numerator and denominator. This is an application of the Fundamental Principle of Fractions.

$$\frac{A}{B} = \frac{AC}{BC}$$

SIX Rational Expressions

We now repeat the problems from Example 1 using Method II. Compare these methods and determine which you prefer.

Example 2

Write each of the following as a simple fraction and reduce to lowest terms:

A. $\dfrac{3\frac{3}{4}}{1\frac{1}{2}}$

$$\dfrac{3\frac{3}{4}}{1\frac{1}{2}} = \dfrac{3 + \frac{3}{4}}{1 + \frac{1}{2}} \qquad \text{LCD} = 4$$

$$= \dfrac{\left(3 + \frac{3}{4}\right) \cdot 4}{\left(1 + \frac{1}{2}\right) \cdot 4} \qquad \text{Fundamental Principle of Fractions}$$

$$= \dfrac{12 + 3}{4 + 2} \qquad \text{Multiply (distributive property)}$$

$$= \dfrac{15}{6} \qquad \text{Simplify}$$

$$= \dfrac{\overset{5}{\cancel{15}}}{\underset{2}{\cancel{6}}} \qquad \text{Reduce}$$

$$= \dfrac{5}{2}$$

B. $\dfrac{y - \frac{1}{x}}{1 - \frac{1}{y}} \qquad \text{LCD} = xy$

$$\dfrac{y - \frac{1}{x}}{1 - \frac{1}{y}} = \dfrac{\left(y - \frac{1}{x}\right) \cdot xy}{\left(1 - \frac{1}{y}\right) \cdot xy} \qquad \text{Fundamental Principle of Fractions}$$

$$= \dfrac{y \cdot xy - \frac{1}{x} \cdot xy}{1 \cdot xy - \frac{1}{y} \cdot xy} \qquad \text{Distributive property}$$

6.5 Complex Fractions

$$= \frac{xy^2 - y}{xy - x} \qquad \text{Multiply}$$

$$= \frac{y(xy - 1)}{x(y - 1)} \qquad \text{Factor; not reducible}$$

C. $\dfrac{3y - 1}{2 + \dfrac{y}{y + 2}} \qquad \text{LCD} = y + 2$

$$\frac{3y - 1}{2 + \dfrac{y}{y + 2}} = \frac{(3y - 1) \cdot (y + 2)}{\left(2 + \dfrac{y}{y + 2}\right) \cdot (y + 2)} \qquad \begin{array}{l}\text{Fundamental}\\ \text{Principle of}\\ \text{Fractions}\end{array}$$

$$= \frac{3y^2 + 6y - y - 2}{2(y + 2) + \dfrac{y}{y+2} \cdot (y+2)} \qquad \text{Multiply}$$

$$= \frac{3y^2 + 5y - 2}{2y + 4 + y}$$

$$= \frac{3y^2 + 5y - 2}{3y + 4}$$

$$= \frac{(3y - 1)(y + 2)}{3y + 4} \qquad \text{Factor; not reducible} \quad \blacktriangleleft$$

Based upon a comparison of Example 1 with Example 2, we observe that Method II is generally faster than Method I. We leave to you the conclusion as to which method is easier. You are encouraged to become proficient with both methods for simplifying complex fractions.

Exercise Set 6.5

Express each complex fraction as a simple fraction and reduce to lowest terms:

1. $\dfrac{1\dfrac{3}{8}}{2\dfrac{1}{2}}$

2. $\dfrac{3\dfrac{3}{4}}{5\dfrac{1}{8}}$

3. $\dfrac{4 + \dfrac{1}{4}}{5 - \dfrac{1}{8}}$

4. $\dfrac{2 + \dfrac{5}{6}}{4 - \dfrac{1}{2}}$

5. $\dfrac{x - \dfrac{x}{y}}{1 - \dfrac{1}{y}}$

6. $\dfrac{\dfrac{x}{y} + 2}{\dfrac{x^2}{y^2} - 4}$

7. $\dfrac{1 + \dfrac{3}{x}}{x - \dfrac{9}{x}}$
8. $\dfrac{1 - \dfrac{a^2}{b^2}}{1 - \dfrac{a}{b}}$
9. $\dfrac{x + \dfrac{1}{4}}{\dfrac{5x}{8}}$

10. $\dfrac{x - \dfrac{2}{3}}{\dfrac{5x}{6}}$
11. $\dfrac{\dfrac{1}{x} + \dfrac{1}{y}}{\dfrac{x}{y} - \dfrac{y}{x}}$
12. $\dfrac{y - \dfrac{x^2}{y}}{\dfrac{1}{x} - \dfrac{1}{y}}$

13. $\dfrac{1 + \dfrac{2}{x} - \dfrac{3}{x^2}}{2 - \dfrac{1}{x} - \dfrac{1}{x^2}}$
14. $\dfrac{1 - \dfrac{9}{x^2}}{2 + \dfrac{5}{x} - \dfrac{3}{x^2}}$
15. $\dfrac{2x + 3}{3 - \dfrac{x}{x - 2}}$

16. $\dfrac{4x - 1}{2 + \dfrac{2x}{x + 1}}$
17. $\dfrac{\dfrac{2}{x + 2} + 1}{2 - \dfrac{1}{x + 2}}$
18. $\dfrac{\dfrac{1}{x - 1} + 2}{1 + \dfrac{3}{x - 1}}$

Perform the following additions or subtractions by first expressing each complex fraction as a simple fraction. Then add or subtract as in Section 6.4.

19. $\dfrac{x + \dfrac{1}{y}}{1 - \dfrac{x}{y}} + \dfrac{\dfrac{x}{y} - 2}{\dfrac{1}{y} - x}$

20. $\dfrac{1 - \dfrac{x}{y}}{1 - \dfrac{x^2}{y^2}} - \dfrac{\dfrac{1}{y} + x}{\dfrac{1}{y} - x}$

21. $\dfrac{1 - \dfrac{9}{x^2}}{1 + \dfrac{6}{x} + \dfrac{9}{x^2}} - \dfrac{1 - \dfrac{4}{x^2}}{1 + \dfrac{5}{x} + \dfrac{6}{x^2}}$

22. $\dfrac{1 + \dfrac{1}{a} - \dfrac{6}{a^2}}{1 - \dfrac{5}{a} + \dfrac{6}{a^2}} - \dfrac{1 - \dfrac{1}{a}}{1 - \dfrac{2}{a} - \dfrac{3}{a^2}}$

23. Why must a complex fraction be simplified before it can be combined with other fractions?

Review

24. Solve each of the following equations:

    a. $5x - 8 = x + 12$ b. $\dfrac{x}{2} - 1 = \dfrac{2x + 1}{5}$

25. Reduce each of the following rational expressions to lowest terms:

    a. $\dfrac{4y}{2y^2 + 6y}$ b. $\dfrac{m^2 + 3m}{2m^2 - m - 21}$ c. $\dfrac{a^3 - 8}{ab + 3a - 2b - 6}$

26. Solve and graph the inequality:
    $6a - (3 + a) < 2a - (3a - 4)$

6.6 Fractional Equations

When solving equations containing rational expressions, we use the same properties of equations and techniques for solving equations that were introduced in Chapter 2. You will recall that if it is necessary, we must first remove the fractions from the equation. The process uses the following properties of equations:

PROPERTIES OF EQUATIONS

A, B, and C are algebraic expressions

1. If $A = B$, then
    a. $A + C = B + C$
    b. $A - C = B - C$
2. If $A = B$ and $C \neq 0$, then
    a. $A \cdot C = B \cdot C$
    b. $\dfrac{A}{C} = \dfrac{B}{C}$

TECHNIQUE FOR SOLVING FRACTIONAL EQUATIONS

1. Note any values of the variable that make a denominator zero.
2. Remove fractions from the equation by multiplying both sides of the equation by the LCD of all fractions involved.
3. Remove parentheses and other symbols of grouping.
4. Simplify both sides.
5. Gather the terms containing the variable on one side of the equation and terms without the variable on the other side.
6. Divide both sides of the equation by the numerical coefficient of the variable (if it is other than $+1$).
7. Check.

Since the equations we solve here have variables in their denominators, we recall that division by (or a denominator of) zero is never permitted. Therefore, we must define as meaningless any value(s) of the variable that make a denominator in any part of the equation equal zero.

SIX Rational Expressions

Example 1

Solve each of the following fractional equations:

A. $\dfrac{x}{4} - \dfrac{x}{5} = 9$

$\dfrac{x}{4} - \dfrac{x}{5} = 9$ LCD = 20

$20\left[\dfrac{x}{4} - \dfrac{x}{5}\right] = 20 \cdot 9$ Multiply both sides by LCD = 20

$20 \cdot \dfrac{x}{4} - 20 \cdot \dfrac{x}{5} = 20 \cdot 9$ Distribute

$5x - 4x = 180$ Clear fractions

$x = 180$ Solve

Check:

$\dfrac{x}{4} - \dfrac{x}{5} = 9$

$\dfrac{180}{4} - \dfrac{180}{5} \stackrel{?}{=} 9$

$45 - 36 = 9$

$9 = 9$ ✓

B. $\dfrac{3}{x} + \dfrac{1}{2} = \dfrac{5}{x}$ Note: $x \neq 0$

$\dfrac{3}{x} + \dfrac{1}{2} = \dfrac{5}{x}$ LCD = $2x$

$2x\left[\dfrac{3}{x} + \dfrac{1}{2}\right] = 2x\left[\dfrac{5}{x}\right]$ Multiply both sides by the LCD

$2x \cdot \dfrac{3}{x} + 2x \cdot \dfrac{1}{2} = 2x \cdot \dfrac{5}{x}$ Distribute and reduce

$6 + x = 10$ Clear fractions

$x = 4$ Solve

Check:

$\dfrac{3}{x} + \dfrac{1}{2} = \dfrac{5}{x}$

$\dfrac{3}{4} + \dfrac{1}{2} \stackrel{?}{=} \dfrac{5}{4}$

6.6 Fractional Equations

$$\frac{3}{4} + \frac{2}{4} = \frac{5}{4}$$

$$\frac{5}{4} = \frac{5}{4} \checkmark$$

C. $1 + \dfrac{2x+2}{x-3} = \dfrac{8}{x-3}$ Note: $x \neq 3$

$$1 + \frac{2x+2}{x-3} = \frac{8}{x-3} \qquad \text{LCD} = x - 3$$

$$(x-3)\left[1 + \frac{2x+2}{x-3}\right] = (x-3) \cdot \frac{8}{x-3} \qquad \text{Multiply both sides by the LCD}$$

$$(x-3) \cdot 1 + \cancel{(x-3)} \cdot \frac{(2x+2)}{\cancel{(x-3)}} = \cancel{(x-3)} \cdot \frac{8}{\cancel{(x-3)}}$$

$$x - 3 + (2x + 2) = 8 \qquad \text{Clear fractions}$$

$$x - 3 + 2x + 2 = 8 \qquad \text{Remove parentheses}$$

$$3x - 1 = 8$$

$$3x = 9$$

$$\cancel{x = 3} \qquad \text{Solve. Discard the solution.}$$

This equation has *no solution* because $x = 3$ makes both denominators zero. The value 3 is called an **extraneous solution**.

Example 2   Solve the following equations:

A. $\dfrac{3x-1}{2x+4} - \dfrac{1}{3} = 2$

$$\frac{3x-1}{2x+4} - \frac{1}{3} = 2$$

$$\frac{3x-1}{2(x+2)} - \frac{1}{3} = 2 \qquad \begin{array}{l}\text{LCD} = 6(x+2)\\ \text{Note: } x \neq -2\end{array}$$

$$6(x+2)\left[\frac{3x-1}{2(x+2)} - \frac{1}{3}\right] = 6(x+2) \cdot 2 \qquad \text{Multiply both sides by the LCD}$$

$$\overset{3}{\cancel{6(x+2)}} \cdot \frac{(3x-1)}{\cancel{2(x+2)}} - \overset{2}{\cancel{6}}(x+2) \cdot \frac{1}{\cancel{3}} = 6(x+2) \cdot 2 \qquad \text{Distributive property}$$

The numerator of the first fraction is enclosed within parentheses to properly apply the distributive property.

$$3(3x - 1) - 2(x + 2) = 12(x + 2)$$
$$9x - 3 - 2x - 4 = 12x + 24 \quad \text{Remove parentheses}$$
$$7x - 7 = 12x + 24 \quad \text{Simplify}$$
$$-5x = 31$$
$$x = -\frac{31}{5} \quad \text{Solve}$$

B. $\dfrac{2x - 1}{x + 2} = \dfrac{2x}{x + 1}$ Note: $x \neq -2, x \neq -1$

$$\dfrac{2x - 1}{x + 2} = \dfrac{2x}{x + 1} \quad \text{LCD} = (x + 2)(x + 1)$$

$$(x + 2)(x + 1) \cdot \dfrac{(2x - 1)}{(x + 2)} = (x + 2)(x + 1) \cdot \dfrac{2x}{(x + 1)} \quad \text{Multiply both sides by the LCD}$$

$$2x^2 + x - 1 = 2x^2 + 4x \quad \text{Remove parentheses}$$

$$x - 1 = 4x \quad \text{Subtract } 2x^2 \text{ from both sides}$$

$$-1 = 3x$$

$$-\dfrac{1}{3} = x$$

$$x = -\dfrac{1}{3} \quad \text{Solve}$$

C. $\dfrac{3}{x + 2} - \dfrac{6}{x + 3} = \dfrac{x - 5}{x^2 + 5x + 6}$

$$\dfrac{3}{x + 2} - \dfrac{6}{x + 3} = \dfrac{x - 5}{x^2 + 5x + 6}$$

$$\dfrac{3}{x + 2} - \dfrac{6}{x + 3} = \dfrac{x - 5}{(x + 3)(x + 2)} \quad \text{Factor the third denominator}$$

LCD $= (x + 3)(x + 2)$
Note: $x \neq -3, x \neq -2$

$$(x + 3)(x + 2)\left[\dfrac{3}{x + 2} - \dfrac{6}{x + 3}\right] = (x + 3)(x + 2) \cdot \dfrac{x - 5}{(x + 3)(x + 2)} \quad \text{Multiply both sides by the LCD}$$

6.6 Fractional Equations

$$(x+3)(x+2) \cdot \frac{3}{(x+2)} - (x+3)(x+2)\frac{6}{(x+3)} = (x+3)(x+2) \cdot \frac{x-5}{(x+3)(x+2)}$$

$3(x+3) - 6(x+2) = x - 5$	Clear fractions
$3x + 9 - 6x - 12 = x - 5$	Remove parentheses
$-3x - 3 = x - 5$	Simplify
$-3 = 4x - 5$	Add $3x$ to both sides
$2 = 4x$	Add 5 to both sides
$\frac{2}{4} = x$	
$x = \frac{1}{2}$	Solve

D. $\dfrac{b}{b-3} = 1 + \dfrac{3}{3-b}$

$\dfrac{b}{b-3} = 1 + \dfrac{3}{3-b}$ Note: $b \neq 3$

$\dfrac{b}{b-3} = 1 + \dfrac{3}{b-3}$ Rewrite last fraction to make like denominators

    Multiply by $\dfrac{-1}{-1}$

LCD $= b - 3$

$(b-3) \cdot \left[\dfrac{b}{b-3}\right] = (b-3)\left[1 - \dfrac{3}{b-3}\right]$ Multiply both sides by the LCD

$(b-3) \cdot \dfrac{b}{b-3} = (b-3) \cdot 1 - (b-3) \cdot \dfrac{3}{b-3}$ Distribute and reduce

$b = b - 3 + 3$ Clear fractions

$b = b$

$0 = 0$ Subtract b from both sides

This is true regardless of the value of b, $b \neq 3$.
Therefore, the solution to this equation consists of *all real numbers, except* $b = 3$.

The equation in Example **2D** represents a special case. Whenever we obtain an obvious equation such as $0 = 0$ or $5 = 5$ as the result of trying to solve an equation, we conclude that the solution to the original equation is the set of all real numbers except for values of the variable that lead to division by zero. We call this type of equation an **identity**. (See Section 2.2.) Similarly, if we obtain an impossible equation such as $0 = 10$ or $-5 = 5$ as the result of trying to solve an equation, we conclude that the original equation has no solution.

Before leaving this section, let us note a very important difference between performing arithmetic operations (addition, subtraction, multiplication, division) on fractions and solving equations that contain fractions. For example, compare the addition/subtraction

$$\frac{3}{2} + \frac{1}{x} - \frac{1}{4}$$

to solving the equation

$$\frac{3}{2} = \frac{1}{x} - \frac{1}{4}$$

In the addition/subtraction process, the denominator *remains*!

$$\frac{3}{2} + \frac{1}{x} - \frac{1}{4} \qquad \text{LCD} = 4x$$

$$= \frac{3}{2} \cdot \frac{2x}{2x} + \frac{1}{x} \cdot \frac{4}{4} - \frac{1}{4} \cdot \frac{x}{x}$$

$$= \frac{6x}{4x} + \frac{4}{4x} - \frac{x}{4x}$$

$$= \frac{6x + 4 - x}{4x}$$

$$= \frac{5x + 4}{4x}$$

However, when solving equations we *remove* the fractions using the property of equations that allows us to multiply both sides by the same non-zero quantity (the LCD).

$$\frac{3}{2} = \frac{1}{x} - \frac{1}{4}$$

$$4x \cdot \frac{3}{2} = 4x \cdot \left(\frac{1}{x} - \frac{1}{4}\right)$$

$$6x = 4 - x$$

$$7x = 4$$

$$x = \frac{4}{7}$$

6.6 Fractional Equations

Unfortunately, some students have a tendency to multiply a fractional expression by the LCD. This is not allowed!

Exercise Set 6.6

Solve each of the following fractional equations:

1. $\dfrac{2x}{3} + \dfrac{5}{2} = 4$
2. $\dfrac{5x}{2} + \dfrac{8}{3} = 3$
3. $\dfrac{5x}{2} + 6 = x$
4. $\dfrac{3x}{2} - 4 = 2x$
5. $\dfrac{x-3}{6} = 2$
6. $\dfrac{y+4}{5} = -1$
7. $\dfrac{b-3}{4} = \dfrac{b+6}{5}$
8. $\dfrac{a+4}{3} = \dfrac{a-2}{5}$
9. $\dfrac{1}{x} - \dfrac{2}{3} = \dfrac{5}{x}$
10. $\dfrac{3}{x} + \dfrac{1}{4} = \dfrac{5}{x}$
11. $\dfrac{2}{b} - \dfrac{1}{8} = \dfrac{1}{2b} + \dfrac{1}{4}$
12. $\dfrac{2}{3y} - \dfrac{1}{9} = \dfrac{4}{9} - \dfrac{1}{y}$
13. $\dfrac{x+6}{x-1} = 8$
14. $\dfrac{x+2}{x-1} = 5$
15. $\dfrac{2x+3}{x-1} = \dfrac{3}{2}$
16. $\dfrac{3x-1}{x+2} = \dfrac{2}{3}$
17. $\dfrac{3}{2x-1} = \dfrac{7}{3x+1}$
18. $\dfrac{4}{3x-1} = \dfrac{2}{2x+3}$
19. $\dfrac{3x-2}{x-2} = \dfrac{3x}{x-1}$
20. $\dfrac{2x-1}{x+2} = \dfrac{2x}{x+1}$
21. $\dfrac{x}{x+4} + 3 = \dfrac{4x}{x-1}$
22. $\dfrac{3x}{x+3} - 1 = \dfrac{2x-1}{x-3}$
23. $2 + \dfrac{2}{a-2} = \dfrac{a}{a-2}$
24. $2 + \dfrac{3}{y-3} = \dfrac{y}{y-3}$
25. $\dfrac{4}{x+3} + \dfrac{3}{2x+6} = \dfrac{1}{x-2}$
26. $\dfrac{3}{x-2} - \dfrac{5}{2x-4} = \dfrac{2}{x+1}$
27. $\left(2 - \dfrac{x-5}{x-2}\right)\left(\dfrac{x+1}{x-2}\right)$
28. $2 - \dfrac{2}{x+1} = \dfrac{2x}{x+1}$
29. $\dfrac{4}{x-1} - \dfrac{7}{x+3} = \dfrac{x+3}{x^2+2x-3}$
30. $\dfrac{2}{x-2} = \dfrac{4}{x-3} + \dfrac{x-7}{x^2-5x+6}$
31. $\dfrac{2x}{2x+2} - \dfrac{5}{x-1} = \dfrac{2x^2-3}{x^2+x-2}$
32. $\dfrac{x+1}{x-5} = \dfrac{2}{x^2-3x-10} + \dfrac{x+3}{x+2}$
33. $\dfrac{1}{2x+1} - 2 = \dfrac{-2x}{2x+1}$
34. $\dfrac{y}{y+3} + 3 = \dfrac{3(y+2)}{y+3}$

SIX Rational Expressions

35. Compare the reasons for finding the LCD when combining fractional expressions or when solving fractional equations.

36. Why are you encouraged to place parentheses around the numerator of any fraction whose numerator has more than one term?

Review

37. Reduce each of the following rational expressions to lowest terms:

    a. $\dfrac{4}{6}$ b. $\dfrac{60}{75}$ c. $\dfrac{6x^2 y}{2xy^2}$ d. $\dfrac{4x+8}{6x+12}$

38. Is $\dfrac{2x^2 - 8}{x+2} = 2x - 4$ for all values of x? Explain.

39. Solve for the variable:

    a. $\dfrac{2}{3} = \dfrac{x}{45}$ b. $\dfrac{5}{8} = \dfrac{10}{3x+1}$

40. Draw the graph of the equation $3x - 5y = 15$.

6.7 Ratio and Proportion

One common way to compare two quantities is by their **ratio**. The ratio of two numbers, a and b, is their quotient, $\dfrac{a}{b}$, which is the first number divided by the second. The ratio of two numbers, a and b, is also frequently expressed as $a:b$. For either notation, a and b are the **terms** of the ratio.

Example 1

A mathematics class contains 17 women and 16 men.

A. Find the ratio of men to women.

   The ratio of men to women is $\dfrac{16}{17}$ or $16:17$.

B. Find the ratio of women to men.

   The ratio of women to men is $\dfrac{17}{16}$ or $17:16$.

C. Find the ratio of men to total class size.

   There are $17 + 16 = 33$ people in the class. The ratio of men to total class size is $\dfrac{16}{33}$ or $16:33$.

6.7 Ratio and Proportion

Example 2

The "intelligence" of students is frequently measured by calculating their IQ or intelligence quotient. IQ is the ratio of 100 times the mental age (MA) of a student (as determined by some standardized test) to the chronological age (CA) of the student. That is,

$$IQ = \frac{100 \cdot MA}{CA}$$

If a child has a mental age of 15 and a chronological age of 12, what is her IQ? We use the formula:

$$IQ = \frac{100 \cdot MA}{CA} \quad \text{where MA} = 15, \text{CA} = 12$$

$$IQ = \frac{100 \cdot 15}{12}$$

$$IQ = 125$$

◀

A statement of equality between two ratios is called a **proportion**. For example, $\frac{a}{b} = \frac{c}{d}$ is a proportion. If one or more of the terms of a proportion contains an unknown, finding its value is called *solving the proportion*.

Example 3

Solve each of the following proportions:

A. $\dfrac{a}{9} = \dfrac{11}{6}$

$$\overset{2}{\cancel{18}} \cdot \frac{a}{\cancel{9}} = \overset{3}{\cancel{18}} \cdot \frac{11}{\cancel{6}} \quad \text{Multiply by LCD} = 18$$

$$2a = 33$$

$$a = \frac{33}{2}$$

B. $\dfrac{x-1}{3} = \dfrac{5x-1}{21}$

$$\overset{7}{\cancel{21}} \cdot \frac{(x-1)}{\cancel{3}} = \cancel{21} \cdot \frac{(5x-1)}{\cancel{21}} \quad \text{Multiply by LCD} = 21$$

$$7(x-1) = 5x - 1$$

$$7x - 7 = 5x - 1$$

$$2x = 6$$

$$x = 3$$

◀

SIX   Rational Expressions

Ratios and proportions are encountered in a wide variety of applications in engineering, education, economics, music, and other areas. The examples and problems will illustrate some of these applications.

Example 4

If a $7\frac{1}{2}$-year-old has an IQ of 120, calculate his mental age.

$$IQ = \frac{100 \cdot MA}{CA} \qquad \text{where CA} = 7.5, IQ = 120$$

$$120 = \frac{100 \cdot MA}{7.5} \qquad \text{Substitute}$$

LCD = 7.5

$$(7.5)(120) = (7.5) \cdot \frac{100 \cdot MA}{7.5} \qquad \text{Multiply both sides by 7.5}$$

$$900 = 100 \text{ MA}$$

$$9 = \text{MA} \qquad \text{Divide both sides by 100}$$

The boy has a mental age of 9 years.   ◀

Example 5

Suppose the ratio of full-time employees to part-time employees in a factory is 3 : 7 and that there are 350 part-time employees. How many full-time employees work in the factory?

We begin by assigning a variable to represent the unknown quantity.

Let x = number of full-time employees.

We use the ratio $\dfrac{\text{full-time}}{\text{part-time}}$ to establish the proportion.

$$\frac{\text{full-time}}{\text{part-time}} = \frac{x}{350} = \frac{3}{7}$$

Now, solve:

$$\frac{x}{350} = \frac{3}{7} \qquad \text{LCD} = 350$$

$$350 \cdot \left(\frac{x}{350}\right) = 350 \cdot \left(\frac{3}{7}\right) \qquad \text{Multiply both sides by 350}$$

$$x = 150 \text{ full-time employees} \qquad \blacktriangleleft$$

6.7 Ratio and Proportion

Example 6

If one dollar ($1) in United States currency can be exchanged for $1.23 in Canadian money, how much Canadian money would a U.S. tourist in Canada receive in exchange for $500 in U.S. money?

Let x = amount of Canadian money.

We use the ratio $\dfrac{\text{Canadian money}}{\text{U.S. money}}$ to establish the proportion.

$$\frac{\text{Canadian money}}{\text{U.S. money}} = \frac{\$1.23}{\$1.00} = \frac{x}{500}$$

$$(500) \cdot \frac{1.23}{1.00} = (500) \cdot \frac{x}{500} \qquad \text{Multiply both sides by 500}$$

$$500(\$1.23) = x$$

$$\$615.00 = x$$

The tourist will receive $615 in Canadian money. ◀

(handwritten note: set up a ratio C money / US money)

Example 7

A beaker contains 135 mL of solution of which 45 mL is sulfuric acid. If another beaker contains 600 mL of the same solution, how much of it is sulfuric acid?

Let x = amount of sulfuric acid in the second beaker.

We use the ratio $\dfrac{\text{mL of sulfuric acid}}{\text{mL of total solution}}$ to create the proportion.

$$\frac{\text{mL of sulfuric acid}}{\text{mL of total solution}} = \frac{45}{135} = \frac{x}{600}$$

$\qquad\qquad\qquad\qquad\qquad$ First beaker $\qquad$ Second beaker

Solve:

$$\frac{45}{135} = \frac{x}{600} \qquad\qquad \begin{aligned} 135 &= 3^3 \cdot 7 \\ 600 &= 2^3 \cdot 3 \cdot 5^2 \\ \text{LCD} &= 2^3 \cdot 3^3 \cdot 5^2 \cdot 7 \\ &= 37{,}800 \end{aligned}$$

$$37{,}800 \cdot \left(\frac{45}{135}\right) = 37{,}800 \cdot \left(\frac{x}{600}\right) \qquad \text{Multiply both sides by the LCD} = 37{,}800$$

$$280 \cdot 45 = 63x$$

$$12{,}600 = 63x \qquad\qquad \text{Divide both sides by 63}$$

$$x = 200 \text{ mL of sulfuric acid} \qquad\qquad\qquad ◀$$

Example 8

Mr. Smith donated his lottery winnings to his favorite college to be divided among the mathematics, music, and athletic departments in the ratio $4:3:2$, respectively. If Mr. Smith's lottery winnings amount to \$180,000, how much money did each department receive?

The ratio $4:3:2$ means that for each \$9 received by the college ($4 + 3 + 2 = 9$), the mathematics department receives \$4, the music department receives \$3, and the athletic department receives \$2.

Let x = one "share" of the lottery winnings.

Then $4x$ = amount received by mathematics.

$3x$ = amount received by music.

$2x$ = amount received by athletics.

$$\begin{bmatrix} \text{Amount} \\ \text{to} \\ \text{math} \end{bmatrix} + \begin{bmatrix} \text{Amount} \\ \text{to} \\ \text{music} \end{bmatrix} + \begin{bmatrix} \text{Amount} \\ \text{to} \\ \text{athletics} \end{bmatrix} = \begin{bmatrix} \text{Total} \\ \text{amount} \end{bmatrix}$$

$$4x + 3x + 2x = 180{,}000$$
$$9x = 180{,}000$$
$$x = 20{,}000$$

$4x = 4(20{,}000) = \$80{,}000$ to mathematics
$3x = 3(20{,}000) = \$60{,}000$ to music
$2x = 2(20{,}000) = \$40{,}000$ to athletics
Total = \$180,000 ◂

Exercise Set 6.7

1. A class has 40 men and 25 women enrolled. Find:
   a. the ratio of men to women students.
   b. the ratio of women to men students.
   c. the ratio of women students to total enrollment.

2. An office has 45 female and 27 male employees. Find:
   a. the ratio of female to male employees.
   b. the ratio of male to female employees.
   c. the ratio of male employees to total employees.

3. If a student has a mental age of 18 and a chronological age of 15, what is his IQ?

6.7 Ratio and Proportion

4. If a student has a chronological age of 6 and a mental age of 9, what is her IQ?

5. If a student who is $12\frac{1}{2}$-years-old has a mental age of 11, find her IQ to the nearest whole number.

6. If a student having mental age 17 is actually 14 years old, find his IQ to the nearest whole number.

7. How old is a student having an IQ of 125 if her tested mental age is 15?

8. How old is a student having an IQ of 120 if his tested mental age is 18?

Solve the proportions in Problems 9–12:

9. $\dfrac{x}{18} = \dfrac{10}{12}$ 10. $\dfrac{y}{16} = \dfrac{15}{12}$ 11. $\dfrac{7}{35} = \dfrac{m}{15}$ 12. $\dfrac{12}{21} = \dfrac{x}{14}$

Solve the proportions in Problems 13–16. Round answers to two decimal places where necessary.

13. $\dfrac{3.6}{7} = \dfrac{x}{9}$ 14. $\dfrac{5.7}{11} = \dfrac{y}{17}$ 15. $\dfrac{L}{25} = \dfrac{12.56}{70}$ 16. $\dfrac{p}{40} = \dfrac{14.71}{50}$

17. A secretary can type a paper containing 11,700 words in 3 hours. At this rate, how many words can she type in a full 8-hour day?

18. A secretary can type a paper containing 13,200 words in 4 hours. At this rate, how many words can he type in a full 8-hour day?

19. If a bottle-capping machine can cap 700 bottles in 5 minutes, how long will it take to cap 8760 bottles?

20. If a duplicating machine can create 320 copies in 8 minutes, how long will it take to create 1960 copies?

21. A designer plans to make a scale drawing of a rectangular building measuring 120-by-72 feet. If she chooses 15 inches to represent the length of the building, what is the width of the building on the drawing?

22. A 5 × 7 inch photograph is to be enlarged into a poster. If the poster is to be 42 inches long, how wide will it be?

23. An 8 × 11 inch photograph is to be enlarged into a poster. If the poster is to be 48 inches wide, how long will it be?

24. A designer plans to make a scale drawing of a rectangular building measuring 150-by-90 feet. If he chooses 25 inches to represent the length of the building, what is the width of the building?

25. A business owned by three partners earns a profit of $96,000 for the year. If the partners agree to share the profits in the ratio 4:3:1, how much will each partner receive?

354   SIX   Rational Expressions

26. Three partners decide to buy land valued at $160,000. They agree to share the cost of the land in the ratio 10:6:4. How much must each partner contribute so they can buy the land together?

27. The angles in a triangle are in the ratio 3:2:1. How large is each angle if the sum of the angles of a triangle is 180 degrees?

28. The angles in a triangle are in the ratio 9:11:16. How large is each angle if the sum of the angles of a triangle is 180 degrees?

29. A beaker contains 250 mL of solution, of which 45 mL is sulfuric acid. If another beaker contains 650 mL of the same concentration, how much of it is sulfuric acid?

30. A punch bowl contains 300 oz of punch, of which 70 oz is grapefruit juice. If another bowl is to contain 720 oz of the same punch, how much grapefruit juice must it contain?

31. A mixture has 35 mL of H_2SO_4 and 150 mL of water. If another mixture is to have the same concentration of H_2SO_4 and water, but is to contain 252 mL of H_2SO_4, how much water must it contain?

32. A mixture has 65 mL of HCl and 250 mL of water. If another mixture is to have the same concentration of HCl and water, but is to contain 850 mL of water, how much HCl must it contain?

In geometry, similar triangles are defined as two (or more) triangles whose corresponding angles have the same measure—regardless of how long the sides are. For example, the following two triangles are similar. They have the same shape but are different in size.

It is proven in geometry that in similar triangles the corresponding sides are proportional. This can be stated symbolically if we label the sides of the preceding triangles.

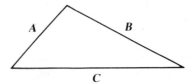

Ratios of corresponding sides:

$$\frac{A}{a} = \frac{B}{b} = \frac{C}{c}$$

6.7 Ratio and Proportion

In the following problems, find the value of the variable side, assuming the triangles are similar:

33.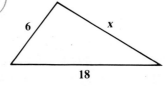

34.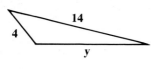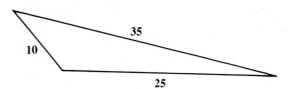

35.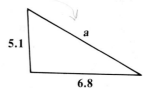

36.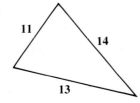

37. How does a ratio differ from a proportion?

38. Describe two ratios you've encountered in your life.

Review

39. For what value of the variable is each of the following expressions undefined?

    a. $\dfrac{3}{x}$

    b. $\dfrac{5}{x^2}$

    c. $\dfrac{c+2}{c-3}$

    d. $\dfrac{u}{u^2 + u - 12}$

    e. $\dfrac{5m}{m^2 - 4}$

40. Find the distance in miles that a vehicle will travel at the given rate R, in miles per hour, and time T, in hours.

	R	T
a.	60	3
b.	x + 2	6
c.	50	t + 3

41. Find the time T, in hours, required to travel the indicated distance D, in miles, at the indicated rate, R, in miles per hour.

	D	R
a.	600	50
b.	x	40
c.	175	x

6.8 Applications

Many types of applied problems lead to equations involving fractions. These problems are solved with the same sequence of steps we used in Section 2.7. Recall:

Approach to applications

1. Read the problem as many times as needed.
2. Draw a sketch (if helpful).
3. Assign a variable to one of the unknown quantities and express the other unknowns in terms of that same variable.
4. Translate the word problem into an equation (which may now involve fractions).
5. Solve the equation for the unknown. Express values for the other unknowns.
6. Check the solution, in the original problem.

Example 1 One number is three times another. The sum of their reciprocals is $\frac{4}{15}$. Find the numbers.

Let x = first number $\left(\text{reciprocal} = \frac{1}{x}\right)$.

Let $3x$ = second number $\left(\text{reciprocal} = \frac{1}{3x}\right)$.

(Sum of their reciprocals) is $\frac{4}{15}$.

6.8 Applications

$$\frac{1}{x} + \frac{1}{3x} = \frac{4}{15} \quad \text{Translate}$$

LCD = 15x
Note: $x \neq 0$

$$15x\left[\frac{1}{x} + \frac{1}{3x}\right] = 15x \cdot \frac{4}{15} \quad \text{Multiply both sides by the LCD}$$

$$15\cancel{x} \cdot \frac{1}{\cancel{x}} + \overset{5}{\cancel{15}}\cancel{x} \cdot \frac{1}{\cancel{3}\cancel{x}} = \cancel{15}x \cdot \frac{4}{\cancel{15}}$$

$$15 + 5 = 4x$$

$$20 = 4x$$

$$x = 5 \quad \text{First number}$$

$$3x = 15 \quad \text{Second number}$$

Check:

$$\frac{1}{5} + \frac{1}{15} = \frac{3}{15} + \frac{1}{15} = \frac{4}{15} \quad \checkmark$$

Example 2

Find a number such that 5 more than two-thirds the number is three-fourths the number.

Let x = the number.

(Two-thirds the number) + (Five) = (Three-fourths the number)

$$\left(\frac{2}{3}\right)(x) \quad + \quad 5 \quad = \quad \left(\frac{3}{4}\right)(x) \quad \text{Translate}$$

$$\frac{2x}{3} + 5 = \frac{3x}{4} \quad \text{Rewrite}$$

LCD = 12

$$12 \cdot \left[\frac{2x}{3} + 5\right] = 12 \cdot \frac{3x}{4} \quad \text{Multiply both sides by the LCD}$$

$$\overset{4}{\cancel{12}} \cdot \frac{2x}{\cancel{3}} + 12 \cdot 5 = \cancel{12} \cdot \frac{3x}{\cancel{4}}$$

$$8x + 60 = 9x \quad \text{Clear fractions}$$

$$60 = x$$

$$x = 60$$

Example 3

$\dfrac{x+1}{x+4+3} = \dfrac{3}{5}$

The denominator of a fraction is 4 more than the numerator. If the numerator is increased by 1 and the denominator is increased by 3, the value of the new fraction is $\frac{3}{5}$. What is the original fraction?

Let x = original numerator.

$x + 4$ = original denominator.

Then,

New numerator = $x + 1$ (increased by 1).

New denominator = $(x + 4) + 3$ (increased by 3)

$\qquad\qquad\qquad\quad = x + 7$

Now,

$$\text{New fraction} = \dfrac{3}{5}$$

$\dfrac{x+1}{x+7} = \dfrac{3}{5}$ \qquad Translate

Note: $x \neq -7$
LCD = $5(x + 7)$

$5(\cancel{x+7}) \cdot \dfrac{x+1}{\cancel{x+7}} = \cancel{5}(x+7) \cdot \dfrac{3}{\cancel{5}}$ \qquad Multiply both sides by the LCD

$5(x + 1) = 3(x + 7)$ \qquad Clear fractions

$5x + 5 = 3x + 21$ \qquad Remove parentheses

$2x = 16$

$x = 8$ \qquad Original numerator

$x + 4 = 12$ \qquad Original denominator

Original fraction is $\frac{8}{12}$.
Check:

$\dfrac{8}{12}$ \qquad Original fraction

$\dfrac{8 + 1}{12 + 3}$ \qquad $\left.\begin{array}{l}\text{Increased by 1}\\ \text{Increased by 3}\end{array}\right\}$ new fraction

$= \dfrac{9}{15}$

$= \dfrac{\cancel{9}}{\cancel{15}_5} = \dfrac{3}{5}$ ✓

◀

6.8 Applications

Example 4

Jason can mow a lawn in 3 hours, working alone. His brother, Rob, can mow the lawn alone in 2 hours. How long will it take the brothers to mow the lawn if they work together?

Note: Problems of this type are worked under the assumption that work is performed at a fixed rate. That is, in this problem we assume the brothers each work at the same fixed rate whether working alone or together.

Let x = time, in hours, required for the brothers to mow the lawn working together. We consider the part of this job that can be done by each brother in one hour. Since Jason requires 3 hours to mow the entire lawn, then,

$$\frac{1}{3} = \text{portion of lawn mowed by Jason in 1 hour (Jason's rate)}$$

Rob requires 2 hours to mow the entire lawn, so

$$\frac{1}{2} = \text{portion of lawn mowed by Rob in 1 hour (Rob's rate)}$$

Working together, the brothers require x hours to mow the entire law, so

$$\frac{1}{x} = \text{portion of lawn mowed in 1 hour by the brothers (combined rate)}$$

Now, the rate at which the lawn is being mowed by Jason and Rob together is equal to the sum of their individual rates. So,

$$\begin{pmatrix}\text{Combined}\\\text{rate}\end{pmatrix} = \begin{pmatrix}\text{Jason's}\\\text{rate}\end{pmatrix} + \begin{pmatrix}\text{Rob's}\\\text{rate}\end{pmatrix}$$

$$\frac{1}{x} = \frac{1}{3} + \frac{1}{2}$$

$$x \neq 0$$

LCD = $6x$

$$6x \cdot \left(\frac{1}{x}\right) = 6x \cdot \left(\frac{1}{3} + \frac{1}{2}\right)$$

$$6x \cdot \frac{1}{x} = \overset{2}{6}x \cdot \frac{1}{3} + \overset{3}{6}x \cdot \frac{1}{2}$$

$$6 = 2x + 3x$$

$$6 = 5x$$

$$x = \frac{6}{5} \text{ hours, working together}$$

$$x = 1\frac{1}{5} \text{ hours}$$

Note: $\frac{1}{5}$ hour is $\frac{1}{5}$ of 60 minutes.
So, $(\frac{1}{5})(60) = 12$ minutes

$x = 1$ hour, 12 minutes ◀

Example 5

A garden hose can fill a swimming pool in 18 hours. With the help of an inlet pipe in the bottom of the pool, the pool can be filled in 6 hours. How long would it take the inlet pipe to fill the pool working alone?

Let $x =$ time, in hours, required for the inlet pipe to fill the pool working alone. Once again, we consider how much of the pool can be filled by each supplier of water in 1 hour. Since the hose requires 18 hours to fill the pool, then

$$\frac{1}{18} = \text{portion (fraction) of the pool filled by the hose in 1 hour}$$
(garden hose rate)

The inlet pipe requires x hours to fill the pool, so

$$\frac{1}{x} = \text{portion (fraction) of the pool filled by the inlet pipe in 1 hour}$$
(inlet pipe rate)

Working together it takes 6 hours to fill the pool.

$$\frac{1}{6} = \text{portion of the pool filled by both the hose and inlet pipe working together for 1 hour (combined rate)}$$

$$\begin{pmatrix} \text{Combined} \\ \text{rate} \end{pmatrix} = \begin{pmatrix} \text{Garden hose} \\ \text{rate} \end{pmatrix} + \begin{pmatrix} \text{Inlet pipe} \\ \text{rate} \end{pmatrix}$$

$$\frac{1}{6} = \frac{1}{18} + \frac{1}{x} \qquad x \neq 0$$

LCD $= 18x$

$$18x \cdot \left(\frac{1}{6}\right) = 18x \cdot \left(\frac{1}{18} + \frac{1}{x}\right)$$

$$\overset{3}{\cancel{18x}} \cdot \frac{1}{\cancel{6}} = \cancel{18}x \cdot \frac{1}{\cancel{18}} + 18\cancel{x} \cdot \frac{1}{\cancel{x}}$$

$$3x = x + 18$$

$$2x = 18$$

$$x = 9 \text{ hours} \qquad ◀$$

6.8 Applications

Example 6

The current in a river measures 4 mph. If it takes a boat the same time to travel 48 miles upstream as it does to travel 72 miles downstream, what is the speed of the boat in still water?

We use the formula:

$$d = r \cdot t$$

(distance) = (rate) · (time)

A convenient way to keep track of the information in a problem such as this is to create a chart containing a place to record distance, rate, and time for *both trips*, upstream and downstream.

	d	r	t
Upstream			
Downstream			

Let x = speed of boat in still water.

Then $x + 4$ = speed of boat downstream (boat is traveling with the current).

$x - 4$ = speed of boat upstream (boat is traveling against the current).

The chart can be partially completed as follows:

	d	r	t
Upstream	48	$x - 4$	
Downstream	72	$x + 4$	

From the equation $d = r \cdot t$, we have

$$t = \frac{d}{r} \qquad \text{Divide both sides by } r$$

We can now complete the chart:

	d	r	t
Upstream	48	$x - 4$	$\dfrac{48}{x - 4}$
Downstream	72	$x + 4$	$\dfrac{72}{x + 4}$

Finally, since

$$\text{time (upstream)} = \text{time (downstream)}$$

$$\frac{48}{x-4} = \frac{72}{x+4}$$

Note: $x \neq 4, x \neq -4$
LCD $= (x-4)(x+4)$

$$(x-4)(x+4) \cdot \left(\frac{48}{x-4}\right) = (x-4)(x+4) \cdot \left(\frac{72}{x+4}\right) \quad \text{Multiply both sides by the LCD}$$

$$48(x+4) = 72(x-4) \quad \text{Clear fractions}$$

$$48x + 192 = 72x - 288$$

$$192 = 24x - 288 \quad \text{Subtract } 48x \text{ from both sides}$$

$$480 = 24x \quad \text{Add 288 to both sides}$$

$$x = 20 \text{ mph}$$

The boat travels at 20 mph in still water. ◀

Example 7

One plane travels 2400 miles in the same time it takes a second plane traveling 50 mph slower than the first to go 2200 miles. What are the speeds of the two planes?

Remember the chart.

	d	r	t
Slow plane			
Fast plane			

Let x = speed of fast plane.
$x - 50$ = speed of slow plane.

	d	r	t
Slow plane	2200	$x-50$	$\dfrac{2200}{x-50}$
Fast plane	2400	x	$\dfrac{2400}{x}$

$t = \dfrac{d}{r}$

6.8 Applications

Since

$$\text{time (slow plane)} = \text{time (fast plane)}$$

$$\frac{2200}{x-50} = \frac{2400}{x}$$

Note: $x \neq 0$, $x \neq 50$
LCD $= x(x-50)$

$$x(x-50)\left(\frac{2200}{x-50}\right) = x(x-50)\left(\frac{2400}{x}\right)$$

Multiply both sides by the LCD

$$2200x = 2400(x-50)$$ Clear fractions

$$2200x = 2400x - 120{,}000$$

$$-200x = -120{,}000$$

$$x = \frac{-120{,}000}{-200}$$

$$x = 600 \text{ mph (fast plane)}$$

$$x - 50 = 550 \text{ mph (slow plane)}$$

◀

Exercise Set 6.8

Solve each of the following application problems:

1. Find a number such that 3 less than one-half the number is one-third the number.

2. Find a number such that 4 more than one-fourth the number is one-third the number.

3. One number is five times another. The sum of their reciprocals is $\frac{3}{10}$. Find the numbers.

4. One number is four times another. The sum of their reciprocals is $\frac{5}{12}$. Find the numbers.

5. If $\frac{3}{4}$ is added to the reciprocal of a number, the result is $\frac{7}{8}$. Find the number.

6. If $\frac{5}{6}$ is subtracted from the reciprocal of a number, the result is $-\frac{19}{30}$. Find the number.

7. If the same number is added to both the numerator and denominator of the fraction $\frac{13}{16}$, the new fraction becomes $\frac{6}{7}$. Find the number.

8. If the same number is added to the numerator and denominator of $\frac{1}{5}$, the new fraction has the value $\frac{1}{2}$. Find the number.

9. If the same number is subtracted from both the numerator and denominator of the fraction $\frac{11}{15}$, the new fraction becomes $\frac{2}{3}$. Find the number.

10. Find the number that can be subtracted from the numerator and denominator of the fraction $\frac{7}{10}$ to give the fraction $\frac{1}{2}$.

11. The numerator of a fraction is 2 more than the denominator. If the numerator is decreased by 1 and the denominator is increased by 5, the value of the new fraction is $\frac{3}{5}$. What is the original fraction?

12. The denominator of a fraction is 3 more than the numerator. If the numerator is increased by 2 and the denominator is decreased by 1, the value of the new fraction is 1. What is the original fraction?

13. Gene can build a table in 6 hours, working alone. His daughter, Megan, can build the same table in 12 hours, working alone. How long will it take to build the table if Gene and his daughter work together?

14. Donna can paint a dormitory room in 4 hours, working alone. Her brother, Mike, requires 6 hours to paint a similar room, working alone. How long will it take to paint a dormitory room if Donna and Mike work together?

15. A large hose can fill a tank in 12 hours. With the help of a small hose, the tank can be filled in 9 hours. How long would it take the small hose to fill the tank, working alone?

16. Ed can paint a car in 3 hours, working alone. With help from Dale, the car can be painted in 2 hours. How long would it take Dale to paint the car, working alone?

17. An inlet pipe can fill a pool in 20 hours, and an outlet pipe can drain the pool in 30 hours. How long will it take to fill the pool if both pipes are left open?

18. An inlet pipe can fill a pool in 14 hours, and an outlet pipe can drain the pool in 35 hours. How long will it take to fill the pool if both pipes are left open?

19. The current in a river measures 5 mph. If it takes a boat the same time to travel 44 miles upstream as it does to travel 84 miles downstream, what is the speed of the boat in still water?

20. The current in a river measures 3 mph. If it takes a boat the same time to travel 42 miles upstream as it does to travel 60 miles downstream, what is the speed of the boat in still water?

21. An airplane flying with the wind flies 780 miles in the same time it takes to fly 660 miles against the wind. If the speed of the wind is 15 mph, what is the speed of the plane in still air?

22. An airplane flying against the wind flies 570 miles in the same time it takes to fly 690 miles with the wind. If the speed of the wind is 20 mph, what is the speed of the plane in still air?

23. One car travels 10 mph faster than another. One car travels 260 miles in the same time it takes the other car to travel 220 miles. What are the speeds of the cars?

24. One train travels 180 miles in the same time it takes a second train to travel 135 miles. If the fast train travels 15 mph faster than the slow train, what are the speeds of the two trains?

Review

25. Multiply each of the following expressions:

    a. $x^2 \cdot x^7$
    b. $(2x^2)(-5x^7)$
    c. $3x(2x^2 - 5x^7)$
    d. $\left(\dfrac{x}{2}\right)\left(\dfrac{x}{2}\right)\left(\dfrac{x}{2}\right)$

26. Complete the indicated divisions:

    a. $\dfrac{8y^8}{2y^2}$
    b. $\dfrac{-24ab^3}{4ab}$
    c. $\dfrac{15x^2y^5z}{-3x^2yz}$

27. Find the slope of each line:

    a. Passing through points $(-3, 5), (1, -1)$
    b. Whose equation is $3x + 4y = 9$
    c. Whose equation is $\dfrac{x}{2} + \dfrac{y}{3} = 1$

6.9 Review and Chapter Test

(6.1) A *rational expression* is an algebraic expression that may be written in fractional form with the numerator and denominator as polynomials. It is defined for all values of the variables that *do not* make the denominator zero.

Example 1

Each of the following is a rational expression:

A. $\dfrac{x^2 - 2x + 4}{x + 2}, \; x \neq -2$

B. $\dfrac{4}{x - x^2}, \; x \neq 0, 1$

C. $\dfrac{2m + 3}{5}$

D. $x^2 - 3x = \dfrac{x^2 - 3x}{1}$

E. $7 = \dfrac{7}{1}$ F. $\dfrac{3}{5}$

FUNDAMENTAL PRINCIPLE OF FRACTIONS

Let A and B represent polynomials that form the rational expression $\dfrac{A}{B}$ where $B \neq 0$. Let C be an expression (polynomial or fractional) where $C \neq 0$. Then,

a. $\dfrac{AC}{BC} = \dfrac{A\cancel{C}}{B\cancel{C}} = \dfrac{A}{B}$ Reduce a fraction to "lower terms" by dividing the numerator and denominator by C. We also call this operation *canceling*.

b. $\dfrac{A}{B} = \dfrac{AC}{BC}$ Raise a fraction to "higher terms" by multiplying the numerator and denominator by the same non-zero factor.

Example 2

Reduce the fraction to lowest terms:

$$\dfrac{9m^3n^2}{12m^2n^3}$$

$$\dfrac{9m^3n^2}{12m^2n^3} = \dfrac{(\cancel{3m^2n^2})(3m)}{(\cancel{3m^2n^2})(4n)} = \dfrac{3m}{4n}$$

Example 3

Raise the fraction $\dfrac{4y}{5x^2}$ to the indicated higher terms by replacing the ? with an appropriate expression.

$$\dfrac{4y}{5x^2} = \dfrac{?}{10x^3y}$$

$$\dfrac{?}{5x^2(2xy)} \quad \text{Factor}$$

therefore

$$\dfrac{4y}{5x^2} = \dfrac{4y(2xy)}{5x^2(2xy)} \quad \text{Multiply numerator and denominator by } 2xy$$

$$= \dfrac{8xy^2}{10x^3y}$$

6.9 Review and Chapter Test

PROPERTIES OF SIGNS

$$\frac{-A}{-B} = -\frac{A}{-B} = -\frac{-A}{B} = \frac{A}{B}$$

$$\frac{-A}{B} = \frac{A}{-B} = -\frac{A}{B}$$

Example 4 Reduce the fraction to lowest terms: $\dfrac{2x^3 - 6x^2}{9y - 3xy}$

$$\frac{2x^3 - 6x^2}{9y - 3xy} = \frac{2x^2(x - 3)}{3y(3 - x)} \quad \text{Factor}$$

$$= \frac{2x^2\overset{-1}{\cancel{(x - 3)}}}{3y\cancel{(3 - x)}} \quad \text{Cancel}$$

$$= \frac{-2x^2}{3y}$$

or $\qquad = -\dfrac{2x^2}{3y}$ ◀

(6.2)

MULTIPLICATION OF RATIONAL EXPRESSIONS

Let $\dfrac{A}{B}$ and $\dfrac{C}{D}$ be rational expressions with $B \neq 0$, $D \neq 0$. Then,

$$\frac{A}{B} \cdot \frac{C}{D} = \frac{AC}{BD}$$

Remember, first factor the numerator and denominator, then reduce, and then complete the multiplication.

Example 5 Multiply and reduce to lowest terms:

$$\frac{a^2 - 1}{a^2 - a} \cdot \frac{2a}{a + 1}$$

$$\frac{a^2 - 1}{a^2 - a} \cdot \frac{2a}{a + 1} = \frac{(a - 1)(a + 1)}{a(a - 1)} \cdot \frac{2a}{a + 1} \quad \text{Factor}$$

$$= \frac{(\cancel{a-1})(a+1)}{\cancel{a}(\cancel{a-1})} \cdot \frac{2\cancel{a}}{\cancel{a+1}} \qquad \text{Cancel}$$

$$= \frac{2}{1} = 2 \qquad \blacktriangleleft$$

To divide one rational expression by another, invert the divisor and multiply.

DIVISION OF RATIONAL EXPRESSIONS

Let $\dfrac{A}{B}$ and $\dfrac{C}{D}$ be rational expressions where $B \neq 0$, $D \neq 0$, $C \neq 0$.

Then,

$$\frac{A}{B} \div \frac{C}{D} = \frac{A}{B} \cdot \frac{D}{C} = \frac{AD}{BC}$$

Example 6

Divide and reduce to lowest terms:

$$\frac{x^2 - 7x}{3x^2 - 48} \div \frac{x^2 - 9}{x^2 - 7x + 12}$$

$$\frac{x^2 - 7x}{3x^2 - 48} \div \frac{x^2 - 9}{x^2 - 7x + 12}$$

$$= \frac{x(x-7)}{3(x^2 - 16)} \cdot \frac{(x-4)(x-3)}{(x-3)(x+3)} \qquad \text{Invert divisor and factor}$$

$$= \frac{x(x-7)}{3(x-4)(x+4)} \cdot \frac{(\cancel{x-4})(\cancel{x-3})}{(\cancel{x-3})(x+3)} \qquad \text{Reduce}$$

$$= \frac{x(x-7)}{3(x+4)(x+3)} \qquad \blacktriangleleft$$

(6.3) Let $a, b, x \in N$ such that $a \cdot b = b \cdot a = x$. Then a and b are *factors* of x, and x is a *multiple* of both a and b.

Example 7

Since $3 \cdot 7 = 21$, then 3 and 7 are factors of 21 while 21 is a multiple of both 3 and 7. $\blacktriangleleft$

A *prime number* is any natural number, except 1, whose only factors are 1 and the number itself. A *composite number* is any natural number, except 1, which is not prime. The number 1 is neither prime nor composite.

FUNDAMENTAL THEOREM OF ARITHMETIC
Every natural number, except 1, is either prime or can be expressed uniquely as a product of primes.

Example 8

Find the prime factorization of the following numbers:

A. 33

$$33 = 3 \cdot 11 \qquad \text{All primes}$$

B. 19

19 is prime

C. 18

$$18 = 6 \cdot 3 \qquad \text{Factor 18}$$
$$= 2 \cdot 3 \cdot 3 \qquad \text{Factor 6}$$
$$= 2 \cdot 3^2 \qquad \text{All primes}$$

◀

The *least common multiple (LCM)* of a collection of natural numbers is the smallest number that contains each of the original numbers as a factor.

Technique for finding the least common multiple (LCM) of a group of numbers

1. Write each number in prime factored form.

2. The LCM is the number composed of factors that are all the different primes from Step 1, each raised to the highest power observed on that prime in any one factorization.

If our task is to find the LCM of a collection of expressions containing variable factors as well as constant factors, we proceed in a similar manner.

Technique for finding the least common multiple (LCM) of a group of expressions containing variable factors

1. Write each expression in completely factored form and each coefficient in prime factored form.

2. The LCM is the expression composed of factors that are all the different prime factors from Step 1, each raised to the highest power appearing on that factor in any one factorization.

Example 9   Find the LCM for each of the following groups of expressions:

A. $6a^2b, 15ab^3, 24a^2b^2$

Step 1: $6a^2b = 2 \cdot 3 \cdot a^2 \cdot b$
$15ab^3 = 3 \cdot 5 \cdot a \cdot b^3$ Prime factors
$24a^2b^2 = 2^3 \cdot 3 \cdot a^2 \cdot b^2$

Step 2:   LCM $(6a^2b, 15ab^3, 24a^2b^2)$

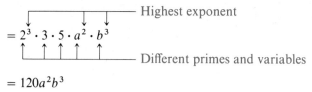

$= 2^3 \cdot 3 \cdot 5 \cdot a^2 \cdot b^3$

$= 120a^2b^3$

B. $x^2 + 2x + 1, x^2 - 1$

Step 1: $x^2 + 2x + 1 = (x + 1)^2$
$x^2 - 1 = (x + 1)(x - 1)$ Prime factors

Step 2:   LCM $(x^2 + 2x + 1, x^2 - 1)$

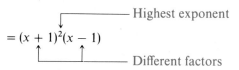

$= (x + 1)^2(x - 1)$

◀

(6.4)

ADDITION AND SUBTRACTION OF FRACTIONS HAVING COMMON (LIKE) DENOMINATORS

Let A, B, and C be polynomials where $C \neq 0$. Then,

$$\frac{A}{C} + \frac{B}{C} = \frac{A + B}{C}$$

$$\frac{A}{C} - \frac{B}{C} = \frac{A - B}{C}$$

When we add or subtract rational expressions that do not have like denominators, we use the Fundamental Principle of Fractions (Section 6.1). We *rewrite* the fractions so they have a common (same) denominator and then proceed according to the rule for adding and subtracting fractions having like denominators.

In summary, to add or subtract rational expressions, we follow these steps:

1. Determine the least common denominator (LCD) of all fractions involved.
2. Use the Fundamental Principle of Fractions to rewrite each fraction so that each of them has the LCD as its denominator.
3. Add or subtract the fractions by combining their numerators, leaving the denominator unchanged as the LCD.
4. Reduce the resulting rational expression to lowest terms if necessary.

Example 10

Subtract as indicated. Reduce answer to lowest terms.

$$\frac{x}{x+2} - \frac{2}{x^2-4}$$

$$\frac{x}{x+2} - \frac{2}{x^2-4} = \frac{x}{x+2} - \frac{2}{(x+2)(x-2)} \qquad \text{Factor denominator} \\ \text{LCD} = (x+2)(x-2)$$

$$= \frac{x}{x+2} \cdot \frac{(x-2)}{(x-2)} - \frac{2}{(x+2)(x-2)} \qquad \text{Fundamental Principle}$$

$$= \frac{x(x-2)}{(x+2)(x-2)} - \frac{2}{(x+2)(x-2)} \qquad \text{Common denominators}$$

$$= \frac{x(x-2) - 2}{(x+2)(x-2)} \qquad \text{Combine numerators}$$

$$= \frac{x^2 - 2x - 2}{(x+2)(x-2)} \qquad \text{Remove parentheses}$$

◀

(6.5) A *complex fraction* is a fractional expression in which a fraction appears in the numerator or the denominator. A complex fraction may be reduced to a simple fraction by either of two methods.

Method I: Write both the numerator and denominator of the complex fraction as single fractions. Then divide the numerator by the denominator using the rules for dividing fractions (invert and multiply).

Method II: Multiply the numerator and denominator of the complex fraction by the LCM of all denominators appearing in the numerator and denominator. This is an application of the Fundamental Principle of Fractions.

Example 11 Express the following complex fraction as a simple fraction and reduce to lowest terms:

$$\frac{\dfrac{1}{x} - \dfrac{1}{y}}{\dfrac{1}{x} + \dfrac{1}{y}}$$

Method I:

$$\frac{\dfrac{1}{x} - \dfrac{1}{y}}{\dfrac{1}{x} + \dfrac{1}{y}} = \frac{\dfrac{1}{x} \cdot \dfrac{y}{y} - \dfrac{1}{y} \cdot \dfrac{x}{x}}{\dfrac{1}{x} \cdot \dfrac{y}{y} + \dfrac{1}{y} \cdot \dfrac{x}{x}} \qquad \text{Fundamental Principle of Fractions}$$

$$= \frac{\dfrac{y - x}{xy}}{\dfrac{y + x}{xy}} \qquad \text{Express numerator and denominator as a single fraction}$$

$$= \frac{y - x}{\cancel{xy}} \cdot \frac{\cancel{xy}}{y + x} \qquad \text{Invert and multiply}$$

$$= \frac{y - x}{y + x}$$

Method II:

$$\frac{\dfrac{1}{x} - \dfrac{1}{y}}{\dfrac{1}{x} + \dfrac{1}{y}} \qquad \text{LCD} = xy$$

$$\frac{\left(\dfrac{1}{x} - \dfrac{1}{y}\right) \cdot xy}{\left(\dfrac{1}{x} + \dfrac{1}{y}\right) \cdot xy} \qquad \text{Fundamental Principle}$$

$$= \frac{\dfrac{1}{\cancel{x}} \cdot \cancel{x}y - \dfrac{1}{\cancel{y}} \cdot x\cancel{y}}{\dfrac{1}{\cancel{x}} \cdot \cancel{x}y + \dfrac{1}{\cancel{y}} \cdot x\cancel{y}} \qquad \text{Distributive property}$$

$$= \frac{y - x}{y + x} \qquad \text{Multiply}$$

◀

6.9 Review and Chapter Test

(6.6) Technique for solving fractional equations

1. Note any values of the variable that make a denominator zero.
2. Remove fractions from the equation by multiplying both sides of the equation by the LCD of all fractions involved.
3. Remove parentheses and other symbols of grouping.
4. Simplify both sides.
5. Gather the terms containing the variable on one side of the equation and terms without the variable on the other side.
6. Divide both sides of the equation by the numerical coefficient of the variable (if it is other than $+1$).
7. Check.

Example 12

Solve the equation:

$$\frac{3}{m+2} = \frac{m-2}{m^2+m-2}$$

$\frac{3}{m+2} = \frac{m-2}{(m+2)(m-1)}$ Factor denominators

Note: $m \neq -2, m \neq 1$
LCD $= (m+2)(m-1)$

$(m+2)(m-1)\left(\frac{3}{m+2}\right) = (m+2)(m-1)\left(\frac{m-2}{(m+2)(m-1)}\right)$ Multiply both sides by the LCD

$3(m-1) = m-2$ Clear fractions

$3m - 3 = m - 2$ Remove parentheses

$2m = 1$

$m = \frac{1}{2}$ ◀

Remember, when we are *solving equations* we can *remove fractions* by multiplying both sides of the equation by the LCD. However, when we are *performing operations* the denominators *remain!*

(6.7) The *ratio* of two numbers, a and b, is their quotient, $\frac{a}{b}$, or $a:b$.

Example 13

A child's bank contains 35 nickels, 30 dimes, and 20 quarters.

A. Find the ratio of number of dimes to the number of quarters:
The ratio of the number of dimes to the number of quarters is

$$\frac{30}{20} = \frac{3}{2} \quad \text{or } 3:2$$

B. Find the ratio of the number of nickels to the total number of coins in the bank:

$$\text{Total coins} = \underset{\text{(Nickels)}}{35} + \underset{\text{(Dimes)}}{30} + \underset{\text{(Quarters)}}{20} = 85$$

The ratio of the number of nickels to the total number of coins is

$$\frac{35}{85} = \frac{7}{13} \quad \text{or } 7:13 \quad \blacktriangleleft$$

A *proportion* is a statement of equality between two ratios. To *solve* a proportion is to find the value of a variable appearing in any of the terms.

Example 14

Solve the proportion:

$$\frac{7}{y} = \frac{4.6}{20}$$

Round the answer to two decimal places if necessary.

$$\frac{7}{y} = \frac{4.6}{20}$$

$$20y \cdot \frac{7}{y} = 20y \cdot \frac{4.6}{20} \qquad \text{Multiply both sides by } 20y = \text{LCD}$$

$$140 = 4.6y$$

$$\frac{140}{4.6} = \frac{4.6y}{4.6} \qquad \text{Divide}$$

$$30.4347 = y$$

$$y = 30.43 \qquad \text{Round} \qquad \blacktriangleleft$$

Example 15

An automatic robot welder can perform 20 welds in 5 minutes. At this rate, how many welds can the robot perform in 8 hours?

Let x = number of welds in 8 hours.

We use the ratio $\dfrac{\text{welds}}{\text{minutes}}$ to establish the proportion.

6.9 Review and Chapter Test

Note: 8 hours = 8 · 60 minutes
= 480 minutes

$$\frac{\text{welds}}{\text{minutes}} = \frac{20}{5} = \frac{x}{480}$$

LCD = 480

$$\overset{96}{\cancel{480}} \cdot \frac{20}{\cancel{5}} = \cancel{480} \cdot \frac{x}{\cancel{480}}$$

$$1920 = x$$

$$x = 1920 \text{ welds in 8 hours} \quad \blacktriangleleft$$

(6.8) Approach to applications

1. Read the problem as many times as needed.
2. Draw a sketch (if helpful).
3. Assign a variable to one of the unknown quantities and express the other unknowns in terms of that same variable.
4. Translate the word problem into an equation (which may now involve fractions).
5. Solve the equation for the unknown. Express values for the other unknowns.
6. Check the solution.

Example 16

If an inlet pipe can fill a pool in 20 hours and an outlet pipe can drain the pool in 32 hours, how long will it take to fill the pool if both pipes are left open?

Let x = time, in hours, required for both pipes to fill the pool.

Since the inlet pipe requires 20 hours to fill the pool, then

$$\frac{1}{20} = \text{portion of pool } \textit{filled} \text{ by the inlet pipe } \textit{in 1 hour} \text{ (inlet rate)}$$

Outlet pipe requires 32 hours to empty the pool, so

$$\frac{1}{32} = \text{portion of pool } \textit{emptied} \text{ by outlet pipe } \textit{in 1 hour} \text{ (outlet rate)}$$

Working together, the pipes require x hours, so

$$\frac{1}{x} = \text{portion of pool } \textit{filled} \text{ in 1 hour by both pipes (combined rate)}$$

$$\begin{pmatrix} \text{Inlet} \\ \text{rate} \end{pmatrix} - \begin{pmatrix} \text{Outlet} \\ \text{rate} \end{pmatrix} = \begin{pmatrix} \text{Combined} \\ \text{rate} \end{pmatrix}$$

The outlet pipe is working against the inlet pipe, so we subtract its rate:

$$\frac{1}{20} - \frac{1}{32} = \frac{1}{x}$$

$$20 = 2^2 \cdot 5$$

$$32 = 2^5$$

$$\text{LCD} = 2^5 \cdot 5 \cdot x$$

$$= 160x$$

$$160x \cdot \left(\frac{1}{20} - \frac{1}{32}\right) = 160x \cdot \frac{1}{x}$$

$$\overset{8}{\cancel{160x}} \cdot \frac{1}{\cancel{20}} - \overset{5}{\cancel{160x}} \cdot \frac{1}{\cancel{32}} = \cancel{160x} \cdot \frac{1}{\cancel{x}}$$

$$8x - 5x = 160$$

$$3x = 160$$

$$x = \frac{160}{3}$$

$$x = 53\frac{1}{3} \quad \text{hours is required to fill the pool} \quad \blacktriangleleft$$

Exercise Set 6.9 Review

Reduce each of the following fractions to lowest terms:

1. $\dfrac{4 - x^2}{x^2 + x - 6}$

2. $\dfrac{12a^3(2x + y)^5}{32a^4(2x + y)^3}$

3. $\dfrac{m^2n^2 - mn^3}{m^3n - m^2n^2}$

4. $\dfrac{9a^2 - 25b^2}{6a^2 - 10ab}$

Raise each of the following fractions to the indicated higher terms. Replace the ? with an appropriate expression.

5. $\dfrac{3}{7y} = \dfrac{?}{35x^2y^2}$

6. $\dfrac{2x}{3y} = \dfrac{?}{3x^2y - 6y^2}$

7. $\dfrac{x - 2}{x + 3} = \dfrac{?}{x^2 - x - 12}$

8. $\dfrac{m + 2}{m - 2} = \dfrac{?}{m^2 - 4}$

6.9 Review and Chapter Test

Multiply or divide as indicated. Write all answers in lowest terms.

9. $\dfrac{6x^2ym}{21xn^2} \cdot \dfrac{14nm^2}{9xy^2}$

10. $\dfrac{6x^2ym}{21xn^2} \div \dfrac{14nm^2}{9xy^2}$

11. $\dfrac{2x^2 - 4x}{x^2 + 3x} \cdot \dfrac{x^2 + 2x - 3}{4xy - 8y}$

12. $\dfrac{x^2 + x - 6}{x^2 + x - 2} \div \dfrac{x^2 + 4x + 3}{x^2 - 1}$

13. $\dfrac{8xy^3}{5a^2b} \div \dfrac{12x^2y}{15a} \cdot \dfrac{yb^2}{6xa}$

Find the prime factorization of each of the following numbers:

14. 27 15. 36 16. 23 17. 500

Find the least common multiple (LCM) of each of the following sets of expressions:

18. $5x - 5, 4x^2 - 4$

19. $12x^2yz^3, 8xy^2, 9x^3y^2$

20. $x^2 - 9, x^2 + x - 6, x - 2$

21. $4m^2 - 2m - 30, 4m^3 - 12m^2, 2m^2 + 5m$

Add or subtract as indicated. Reduce answers to lowest terms.

22. $\dfrac{5}{18} + \dfrac{7}{12}$

23. $\dfrac{8}{35} - \dfrac{3}{14}$

24. $\dfrac{3}{x+2} + \dfrac{5}{x-1}$

25. $\dfrac{2}{ab^2c} - \dfrac{4}{5ac^2} + \dfrac{1}{2a^2}$

26. $\dfrac{3}{x+4} + \dfrac{1}{x-4} - \dfrac{3x}{x^2 - 16}$

27. $\dfrac{4}{x^2 - 3x + 2} - \dfrac{x}{x - 2} + \dfrac{2x}{x - 1}$

28. $\dfrac{x+2}{2x^2 - x - 6} - \dfrac{x-1}{2x^2 + 5x + 3}$

Express each of the following complex fractions as a simple fraction and reduce to lowest terms:

29. $\dfrac{\dfrac{4}{5x^2}}{\dfrac{3}{7x}}$

30. $\dfrac{\dfrac{1}{3} + \dfrac{2}{5}}{\dfrac{3}{10} + \dfrac{2}{3}}$

31. $\dfrac{1 + \dfrac{2}{a}}{a - \dfrac{4}{a}}$

32. $\dfrac{3 + \dfrac{5}{x} - \dfrac{2}{x^2}}{3 - \dfrac{10}{x} + \dfrac{3}{x^2}}$

Solve each of the following equations:

33. $\dfrac{3}{4y} - \dfrac{1}{y} = \dfrac{1}{4}$

34. $\dfrac{6}{5y - 3} = \dfrac{3}{2y + 1}$

35. $\dfrac{4a}{a - 3} - \dfrac{7}{3a} = 4$

36. $\dfrac{2x - 1}{x^2 - x - 6} = \dfrac{4}{x - 3}$

37. $\dfrac{5}{2 - a} = \dfrac{3a}{a - 2} + 1$

Solve the following proportions. Round answers to two decimal places if necessary.

38. $\dfrac{m}{3.2} = \dfrac{11}{20}$

39. $\dfrac{19}{15} = \dfrac{8}{6x}$

40. A clerk waits on 7 people in 15 minutes. At this rate, how long will it take the clerk to wait on 105 people?

41. The angles in a triangle are in the ratio 5:6:7. How large is each angle if the sum of the angles of a triangle is 180 degrees?

42. Mike can paint a house in 12 days. Sally can paint the same house in 15 days. How long would it take to paint the house if Mike and Sally work together?

43. The current in a river measures 7 mph. If it takes a boat the same time to travel 40 miles upstream as it does to travel 110 miles downstream, what is the speed of the boat in still water?

Chapter 6 Review Test

1. Reduce to lowest terms:

   $$\dfrac{x^3 y - xy^3}{x^3 y + 3x^2 y^2 + 2xy^3}$$

2. Raise the following fractions to the indicated higher terms by replacing the ? with an appropriate expression:

   a. $\dfrac{4x}{3y} = \dfrac{?}{15x^2 y^2}$

   b. $\dfrac{2y + 3}{y - 2} = \dfrac{?}{3y^2 - 4y - 4}$

3. Perform the following operations, as indicated, and write all answers in lowest terms:

   a. $\dfrac{3a^2 bc^3}{5xm^2} \div \dfrac{9ab^2 c^2}{10x^2 m}$

   b. $\dfrac{x^2 - 4}{x^2 + x - 2} \cdot \dfrac{x^2 + 8x + 15}{x^2 + x - 6}$

6.9 Review and Chapter Test

c. $\dfrac{y^2 + y - 6}{y^2 + 6y + 9} \cdot \dfrac{y + 2}{y - 3} \div \dfrac{y^2 - y - 2}{y^2 - 7y + 12}$

d. $\dfrac{4}{m - 3} - \dfrac{5}{m + 2}$

e. $\dfrac{3}{4x^2 y} - \dfrac{2}{3xy} + \dfrac{1}{y^2}$

f. $\dfrac{x + 3}{x^2 - x - 6} - \dfrac{x - 4}{x^2 - 3x - 10}$

4. Express the complex fraction as a simple fraction reduced to lowest terms:

$$\dfrac{1 - \dfrac{1}{x} - \dfrac{6}{x^2}}{1 + \dfrac{6}{x} + \dfrac{8}{x^2}}$$

5. Solve each equation:

   a. $\dfrac{2}{x} = \dfrac{6}{x^2} - \dfrac{1}{x}$

   b. $\dfrac{3}{x^2 - 2x} = \dfrac{10}{x^2 - 4}$

6. A math student solves 4 problems in 20 minutes. At this rate, how long will it take her to solve 50 problems?

7. Larry can paint a house in 5 days. Ed and Larry, working together, can paint the house in 3 days. How long would it take Ed to paint the house, working alone?

CHAPTER SEVEN

EXPONENTS AND RADICALS

Working with exponents is not new to us. Exponents were first introduced in Chapter 1, and then exponential forms were used extensively in Chapter 5 to operate on polynomials. In this chapter we develop the operations on exponential forms in more detail, and then study the closely related topics of roots and radicals.

7.1 Natural Number Exponents

Recall that b^n is a symbolic form for the product of n factors of the number b. The real number b is the **base**, and the natural number n is the **exponent**, while the expression b^n is called the nth **power** of b. An exponent of 1 is generally not written, and we define $b^1 = b$.

Exponential form: For $b \in R$, $n \in N$:

$$b^n = \underbrace{b \cdot b \cdots b}_{n \text{ factors}}$$

$$b^1 = b$$

From the definition, we developed the product rule of exponents in Chapter 5.

7.1 Natural Number Exponents

PRODUCT RULE

$$b^m \cdot b^n = b^{m+n}$$

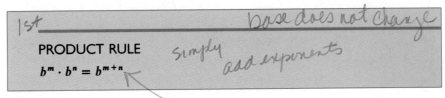

Example 1

Simplify each expression:

A. $x^3 \cdot x^6$

$$x^3 \cdot x^6 = x^{3+6} = x^9$$

B. $a^4 \cdot a^5 \cdot a$

$$a^4 \cdot a^5 \cdot a = a^{4+5+1} = a^{10}$$

Notice that the product rule applies to two *or more* factors of the *same base*, and also note the addition of the *understood* exponent of 1 ($a = a^1$).

Power of a Product Rule

A product of two or more factors may be raised to a power, such as $(2x)^4$.

$(2x)^4 = (2x)(2x)(2x)(2x)$ Definition

$ = (2 \cdot 2 \cdot 2 \cdot 2)(xxxx)$ Commutative and associative properties

$ = 2^4 x^4$ Definition

$ = 16x^4$

Observe that *each factor* is raised to the indicated power. In general,

POWER OF A PRODUCT RULE

$$(ab)^m = a^m b^m$$

Example 2

Use one or both of the preceding rules to simplify each expression:

A. $(3x)^2$

$$(3x)^2 = 3^2 x^2$$

$$ = 9x^2$$

B. $(3x)^3 \cdot x^4$

$$(3x)^3 x^4 = 3^3 x^3 x^4$$
$$= 3^3 x^{3+4}$$
$$= 27 x^7$$

C. $(2x)^3 \cdot (xy)^4$

$$(2x)^3 (xy)^4 = 2^3 x^3 x^4 y^4$$
$$= 2^3 \cdot x^{3+4} y^4$$
$$= 8 x^7 y^4$$

◀

Power of a Quotient Rule

We show that when a quotient or fraction (such as $\frac{2}{x}$) is raised to a power, both the numerator and the denominator are raised to that power. Consider:

$$\left(\frac{2}{x}\right)^4 = \left(\frac{2}{x}\right)\left(\frac{2}{x}\right)\left(\frac{2}{x}\right)\left(\frac{2}{x}\right) \quad \text{Definition}$$

$$= \frac{2 \cdot 2 \cdot 2 \cdot 2}{xxxx} \quad \text{Multiplication of fractions}$$

$$= \frac{2^4}{x^4} = \frac{16}{x^4} \quad \text{Definition}$$

In general,

POWER OF A QUOTIENT RULE

$$\left(\frac{a}{b}\right)^m = \frac{a^m}{b^m}, \ b \neq 0$$

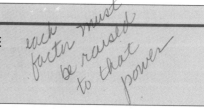
each factor must be raised to that power

Example 3

Use any of the preceding rules to simplify each expression:

A. $\left(\dfrac{x}{y}\right)^3$

$$\left(\frac{x}{y}\right)^3 = \frac{x^3}{y^3}$$

7.1 Natural Number Exponents

B. $\left(\dfrac{2x}{y}\right)^3$

$$\left(\dfrac{2x}{y}\right)^3 = \dfrac{(2x)^3}{y^3} \quad \text{Power of a quotient}$$

$$= \dfrac{2^3 x^3}{y^3} \quad \text{Power of a product}$$

$$= \dfrac{8x^3}{y^3} \quad \text{Simplify } (2^3 = 8)$$

C. $\left(\dfrac{3a}{b}\right)^2 \cdot \left(\dfrac{a}{b}\right)^3$

$$\left(\dfrac{3a}{b}\right)^2 \cdot \left(\dfrac{a}{b}\right)^3 = \dfrac{(3a)^2}{b^2} \cdot \dfrac{a^3}{b^3} \quad \text{Power of a quotient}$$

$$= \dfrac{3^2 a^2}{b^2} \cdot \dfrac{a^3}{b^3} \quad \text{Power of a product}$$

$$= \dfrac{9a^2}{b^2} \cdot \dfrac{a^3}{b^3} \quad \text{Simplify } (3^2 = 9)$$

$$= \dfrac{9a^2 a^3}{b^2 b^3} \quad \text{Product of fractions}$$

$$= \dfrac{9a^5}{b^5} \quad \text{Product rule}$$

Power to a Power Rule

Consider a power raised to a power, such as $(b^2)^4$.

$(b^2)^4 = (b^2)(b^2)(b^2)(b^2)$ Definition

$ = b^{2+2+2+2}$ Product rule

$ = b^{4 \cdot 2}$ Definition of multiplication as successive addition

$ = b^8$

In general,

POWER TO A POWER RULE

$(b^m)^n = b^{mn}$

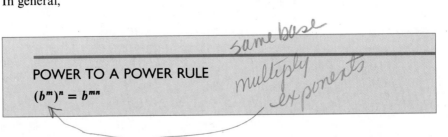

same base
multiply exponents

Example 4 Use the preceding rules to simplify each expression:

A. $(-4x^3)^2$

$$(-4x^3)^2 = (-4)^2(x^3)^2 \quad \text{Power of a product}$$
$$= 16x^{3 \cdot 2} \quad \text{Power to a power}$$
$$= 16x^6$$

B. $x^2 \cdot (y^2)^3$

$$x^2 \cdot (y^2)^3 = x^2 y^{2 \cdot 3} = x^2 y^6$$

C. $(y^3)^2 \cdot (y^2)^4$

$$(y^3)^2 \cdot (y^2)^4 = y^6 \cdot y^8 \quad \text{Power to a power}$$
$$= y^{14}$$

D. $(2x^3 y^2)^4$

$$(2x^3 y^2)^4 = 2^4 (x^3)^4 (y^2)^4 \quad \text{Power of a product}$$
$$= 16 x^{12} y^8 \quad \text{Power to a power}$$

E. $\left(\dfrac{3}{x^3}\right)^2$

$$\left(\dfrac{3}{x^3}\right)^2 = \dfrac{3^2}{(x^3)^2} \quad \text{Power to a quotient}$$
$$= \dfrac{9}{x^6} \quad \text{Power to a power}$$

Quotient Rule

Finally, we review the quotient rule, which was also developed in Chapter 5 (see Section 5.5). In general:

QUOTIENT RULE, $b \neq 0$

Case 1: $\dfrac{b^m}{b^n} = b^{m-n}$ if $m > n$

Case 2: $\dfrac{b^m}{b^n} = \dfrac{1}{b^{n-m}}$ if $m < n$

Case 3: $\dfrac{b^m}{b^n} = 1$ if $m = n$

7.1 Natural Number Exponents

At this time, only natural number exponents are meaningful, so the quotient rule was written with three parts, depending on the size of the exponents, m and n. In later sections, the restrictions will be removed, and the quotient rule will be simplified. In the meantime, rather than a strict memorization of this rule, you might consider such expressions as applications of the Fundamental Property of Fractions. For example, to simplify

$$\frac{5x^2y^9}{4x^3y^4}$$

think: "let's cancel two factors of x and four factors of y," giving

$$\frac{5y^5}{4x}$$

Example 5

Simplify each expression using the preceding five rules of exponents:

A. $\dfrac{x^8}{x^2}$

$\dfrac{x^8}{x^2} = x^{8-2}$ Quotient rule

$= x^6$

B. $\dfrac{5m^3}{15m^4}$

$\dfrac{5m^3}{15m^4} = \dfrac{\cancel{5}}{\underset{3}{\cancel{15}}m^{4-3}}$ Fundamental Property of Fractions and the quotient rule

$= \dfrac{1}{3m}$

C. $\dfrac{(4m^2)^3}{(2m^2)^4}$

$\dfrac{(4m^2)^3}{(2m^2)^4} = \dfrac{4^3(m^2)^3}{2^4(m^2)^4}$ Power of a product

$= \dfrac{64m^6}{16m^8}$ Power of a power

$= \dfrac{\underset{}{\cancel{64}}^{4}}{\cancel{16}m^{8-6}}$ Cancel 16 and apply the quotient rule

$= \dfrac{4}{m^2}$

◀

SEVEN   Exponents and Radicals

Two generalizations can be made regarding the preceding rules of exponents:

1. The rules may be used in any sequence.
2. The rules apply only to quantities that are multiplied or divided.

For example, $(ab)^2 = a^2 b^2$ by the power of a product rule. However, in general,
$$(a + b)^2 \neq a^2 + b^2$$
Recall $(a + b)^2 = (a + b)(a + b) = a^2 + 2ab + b^2$ as was shown by FOIL.

Example 6

Simplify each expression:

A. $\dfrac{a^7}{(a^2)^4}$

$$\dfrac{a^7}{(a^2)^4} = \dfrac{a^7}{a^8} \qquad \text{Power to a power}$$

$$= \dfrac{1}{a^{8-7}} \qquad \text{Quotient rule}$$

$$= \dfrac{1}{a}$$

B. $\left(\dfrac{x^3}{c^2}\right)^4$

$$\left(\dfrac{x^3}{c^2}\right)^4 = \dfrac{(x^3)^4}{(c^2)^4} \qquad \text{Power of a quotient}$$

$$= \dfrac{x^{12}}{c^8} \qquad \text{Power to a power}$$

Note that no further simplification is possible since the bases are not the same.

C. $\dfrac{4x^2 y^6 z^7}{6x^8 y^6 z}$

$$\dfrac{4x^2 y^6 z^7}{6x^8 y^6 z} = \dfrac{4 y^6 z^{7-1}}{6 x^{8-2} y^6} \qquad \text{Quotient rule}$$

$$= \dfrac{2 z^6}{3 x^6}$$

D. $\left(\dfrac{2x^2}{5y}\right)^3 \cdot \left(\dfrac{y}{4x}\right)^4$

$$\left(\dfrac{2x^2}{5y}\right)^3 \cdot \left(\dfrac{y}{4x}\right)^4 = \dfrac{2^3 (x^2)^3}{5^3 y^3} \cdot \dfrac{y^4}{4^4 x^4} \qquad \begin{array}{l}\text{Power of a product and} \\ \text{of a quotient}\end{array}$$

7.1 Natural Number Exponents

$$= \frac{8x^6y^4}{125 \cdot 256x^4y^3} \quad \text{Power to a power}$$

$$= \frac{x^{6-4}y^{4-3}}{125 \cdot 32} \quad \text{Quotient rule}$$

$$= \frac{x^2y}{4000}$$

◀

Exercise Set 7.1

Use the rules of exponents to write each expression in simplified form. Each variable should appear only once, and constants should be evaluated.

1. $x^3 \cdot x$
2. $y \cdot y^4$
3. $(2m)^2$
4. $(3y)^2$
5. $(ab)^5$
6. $(4a)^3$
7. $\left(\dfrac{2}{b}\right)^3$
8. $\left(\dfrac{m}{3}\right)^4$
9. $\dfrac{c^7}{c}$
10. $\dfrac{q^{11}}{q}$
11. $\dfrac{8x^2}{2x^2}$
12. $\dfrac{4m^3}{12m^3}$
13. $\dfrac{5a}{10a^2}$
14. $\dfrac{6p}{18p^4}$
15. $\left(\dfrac{3u}{5t}\right)^2$
16. $\left(\dfrac{6m}{5n}\right)^2$
17. $\dfrac{(4ab)^2}{8}$
18. $\dfrac{(6a^2)^2}{9}$
19. $\dfrac{6x^4}{9x}$
20. $\dfrac{8x}{12x^4}$
21. $\dfrac{3p^8}{12p^2}$
22. $\dfrac{20L^3}{6L^2}$
23. $(3xy^2)^3$
24. $(2x^3y)^2$
25. $\left(\dfrac{3a^2b}{6a}\right)^2$
26. $\left(\dfrac{4ab^2}{6a}\right)^2$
27. $\dfrac{(2m^2n)^2}{4mn^2p}$
28. $\dfrac{(2p^2q)^3}{16pq^3}$
29. $\dfrac{-6u^2v^3w}{(2uv^2)^3}$
30. $\dfrac{-16uvw^4}{(-2uw^2)^2}$
31. $\dfrac{-2^3}{8}$
32. $\dfrac{(-2)^4}{8}$
33. $\left(\dfrac{mn}{3}\right)^2 \cdot \dfrac{6m}{5n}$
34. $\dfrac{(2x)^2}{(3y)^3} \cdot \left(\dfrac{6y}{x^2}\right)^2$
35. $\dfrac{(9a^3)^2}{-3a^2}$
36. $\dfrac{-3^2}{9m}$
37. $\dfrac{8ab}{5} \cdot \left(\dfrac{-3}{2a}\right)^3$
38. $\dfrac{6x}{7y} \cdot \left(\dfrac{3y}{4x}\right)^2$

Consider the rectangle having length L and width W as indicated in the sketch.

$W = 3ab$

$L = 2a^2b^3$

Recall the formulas:

$A = L \cdot W$ (area)

$P = 2L + 2W$ (perimeter)

For Problems 39–44, refer to this sketch:

39. Find an expression in a and b for the area of the rectangle.
40. Find an expression in a and b for the perimeter of the rectangle.
41. Divide the length by the width.
42. Raise the length to the third power.
43. Raise the width to the third power.
44. Divide the width by the length.

Review

45. Simplify each expression:

   a. $a(a - 1) - (2a - 4)$

   b. $x^2(x + 1) - 2x(x + 3) + 4(x - 1)$

   c. $\dfrac{3x^2 + 15x}{xy - 3x + 5y - 15}$

46. Solve the following systems of equations:

   a. $3x + y = -1$
      $2x - 3y = 14$

   b. $y = 2x - 6$
      $3x + 5y = 9$

47. Divide: $(2y^3 - y^2 + 5y + 3) \div (2y + 1)$

7.2 Integer Exponents

In Section 7.1, five rules for exponents were developed that allow us to simplify expressions with *natural number* exponents. We now use these rules of exponents together with the properties of real numbers to give meaning to expressions with *integer* exponents such as a^0 and x^{-4}.

7.2 Integer Exponents

Consider the expression a^0. If the product rule is assumed to remain valid, we could write

$$a^0 \cdot a^n = a^{0+n} = a^n$$

Here the expression a^n is multiplied by some factor (a^0) resulting in the product a^n. This is only possible if that other factor is the identity element for multiplication, 1. Therefore, we define:

ZERO EXPONENT

$a^0 = 1, a \neq 0$

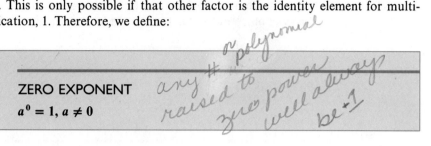

any # "polynomial" raised to zero power will always be 1

Note: The expression 0^0 is undefined.

Example 1

Evaluate or simplify each expression:

A. 2^0

$2^0 = 1$

B. x^0

$x^0 = 1$

C. $(3x)^0$

$(3x)^0 = 1$

Note that the 3 is part of the base for the exponent 0 (as indicated by the parentheses) so the entire quantity $3x$ is raised to the zero power.

D. $3x^0$

$3x^0 = 3(x^0) = 3(1) = 3$

Note that the zero exponent here applies only to the base x.

E. $(-2)^0$

$(-2)^0 = 1$

F. -2^0

$-2^0 = -(2^0) = -1$

G. $3x^2y^0$

$3x^2y^0 = 3x^2 \cdot 1 = 3x^2$

H. $(3x^2y)^0$

$(3x^2y)^0 = 1$

Negative Integer Exponents

To define an exponent such as -4, we assume the product rule is valid and write

$$x^{-4} \cdot x^4 = x^{-4+4} = x^0 = 1$$

Since two numbers whose product is 1 are *reciprocals* of each other, and the product of x^4 and x^{-4} is 1, x^{-4} must be the reciprocal of x^4. We define

$$x^{-4} = \frac{1}{x^4}$$

In general,

NEGATIVE EXPONENT

$$a^{-n} = \frac{1}{a^n}, \quad a \neq 0$$

Also,

$$\frac{1}{a^{-n}} = a^n$$

Note that n and $-n$ are negatives of each other. It is not necessary that we restrict n to be a positive integer. In fact, for *any* integer n, positive or negative, and a a real number other than zero,

$$\frac{1}{a^{-n}} = a^n \quad \text{and} \quad a^{-n} = \frac{1}{a^n}, \quad a \neq 0$$

Example 2

Write each expression with positive exponents:

A. a^{-3}

$$a^{-3} = \frac{1}{a^3}$$

B. m^{-1}

$$m^{-1} = \frac{1}{m}$$

C. 10^{-2}

$$10^{-2} = \frac{1}{10^2} = \frac{1}{100} = 0.01$$

7.2 Integer Exponents

D. $3x^{-4}$ *(handwritten: $\frac{3}{x^4}$)*

$$3x^{-4} = 3\left(\frac{1}{x^4}\right) = \frac{3}{x^4}$$

◀

Be careful in noting the difference between a negative base and a negative exponent!

Example 3 Evaluate each expression:

A. $(-2)^3 = (-2)(-2)(-2) = -8$

B. $(2)^{-3} = \dfrac{1}{2^3} = \dfrac{1}{2 \cdot 2 \cdot 2} = \dfrac{1}{8}$

C. $(-2)^{-3} = \dfrac{1}{(-2)^3} = \dfrac{1}{(-2)(-2)(-2)} = \dfrac{1}{-8} = -\dfrac{1}{8}$

D. $(-x)^2 = (-x)(-x) = x^2$

E. $x^{-2} = \dfrac{1}{x^2}$

◀

With integer exponents now defined, the quotient rule of exponents may be simplified. We remove the restrictions on the relative size of m and n and therefore eliminate the necessity of having three different cases.

QUOTIENT RULE

$$\frac{b^m}{b^n} = b^{m-n} = \frac{1}{b^{n-m}}, \; b \neq 0$$

(handwritten: $\dfrac{x^2}{x^6} = x^{2-6} = x^{-4} = \dfrac{1}{x^4}$)

Example 4 Write each expression with positive exponents only:

A. $\dfrac{x^{-2}}{x^3}$ *(handwritten: $-2-3 = x^{-5} = \dfrac{1}{x^5}$)*

$$\dfrac{x^{-2}}{x^3} = x^{-2-3} = x^{-5} = \dfrac{1}{x^5}$$

B. $\dfrac{y^4}{y^{-3}}$ *(handwritten: $4-(-3)$ $4+3$)*

$$\dfrac{y^4}{y^{-3}} = y^{4-(-3)} = y^{4+3} = y^7$$

SEVEN   Exponents and Radicals

C. $\dfrac{z^{-5}\ -(-7)}{z^{-7}}$

$$\dfrac{z^{-5}}{z^{-7}} = z^{-5-(-7)} = z^{-5+7} = z^2$$

D. $(-2x)(6x^{-4})$

$$(-2x)(6x^{-4}) = -12x^{1-4}$$
$$= -12x^{-3}$$
$$= \dfrac{-12}{x^3}$$

Note, in Example **4D**, the difference between a negative *coefficient*, (-12), and a negative *exponent*, (x^{-3}).

E. $\dfrac{5x^{-3}}{2y^{-5}}$ $\dfrac{5y^5}{2x^3}$

$$\dfrac{5x^{-3}}{2y^{-5}} = \dfrac{5 \cdot \left(\dfrac{1}{x^3}\right)}{2 \cdot \left(\dfrac{1}{y^5}\right)} \qquad \text{Negative exponent}$$

$$= \dfrac{\dfrac{5}{x^3}}{\dfrac{2}{y^5}} \qquad \text{Multiply}$$

$$= \dfrac{5}{x^3} \cdot \dfrac{y^5}{2} \qquad \text{Invert and multiply}$$

$$= \dfrac{5y^5}{2x^3} \qquad \text{Simplify} \qquad \blacktriangleleft$$

Expressions such as the one in Example **4E** occur often enough that a shortcut would be helpful to avoid the long sequence of steps shown previously. Consider the expression

$$\dfrac{ab^{-m}}{cd^{-n}}$$

which we will simplify:

$$\dfrac{ab^{-m}}{cd^{-n}} = \dfrac{a\left(\dfrac{1}{b^m}\right)}{c\left(\dfrac{1}{d^n}\right)}$$

7.2 Integer Exponents

$$\frac{\dfrac{a}{b^m}}{\dfrac{c}{d^n}} = \frac{a}{b^m} \cdot \frac{d^n}{c}$$

$$= \frac{ad^n}{cb^m}$$

Observe that b^{-m} of the numerator became b^m in the denominator and that d^{-n} of the denominator became d^n in the numerator. We summarize:

A *factor* may be moved from the numerator to the denominator or from the denominator to the numerator of an expression by *changing the sign of the exponent.*

Repeating Example **4E** using this observation:

$$\frac{5x^{-3}}{2y^{-5}} = \frac{5y^5}{2x^3}$$

Example 5 Write each of the following expressions with positive exponents only:

A. $\dfrac{3x^{-2}}{y^{-4}}$

$$\frac{3x^{-2}}{y^{-4}} = \frac{3y^4}{x^2}$$

B. $\dfrac{2x^{-3}}{7x^{-5}}$

$$\frac{2x^{-3}}{7x^{-5}} = \frac{2x^5}{7x^3}$$

$$= \frac{2x^2}{7}$$

C. $\dfrac{5x^{-3}y^2}{-6xy^{-1}}$

$$\frac{5x^{-3}y^2}{-6xy^{-1}} = \frac{5y^2 y}{-6xx^3}$$

$$= -\frac{5y^3}{6x^4}$$

SEVEN Exponents and Radicals

In Example 5, all exponential expressions were *factors* of the numerator or denominator. When an expression contains a negative power on a term in a sum or difference, we must use the negative exponent definition and then simplify the resulting expression.

Example 6

Simplify each expression:

A. $2^{-1} + 3^{-1}$

$$2^{-1} + 3^{-1} = \frac{1}{2^1} + \frac{1}{3^1} \qquad \text{Negative exponent}$$

Because 2^{-1} and 3^{-1} are terms (not factors), they must be dealt with individually.

$$= \frac{1}{2} + \frac{1}{3}$$

$$= \frac{1 \cdot 3}{2 \cdot 3} + \frac{1 \cdot 2}{3 \cdot 2} \qquad \text{LCD} = 6$$

$$= \frac{3 + 2}{6}$$

$$= \frac{5}{6}$$

B. $\left(\dfrac{x^{-1} + y^{-1}}{x + y} \right)$

$$\frac{x^{-1} + y^{-1}}{x + y} = \frac{\frac{1}{x} + \frac{1}{y}}{x + y} \qquad \text{Negative exponent}$$

$$= \frac{\left(\frac{1}{x} + \frac{1}{y}\right)xy}{(x + y)xy} \qquad \text{Complex fraction}$$

$$= \frac{(y + x)}{(x + y)xy} \qquad \text{Cancel}$$

$$= \frac{1}{xy}$$

One last situation involving negative exponents occurs when the base is a fraction. Consider $\left(\dfrac{a}{b}\right)^{-2}$ for example.

$$\left(\frac{a}{b}\right)^{-2} = \frac{a^{-2}}{b^{-2}} \qquad \text{Power of a quotient rule}$$

7.2 Integer Exponents

$$= \frac{b^2}{a^2}$$ By techniques of Example 5

$$= \left(\frac{b}{a}\right)^2$$ Power of a quotient rule

That is, any fraction raised to a negative exponent is equal to the reciprocal of the fraction raised to the corresponding positive power.

$$\left(\frac{a}{b}\right)^{-m} = \left(\frac{b}{a}\right)^m$$

Example 7

Simplify each expression:

A. $\left(\dfrac{3}{4}\right)^{-3}$

$$\left(\frac{3}{4}\right)^{-3} = \left(\frac{4}{3}\right)^3 = \frac{4^3}{3^3} = \frac{64}{27}$$

B. $\left(\dfrac{2x^2}{y^3}\right)^{-2}$

$$\left(\frac{2x^2}{y^3}\right)^{-2} = \left(\frac{y^3}{2x^2}\right)^2$$

$$= \frac{(y^3)^2}{2^2(x^2)^2} = \frac{y^6}{4x^4}$$

C. $\left(\dfrac{3a^{-3}}{2b^{-2}}\right)^{-4}$

$$\left(\frac{3a^{-3}}{2b^{-2}}\right)^{-4} = \left(\frac{2b^{-2}}{3a^{-3}}\right)^4$$

$$= \left(\frac{2a^3}{3b^2}\right)^4 \qquad \text{Technique of Example 5}$$

$$= \frac{2^4(a^3)^4}{3^4(b^2)^4}$$

$$= \frac{16a^{12}}{81b^8}$$

The five rules of exponents are still valid with integer exponents.

SUMMARY OF THE RULES OF EXPONENTS

1. **Product Rule:** $b^m b^n = b^{m+n}$ *add*
2. **Power of a Product Rule:** $(ab)^m = a^m b^m$ *distributive*
3. **Power of a Quotient Rule:** $\left(\dfrac{a}{b}\right)^m = \dfrac{a^m}{b^m}, b \neq 0$ *distributive*
4. **Power to a Power Rule:** $(b^m)^n = b^{mn}$ *multiply*
5. **Quotient Rule:** $\dfrac{b^m}{b^n} = b^{m-n}, b \neq 0$
6. $a^0 = 1, a \neq 0$
7. $a^{-n} = \dfrac{1}{a^n}, a \neq 0$ *negative exponent (reciprocal principle)*
8. **"Shortcuts":** $\dfrac{ab^{-m}}{cd^{-n}} = \dfrac{ad^n}{cb^m}, \left(\dfrac{a}{b}\right)^{-m} = \left(\dfrac{b}{a}\right)^m$

With integer exponents defined, note that the quotient rule may be stated with no restrictions on m and n (as was necessary in Section 7.1).

Example 8

Use the rules of exponents to simplify each of the following expressions. Write the final answer with positive exponents only.

A. $(3x^{-2})(-5x^{-7})$

$$(3x^{-2})(-5x^{-7}) = (3)(-5)(x^{-2}x^{-7}) \quad \text{Commutative and associative properties}$$

$$= -15x^{-9} \quad \text{Product rule}$$

$$= -\dfrac{15}{x^9} \quad \text{Negative exponent}$$

Note the -15 is a *coefficient*, not an exponent. It *does not* move to the denominator.

B. $(2x^{-2}y)^{-2}$

$$(2x^{-2}y)^{-2} = 2^{-2}(x^{-2})^{-2}y^{-2} \quad \text{Power of a product}$$

$$= 2^{-2}x^4 y^{-2} \quad \text{Power to a power}$$

$$= \dfrac{x^4}{2^2 y^2} \quad \text{Negative exponent}$$

$$= \dfrac{x^4}{4y^2}$$

7.2 Integer Exponents

C. $\dfrac{(6a^{-2})^{-1}}{(2ab^2)^{-2}}$

$\dfrac{(6a^{-2}b)^{-1}}{(2ab^2)^{-2}} = \dfrac{6^{-1}a^2b^{-1}}{2^{-2}a^{-2}b^{-4}}$ Power of a product and power to a power

$= \dfrac{2^2 a^2 a^2 b^4}{6^1 b^1}$ Negative exponents

$= \dfrac{4a^4 b^4}{6b}$ Product

$= \dfrac{2a^4 b^3}{3}$ Simplify

D. $\left(\dfrac{4m^{-2}}{5n}\right)^{-3}$

$\left(\dfrac{4m^{-2}}{5n}\right)^{-3} = \dfrac{4^{-3} m^6}{5^{-3} n^{-3}}$ Power of a quotient

$= \dfrac{5^3 m^6 n^3}{4^3}$ Negative exponent

$= \dfrac{125 m^6 n^3}{64}$ ◀

Although exponential expressions are generally simplified by writing the variables with positive exponents only, occasionally we prefer to write them with no factors in the denominator. For example,

$$\dfrac{2}{x^5}$$

may be written as $2x^{-5}$.

Example 9

Write each of the following expressions with no variable in the denominator:

A. $\dfrac{2y}{x}$

$\dfrac{2y}{x} = 2yx^{-1}$

B. $\dfrac{3yx^{-3}}{xy^{-2}}$

$\dfrac{3yx^{-3}}{xy^{-2}} = 3yx^{-3}x^{-1}y^2$

$= 3x^{-4} y^3$ ◀

Exercise Set 7.2

Simplify each expression. Express the answer with positive exponents only and reduce each fraction to lowest terms.

1. $3x^0$
2. $5y^0$
3. $(3x)^0$
4. $(5y)^0$
5. -2^0
6. -8^0
7. $(-2)^0$
8. $(-8)^0$
9. $3x^0 y$
10. $5ab^0$
11. $(3x)^0 y$
12. $5(ab)^0$
13. $2x^2 y^0$
14. $6xy^0$
15. $(x+y)^0$
16. $(3x+2y)^0$
17. $x + y^0$
18. $3x + 2y^0$
19. x^{-3}
20. a^{-1}
21. $2x^{-2}$
22. $3a^{-1}$
23. $(2x)^{-2}$
24. $(3a)^{-1}$
25. $x^{-4} y$
26. $3xy^{-2}$
27. $5x^{-2} y$
28. $6x^{-2} y^{-1}$
29. $\dfrac{3x^{-2}}{x}$
30. $\dfrac{5x \cdot x^2}{x^{-2}}$
31. $\dfrac{3x^{-3}}{y^{-1}}$
32. $\dfrac{x^{-2}}{3y^{-2}}$
33. $\left(\dfrac{2x^{-1}}{y}\right)^2$
34. $\left(\dfrac{x}{2y}\right)^{-2}$
35. $\left(\dfrac{2x}{y}\right)^{-3}$
36. $(2xy^2)^{-2}$
37. $(3x^{-2} y)^{-1}$
38. $(5x^{-1} y)^{-2}$
39. $\left(\dfrac{2x^{-1} y}{3x}\right)^{-2}$
40. $(2x^{-2})^{-3}$
41. $\left(\dfrac{3x^2 y}{x^{-2}}\right)^{-1}$
42. $\left(\dfrac{2x^2 y}{z^{-2}}\right)^{2}$
43. $\left(\dfrac{3xy^{-2}}{x^{-2} y}\right)^{-3}$
44. $\left(\dfrac{5x^2 y}{x^{-2} y^2}\right)^{-2}$

Simplify each of the following expressions. Write answers with positive exponents only.

45. $3^{-1} + 3^{-2}$
46. $2^{-1} + 2^{-2}$
47. $a^{-1} + a^{-2}$
48. $c^{-2} + c^{-3}$
49. $\dfrac{x-y}{x^{-1} - y^{-1}}$
50. $(x^{-1} + y^{-1})^{-1}$
51. $(a^{-1} - b^{-1})^2$
52. $(a+b)^{-2}$

Write each of the following expressions with no variables in the denominator. If the denominator is 1, it need not be written.

53. $\dfrac{3x}{y}$
54. $\dfrac{2xy}{z^2}$
55. $\dfrac{3x^2 y^{-2}}{xy}$
56. $\dfrac{3xy^{-2}}{x^{-2} y}$
57. $\dfrac{5x^2 y}{3x^4}$
58. $\dfrac{5x^2 y}{2x^{-5}}$
59. $(2x^2 y)^{-2}$
60. $(3x^{-2} y)^{-2}$

61. Why is the definition for exponential form on the first page of this chapter *not* appropriate for any exponent that is not a natural number?

Review

62. Multiply and divide as indicated:

    a. $(1000) \cdot (10{,}000)$ b. $(4000)(500{,}000)$

    c. $(0.00006)(0.003)$ d. $(0.0002)(8{,}000{,}000)$

    e. $\dfrac{8{,}000{,}000}{20{,}000}$ f. $\dfrac{4000}{0.002}$

63. Graph the solution of the system:

    $2x - 3y < 12$

    $x > -1$

    $y > -1$

64. Divide as indicated:

    a. $\dfrac{12x^9}{-2x^3}$ b. $\dfrac{8x^5 - 6x^4 + 2x^3}{2x^3}$

    c. $(2a^3 + 6a^2 + 5a + 15) \div (a + 3)$

65. Simplify the complex fractions:

    a. $\dfrac{3 + \frac{1}{2}}{2 - \frac{3}{4}}$ b. $\dfrac{x - \frac{1}{x - 1}}{\frac{x^2 + 2}{x - 1} + 2}$

7.3 Scientific Notation   *Number between 1 & 10*

In some applications, we compute with very large and very small numbers. Normally, performing such computations can be quite tedious. However, such computations can be significantly simplified by writing each of the numbers in a form called **scientific notation**.

A number, N, is in *scientific notation* when it is written as a number between 1 and 10 multiplied by an integer power of 10. In symbols,

$$N = s \cdot 10^p$$

where $1 \leq s < 10$ and $p \in I$.

SEVEN   Exponents and Radicals

First, we express some integer powers of 10:

$$10^4 = 10 \cdot 10 \cdot 10 \cdot 10 = 10{,}000$$
$$10^3 = 10 \cdot 10 \cdot 10 = 1000$$
$$10^2 = 10 \cdot 10 = 100$$
$$10^1 = 10$$
$$10^0 = 1$$
$$10^{-1} = \frac{1}{10} \phantom{= \frac{1}{100}} = 0.1$$
$$10^{-2} = \frac{1}{10^2} = \frac{1}{100} = 0.01$$
$$10^{-3} = \frac{1}{10^3} = \frac{1}{1000} = 0.001$$

Now, using these powers of 10 and your knowledge of multiplication, we convert some numbers from scientific notation into standard form:

$$7.25 \times 10^3 = 7.25 \times 1000 = 7250$$
$$8.03 \times 10^1 = 8.03 \times 10 = 80.3$$
$$6.91 \times 10^0 = 6.91 \times 1 = 6.91$$
$$3.84 \times 10^{-2} = 3.84 \times 0.01 = 0.0384$$
$$1.48 \times 10^{-3} = 1.48 \times 0.001 = 0.00148$$

In the preceding examples notice that numbers written in scientific notation that have positive powers of 10 yield numbers in standard form that are greater than 10 (which we refer to informally as *large* numbers).

$$7.25 \times 10^3 = 7250 \text{ (greater than 10)}$$

(Positive power)

Similarly, numbers written in scientific notation that have negative powers of 10 yield numbers less than 1 (which we refer to informally as *small* numbers) when written in standard form.

$$3.84 \times 10^{-2} = 0.0384 \text{ (less than 1)}$$

(Negative power)

 When writing a number in scientific notation, the power of 10 indicates the number of positions that the decimal point is moved to place it to the right of the first non-zero digit of the number. For large numbers, the power of 10 is positive;

7.3 Scientific Notation

for small numbers, the power of 10 is negative. Using these observations, a number can be easily changed from standard form to scientific notation or from scientific notation to standard form, as is illustrated in the next two examples.

Example 1

Write each of the following numbers in scientific notation:

A. 62,800

$$62,800. = 6.2800 \times 10^4$$

4 positions $= 6.28 \times 10^4$ Trailing zeros are not needed

B. 0.000316

$$0.000316 = 3.16 \times 10^{-4}$$

4 positions

C. 78,000,000

$$78,000,000. = 7.8 \times 10^7$$

7 positions

D. 0.611

$$0.611 = 6.11 \times 10^{-1}$$

1 position

◀

Example 2

Write each of the following numbers in standard form:

A. 6.86×10^{-5}

To produce a small number, the decimal point must be moved to the left.

$$6.86 \times 10^{-5} = 0.0000686$$

5 positions

B. 4.03×10^9

To produce a large number, the decimal point must be moved to the right.

$$4.03 \times 10^9 = 4,030,000,000$$

9 positions

C. 2.84×10^2

$$2.84 \times 10^2 = 284$$

2 positions

D. 3.18×10^{-3}

$$3.18 \times 10^{-3} = 0.00318$$
$$\text{3 positions}$$

◀

Negative numbers can also be written in scientific notation.

$$-2.7 \times 10^{-3} = -0.0027$$
$$\text{3 positions}$$

$$-7840 = -7.84 \times 10^3$$
$$\text{3 positions}$$

Scientific Notation in Computation

One advantage of scientific notation is that it makes computation with large and small numbers relatively simple. With each number written in scientific notation, multiplication, division, and powers can be performed using the properties of real numbers and the rules of exponents. In the examples that follow and in the exercise set, we will illustrate the method of computation using simplified numbers written in scientific notation.

Example 3

Write each number in scientific notation, complete the computation, and then write the final result in standard form:

A. $\dfrac{800}{0.02}$

$$\frac{800}{0.02} = \frac{8 \times 10^2}{2 \times 10^{-2}}$$

$$= \frac{8}{2} \times 10^2 \times 10^2 \qquad \text{Negative exponent}$$

$$= \frac{8}{2} \times 10^4 \qquad \text{Associative property and product rule}$$

$$= 4 \times 10^4$$

$$= 40{,}000 \qquad \text{Standard form}$$

B. $(200)(0.000003)$

$$(200)(0.000003) = (2 \times 10^2)(3 \times 10^{-6})$$
$$= (2 \cdot 3)(10^{2-6})$$
$$= 6 \times 10^{-4} = 0.0006$$

7.3 Scientific Notation

C. $(8000)^3$

$$(8000)^3 = (8 \times 10^3)^3$$
$$= 8^3 \times (10^3)^3$$
$$= 8^3 \times 10^9$$
$$= 512 \times 10^9$$
$$= 512,000,000,000$$

Note: 512×10^9 is *not* in scientific notation since 512 is not between 1 and 10. Consider the following steps to express 512×10^9 in scientific notation:

$$512 \times 10^9 = (5.12 \times 10^2) \times 10^9$$
$$= 5.12 \cdot 10^2 \cdot 10^9$$
$$= 5.12 \cdot 10^{11}$$
$$= 512,000,000,000$$

Most calculations today that involve very large or very small numbers can be performed on hand-held calculators. Even with calculators, however, it is frequently more convenient and quicker to enter numbers in scientific notation. The calculator is quite capable of displaying the results of a calculation in scientific notation. In fact, the calculator must use scientific notation to display results containing more digits than the calculator can display. For this reason, we encourage you to study scientific notation carefully.

Exercise Set 7.3

Write each of the following numbers in scientific notation:

1. 325
2. 2810
3. 5,700,000
4. 67,200
5. 0.00925
6. 0.000388
7. $-21,000$
8. $-385,000$
9. -0.0666
10. -0.0412
11. 26,800,000
12. 419,000,000
13. -0.0000000606
14. -0.0000000421

Write each of the following numbers in standard form:

15. 2.71×10^3
16. 3.15×10^2
17. 9.26×10^0
18. 3.06×10^0
19. 4.11×10
20. 3.17×10
21. -4.61×10^8
22. -9.24×10^7
23. 6.42×10^{-3}
24. 5.82×10^{-4}
25. -4.42×10^{-3}
26. -7.08×10^{-5}

27. 1.38×10^8
28. 4.11×10^9
29. 3.16×10^{-8}
30. 8.29×10^{-7}

Complete each of the following computations by writing the numbers in scientific notation. Write the final result in standard form.

31. $(300)(2000)$
32. $(4000)(200)$
33. $\dfrac{90{,}000}{0.03}$
34. $\dfrac{80{,}000}{0.02}$
35. $\dfrac{0.06}{300}$
36. $\dfrac{0.004}{2000}$
37. $\dfrac{(8000)(-300)}{(0.004)}$
38. $\dfrac{(-0.09)(500)}{(3)(0.003)}$

Express the area of each of the following rectangles and leave the answer in scientific notation:

39.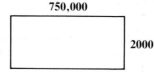

40.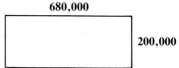

41.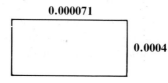

42.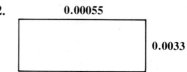

The area of a triangle is given by the formula $A = \tfrac{1}{2}bh$ where b is the *base* of the triangle and h is the *height* (the perpendicular distance from the base to the opposite vertex).

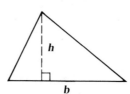

$$A = \frac{1}{2}bh$$

Find the area of each of the following triangles. Express areas in scientific notation:

43.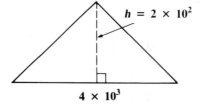

7.3 Scientific Notation

44.
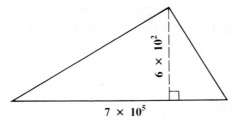

45.
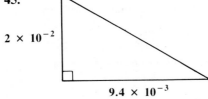

46.
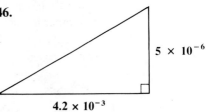

47.
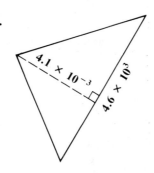

48.
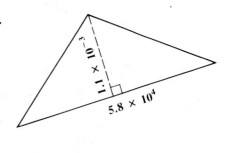

Review

49. Evaluate each of the following expressions:

   a. $(2)(2)(2)$
   b. $(-2)(-2)(-2)$
   c. $(3)(3)(3)(3)$
   d. $(-3)(-3)(-3)(-3)$
   e. $(5)(5)(5)$
   f. $(-5)(-5)(-5)$

50. Simplify:

   a. $p^3 \cdot p^4$ b. $\dfrac{q^7}{q^2}$ c. $(2m)^4$ d. $\left(\dfrac{3}{x}\right)^4$ e. $(s^4)^3$

51. Add or subtract as indicated:

   a. $\dfrac{6x}{x-2} - \dfrac{12}{x-2}$ b. $\dfrac{2}{m+1} + \dfrac{3}{m+2}$

   c. $\dfrac{3}{x-1} + \dfrac{x}{x+1}$

7.4 Roots and Rational Number Exponents

Roots

We have seen and worked with numbers written in exponential form, such as $5^3 = 5 \cdot 5 \cdot 5 = 125$. This statement answers the question "What number do we get when we cube the number 5?" In this section, however, we turn our attention to related questions such as "What number do we cube to get 125?" Such a number is called a **root**, and we define it as follows:

> **DEFINITION**
>
> The number a is an nth root of b if $a^n = b$.

A second root is called a *square* root, and a third root is called a *cube* root.

Example 1

A. 3 is a square root of 9 since $3^2 = 9$.
B. -3 is a square root of 9 since $(-3)^2 = 9$.
C. 2 is a fourth root of 16 since $2^4 = 16$.
D. -4 is a cube root of -64 since $(-4)^3 = -64$.
E. $\frac{2}{3}$ is a square root of $\frac{4}{9}$ since $(\frac{2}{3})^2 = \frac{4}{9}$.
F. $(-\frac{1}{2})$ is a cube root of $-\frac{1}{8}$ since $(-\frac{1}{2})^3 = -\frac{1}{8}$. ◀

Rational Number Exponents

We show that an nth root of b can be written as $b^{1/n}$. From the definition, if $b^{1/n}$ is an nth root of b, then $(b^{1/n})^n$ must equal b. That is,

$$(b^{1/n})^n = b^{(1/n)n} = b^1 = b$$

So,

> An nth root of b is written $b^{1/n}$

Example 2

List the positive roots indicated:

A. $25^{1/2}$

$$25^{1/2} = 5$$

B. $8^{1/3}$

$$8^{1/3} = 2$$

7.4 Roots and Rational Number Exponents

C. $256^{1/4}$

$$256^{1/4} = 4$$

D. $121^{1/2}$

$$121^{1/2} = 11$$ ◄

The number and nature (type) of the real nth root(s) of b are summarized in the following chart:

$b^{1/n}$	b Is Positive	b Is Negative
n is even	There are two real roots: one positive and one negative.	There is no real root.
n is odd	There is one positive root.	There is one negative root.

Here are some examples of the possibilities:

$b^{1/n}$ (power)	b Is Positive (base)	b Is Negative (base)
n is even ($n = 2$)	There are two real square roots of 9: 3 and -3.	There is no real square root of -25.
n is odd ($n = 3$)	There is one positive cube root of 64: 4.	There is one negative cube root of -27: -3.

In order for the symbol $b^{1/n}$ to have a unique meaning for each real number b and each natural number n, we define that when b is positive and n is even, only the positive root is taken and it is called the **principal root**. This principal root is the value of $b^{1/n}$.

Example 3

Find the principal root for each of the following:

A. $25^{1/2}$

$$25^{1/2} = 5$$

B. $36^{1/2}$

$$36^{1/2} = 6$$

C. $16^{1/4}$

$$16^{1/4} = 2$$ ◄

To indicate the negative (non-principal) root of an expression, we write a preceding negative sign.

Example 4

Find each of the following roots:

A. $-25^{1/2}$

$$-25^{1/2} = -5$$

B. $-36^{1/2}$

$$-36^{1/2} = -6$$

C. $-16^{1/4}$

$$-16^{1/4} = -2$$ ◀

It is very important to note the difference between $-25^{1/2}$ and $(-25)^{1/2}$. The first expression $-25^{1/2}$ means $-(25^{1/2}) = -(5) = -5$. The second expression $(-25)^{1/2}$ indicates an even root of a negative number, which has no meaning in the real numbers.

When n is odd, $b^{1/n}$ has only one root, and that root is taken to be the principal root.

Example 5

Find each root:

A. $(-1)^{1/3}$

$$(-1)^{1/3} = -1$$

B. $(1)^{1/3}$

$$(1)^{1/3} = 1$$

C. $(-27)^{1/3}$

$$(-27)^{1/3} = -3$$

D. $64^{1/3}$

$$64^{1/3} = 4$$

E. $(-32)^{1/5}$

$$(-32)^{1/5} = -2$$ ◀

Remember, the expression $b^{1/n}$ is undefined when b is negative and n is even. So, be especially careful with the use and placement of the "$-$" sign. Note that $(-4)^{1/2}$ is undefined, whereas $-4^{1/2}$ means $-(4^{1/2}) = -2$.

The preceding sections have given meaning to expressions written in the form $b^{1/n}$. However, to give meaning to rational number exponents, we define $b^{m/n}$

7.4 Roots and Rational Number Exponents

using our knowledge of the rules of exponents:

DEFINITION

$b^{m/n} = (b^{1/n})^m$

As before, if m/n is in lowest terms, the expression $b^{m/n}$ is undefined when b is negative and n is even.

Example 6

Use the definition to evaluate the following expressions:

A. $25^{3/2}$

$$25^{3/2} = (25^{1/2})^3 = 5^3 = 125$$

B. $16^{5/4}$

$$16^{5/4} = (16^{1/4})^5 = 2^5 = 32$$

C. $(-8)^{2/3} = [(-8)^{1/3}]^2 = (-2)^2 = 4$ ◀

Negative rational exponents are manipulated in the same manner as negative integer exponents.

Example 7

Evaluate or simplify the following expressions:

A. $25^{-3/2}$

$$25^{-3/2} = \frac{1}{25^{3/2}} = \frac{1}{(25^{1/2})^3} = \frac{1}{5^3} = \frac{1}{125}$$

B. $\dfrac{3x^{-1/2}}{y^{-2/3}}$

$$\frac{3x^{-1/2}}{y^{-2/3}} = \frac{3y^{2/3}}{x^{1/2}}$$ ◀

Since meaning was given for an expression of the form $b^{m/n}$ by the use of the properties of exponents, all five rules of exponents still apply.

Example 8

Use the five rules of exponents to simplify each expression. Assume variables represent positive numbers.

A. $x^{1/2} \cdot x^{1/3}$

$$x^{1/2} \cdot x^{1/3} = x^{1/2 + 1/3} = x^{3/6 + 2/6} = x^{5/6}$$

B. $\dfrac{4y}{y^{1/3}}$

$$\dfrac{4y}{y^{1/3}} = 4y^{1-1/3} = 4y^{3/3-1/3} = 4y^{2/3}$$

C. $\dfrac{x^{1/2}y^{-1/2}}{3x^{-1/2}y}$

$$\dfrac{x^{1/2}y^{-1/2}}{3x^{-1/2}y} = \dfrac{x^{1/2}x^{1/2}}{3yy^{1/2}} = \dfrac{x}{3y^{3/2}}$$

D. $(8x^6)^{1/3}$

$$(8x^6)^{1/3} = 8^{1/3}(x^6)^{1/3} = 8^{1/3}x^2 = 2x^2$$

E. $\left(\dfrac{25}{y^6}\right)^{1/2}$

$$\left(\dfrac{25}{y^6}\right)^{1/2} = \dfrac{25^{1/2}}{(y^6)^{1/2}} = \dfrac{5}{y^3}$$

◀

Exercise Set 7.4

Evaluate each expression, where possible:

1. $81^{1/2}$
2. $49^{1/2}$
3. $(-8)^{1/3}$
4. $(-1000)^{1/3}$
5. $64^{1/2}$
6. $100^{1/2}$
7. $27^{1/3}$
8. $64^{1/3}$
9. $(1/16)^{1/4}$
10. $(1/81)^{1/4}$
11. $-16^{1/2}$
12. $-100^{1/2}$
13. $(-16)^{1/2}$
14. $(-100)^{1/2}$
15. $4^{5/2}$
16. $100^{3/2}$
17. $(-8)^{5/3}$
18. $(-27)^{2/3}$
19. $(16)^{3/4}$
20. $(81)^{3/4}$
21. $(25)^{3/2}$
22. $(36)^{3/2}$
23. $-4^{3/2}$
24. $-25^{3/2}$
25. $(-4)^{3/2}$
26. $(-25)^{3/2}$
27. $4^{-1/2}$
28. $100^{-1/2}$
29. $36^{-3/2}$
30. $8^{-5/3}$

Use the rules of exponents to simplify each of the following expressions. Write results with positive exponents only. Assume all variables represent positive real numbers:

31. $(2a^{1/2})^2$
32. $(3m^{1/3})^3$
33. $n^{1/2}n^{1/2}$
34. $y^{1/2}y^{3/2}$
35. $x^2 \cdot x^{1/2}$
36. $a \cdot a^{1/3}$
37. $\dfrac{y^{3/4}}{y^{1/4}}$
38. $\dfrac{x^{3/2}}{x^{1/2}}$
39. $\dfrac{a}{2a^{1/3}}$

40. $\dfrac{4m^2}{m^{1/2}}$ 41. $(x^{2/3})^6$ 42. $(2x^{1/3})^6$

43. $(3a^{3/2})^2$ 44. $\dfrac{a}{a^{3/2}}$ 45. $(x^{1/2}y^{3/4})^4$

46. $\dfrac{u}{u^{3/2}}$ 47. $\dfrac{x^{-1/2}}{x^{1/2}}$ 48. $\dfrac{x^{-2/3}}{x^{1/3}}$

49. $\dfrac{a^{1/2}b^{-1/2}}{a^{1/2}b}$ 50. $\dfrac{3y^{-1/3}}{5y^{2/3}}$ 51. $(3x^{-1/2}y^{1/2})^{-2}$

52. $(2xy^{-1/2})^{-2}$ 53. $(-5xy^{1/4})^{-2}$ 54. $(-2x^{1/2})^{-3}$

55. $\dfrac{4a^{-1/2}}{6a^{-3/2}}$ 56. $\dfrac{8x^{-3/2}}{10x^{-1/2}}$

57. Explain why $(-36)^{1/2}$ and $-36^{1/2}$ do not represent the same quantity.
58. Explain how to evaluate $64^{2/3}$.
59. Why is $b^{1/4}$ undefined when b is negative?
60. What is the difference between "finding the cube of 8" and "finding the cube root of 8"?

Review

61. Add $y^2 + 5y$, $2y^2 - 3y + 6$, and $y^2 + y - 8$.
62. Subtract the sum of $b^2 + 3b - 8$ and $2b + 5$ from the sum of $b^2 + 7$ and $4b - 2$.
63. Write each of the following products directly:
    a. $(3r + 2)(2r + 7)$ b. $(3a + 5)(3a - 5)$
    c. $(u + 7)^2$ d. $(3v - 1)^2$
64. Divide: $(x^2 + 10x + 15) \div (x + 6)$
65. Multiply and divide as indicated:
    a. $\dfrac{2x}{x+3} \cdot \dfrac{x^2 + 3x}{-6}$ b. $\dfrac{4x^2}{x-2} \div \dfrac{-6x}{2-x}$

7.5 Radical Form and Properties of Radicals

An important alternate form in which the nth root of b, or $b^{1/n}$, may be written is **radical form** and looks like $\sqrt[n]{b}$. In this form, b is called the **radicand**, n is the **index** (generally not written when $n = 2$), and "$\sqrt{}$" is the **radical sign**. The same restrictions apply as in the previous section:

a. $\sqrt[n]{b}$ is undefined when n is even and b is negative.

b. Since there are two even roots of a positive number, $\sqrt[n]{b}$ represents the positive, or *principal*, root; $-\sqrt[n]{b}$ represents the negative root.

In the following example, each root is changed from exponential form to radical form and evaluated, if defined.

Example 1

A. $16^{1/2}$

$$16^{1/2} = \sqrt{16} = 4 \qquad \text{Note the index of 2 is not written}$$

B. $64^{1/3}$

$$64^{1/3} = \sqrt[3]{64} = 4$$

C. $625^{1/4}$

$$625^{1/4} = \sqrt[4]{625} = 5$$

D. $-625^{1/4}$

$$-625^{1/4} = -\sqrt[4]{625} = -5$$

E. $(-625)^{1/4}$

$$(-625)^{1/4} = \sqrt[4]{-625} \qquad \text{Not a real number}$$

F. $-1000^{1/3}$

$$-1000^{1/3} = -\sqrt[3]{1000}$$
$$= -10$$

G. $(-1000)^{1/3}$

$$(-1000)^{1/3} = \sqrt[3]{-1000}$$
$$= -10$$

H. $36^{1/2}$

$$36^{1/2} = \sqrt{36} = 6$$

I. $-36^{1/2} = -\sqrt{36} = -6$

J. $(-36)^{1/2} = \sqrt{-36} \qquad \text{Not a real number}$ ◀

The more general exponential form $b^{m/n}$ is equivalent to the following radical form:

$$b^{m/n} = b^{(1/n)m} = (b^{1/n})^m = \sqrt[n]{b}^{\,m}$$

7.5 Radical Form and Properties of Radicals

In the radical form, observe that n (the denominator) indicates the *root*, and m (the numerator) indicates the power:

$$b^{m/n} = \sqrt[n]{b^m} \quad \text{Power/Root}$$

As always, $b^{m/n}$ is not a real number when b is negative and n is even.

Example 2 Write each expression in radical form and evaluate:

A. $8^{5/3}$

$$8^{5/3} = \sqrt[3]{8}^{\,5} = (2)^5 = 32$$

B. $(-8)^{5/3}$

$$(-8)^{5/3} = \sqrt[3]{-8}^{\,5} = (-2)^5 = -32$$

C. $4^{3/2}$

$$4^{3/2} = \sqrt{4^3} = 2^3 = 8$$

D. $-4^{3/2}$

$$-4^{3/2} = -\sqrt{4^3} = -2^3 = -8$$

E. $(-4)^{3/2}$

$$(-4)^{3/2} = \sqrt{-4^3} \quad \text{Not a real number}$$

F. $100^{5/2}$

$$100^{5/2} = \sqrt{100^5} = 10^5 = 100{,}000$$

◀

An algebraic expression containing a radical is called a **radical expression**. Each of the following expressions is a radical expression.

$$\sqrt{x};\ 5\sqrt{x};\ \frac{16}{\sqrt{3}};\ \sqrt{x^2+25};\ \frac{5}{\sqrt{x+3}};\ 2x^2+3x-5\sqrt{x};\ \sqrt{x^2-5x+2}$$

Properties of Radicals

Since radical form is closely related to exponential form, we state and then prove the properties of radicals as follows:

1. Convert the radical expression to exponential form.
2. Change the expression using the properties of exponents.
3. Rewrite the resulting expression in radical form.

For simplicity, we will restrict our attention to only those expressions that have positive factors in the radicand.

PROPERTIES OF RADICALS

Let a and b be positive real numbers; k, n, and m natural numbers, $n \geq 2$.

1. $\sqrt[n]{ab} = \sqrt[n]{a}\sqrt[n]{b}$
2. $\sqrt[n]{\dfrac{a}{b}} = \dfrac{\sqrt[n]{a}}{\sqrt[n]{b}}$
3. $\sqrt[n]{a^n} = a$
4. $\sqrt[kn]{a^{km}} = \sqrt[n]{a^m}$

Proof of 1: $\sqrt[n]{ab} = (ab)^{1/n} = a^{1/n}b^{1/n} = \sqrt[n]{a}\sqrt[n]{b}$

Proof of 2: $\sqrt[n]{\dfrac{a}{b}} = \left(\dfrac{a}{b}\right)^{1/n} = \dfrac{a^{1/n}}{b^{1/n}} = \dfrac{\sqrt[n]{a}}{\sqrt[n]{b}}, \quad b \neq 0$

Proof of 3: $\sqrt[n]{a^n} = a^{n/n} = a^1 = a$

Proof of 4: $\sqrt[kn]{a^{km}} = a^{km/kn} = a^{m/n} = \sqrt[n]{a^m}$

Example 3

Use the properties of radicals to rewrite each expression. Assume the variables represent positive real numbers:

A. $\sqrt{9m}$

$\qquad \sqrt{9m} = \sqrt{9}\sqrt{m} = 3\sqrt{m}$ Property 1

B. $\sqrt[6]{a^4}$

$\qquad \sqrt[6]{a^4} = \sqrt[3 \cdot 2]{a^{2 \cdot 2}} = \sqrt[3]{a^2}$ Property 4

C. $\sqrt[3]{\dfrac{x}{7}}$

$\qquad \sqrt[3]{\dfrac{x}{7}} = \dfrac{\sqrt[3]{x}}{\sqrt[3]{7}}$ Property 2

D. $\sqrt[10]{(r+s)^{10}}$

$\qquad \sqrt[10]{(r+s)^{10}} = (r+s)$ Property 3

$\qquad \qquad \qquad \quad = r + s$

◀

Exercise Set 7.5

Evaluate each expression, where possible:

1. $\sqrt{49}$
2. $\sqrt{81}$
3. $-\sqrt{49}$
4. $-\sqrt{81}$
5. $\sqrt{-49}$
6. $\sqrt{-81}$
7. $\sqrt[3]{8}$
8. $\sqrt[3]{125}$
9. $-\sqrt[3]{8}$
10. $-\sqrt[3]{125}$
11. $\sqrt[3]{-8}$
12. $\sqrt[3]{-125}$

Write each expression in exponential form. Assume all variables represent positive numbers.

13. $\sqrt{y}$
14. $\sqrt{m}$
15. $\sqrt[3]{y}$
16. $\sqrt[3]{m}$
17. $\sqrt{xy}$
18. $\sqrt{6m}$
19. $\sqrt[3]{3u}$
20. $\sqrt[3]{6y}$
21. $\sqrt{a^3}$
22. $\sqrt{11^3}$
23. $\sqrt[3]{a^2}$
24. $\sqrt[3]{x^4}$
25. $\sqrt{x^2 + y^2}$
26. $\sqrt{c^2 - a^2}$

Write each expression in radical form. Assume all variables represent positive numbers.

27. $x^{5/4}$
28. $a^{3/4}$
29. $c^{5/2}$
30. $m^{5/2}$
31. $2x^{1/2}$
32. $5a^{1/3}$
33. $(2x)^{1/2}$
34. $(5a)^{1/3}$
35. $6a^{1/2}b^{1/3}$
36. $4x^{1/3}y^{2/3}$

Use the properties of radicals to rewrite each expression. Assume variables represent positive numbers.

37. $\sqrt{3x}$
38. $\sqrt{5u}$
39. $\sqrt[3]{9y}$
40. $\sqrt[3]{2x}$
41. $\sqrt{\dfrac{5}{m}}$
42. $\sqrt{\dfrac{u}{7}}$
43. $\sqrt[3]{\dfrac{x}{t}}$
44. $\sqrt[3]{\dfrac{t}{11}}$
45. $\sqrt[6]{x^4}$
46. $\sqrt[4]{y^6}$
47. $\sqrt[8]{m^6}$
48. $\sqrt[6]{5^8}$
49. $\sqrt[3]{11^3}$
50. $\sqrt[3]{6^3}$
51. $\sqrt{2x^2}$
52. $\sqrt{3m^2}$

Review

53. Name the property of real numbers illustrated:

    a. $a(b + c) = ab + ac$
    b. $(-7)(2) = 2(-7)$
    c. $(x + 3) + 8 = x + (3 + 8)$

54. Given the numbers $-\frac{1}{2}, 6, \sqrt{3}, 0, 1.7, -3, -8.33\overline{3}$, which are:

    a. whole numbers?
    b. rational numbers?
    c. real numbers?
    d. integers?

55. Solve each of the following equations:

   a. $2.5A + 3.9 = 247.9 - 3.6A$ b. $\dfrac{x+1}{2} = 2x - 10$

56. Solve and graph the inequality:

   $2u - 3 \geq 4 - 5(u + 2)$

7.6 Simplest Radical Form

The properties of radicals are used to write any radical expression in a form called *simplest radical form*. There are several advantages to this form, including standardization and the ease with which numerical values of radical expressions are computed. Although there may be situations in which simplest radical form as defined here is not the most convenient form, each particular situation will dictate the most convenient form. We continue to assume that all variables represent positive numbers.

A radical expression is said to be in *simplest radical form* when the following four conditions are satisfied:

1. The radicand contains no factor whose power is greater than or equal to the index.
2. There are no radicals in the denominator.
3. There are no fractions in the radicand.
4. The index and the power of the radicand contain no common factor.

1. The radicand contains no *factor* whose power is greater than or equal to the index. When this condition is violated, we simplify by using Properties 1 and 3 of radicals, as illustrated in Example 1.

Write each radical expression in simplest radical form:

Example 1

A. $\sqrt{8x^3}$

$\sqrt{8x^3} = \sqrt{4 \cdot 2 \cdot x^2 \cdot x}$

$\phantom{\sqrt{8x^3}} = \sqrt{(4x^2) \cdot (2x)}$ Group perfect squares and non-perfect squares

$\phantom{\sqrt{8x^3}} = \sqrt{4x^2}\sqrt{2x}$ Property 1

$\phantom{\sqrt{8x^3}} = 2x\sqrt{2x}$ Simplest radical form

In the preceding example, the key is to write the radicand as a product in

7.6 Simplest Radical Form

which one factor contains only factors whose exponents are evenly divisible by the index (in Example 1A, perfect *squares*).

B. $\sqrt[3]{16}$

$$\sqrt[3]{16} = \sqrt[3]{8 \cdot 2} \quad \text{Group perfect cubes and non-perfect cubes}$$
$$= \sqrt[3]{8}\sqrt[3]{2} \quad \text{Property 1}$$
$$= 2\sqrt[3]{2}$$

C. $\sqrt{12x^3y^7}$

$$\sqrt{12x^3y^7} = \sqrt{(4x^2y^6)(3xy)} \quad \text{Powers divisible by the index, 2}$$
$$= \sqrt{4x^2y^6}\sqrt{3xy}$$
$$= 2xy^3\sqrt{3xy}$$

D. $\sqrt[3]{32a^5b^7}$

Here, it may be helpful to write 32 in prime factored form.

$$\sqrt[3]{32a^5b^7} = \sqrt[3]{2^5a^5b^7}$$
$$= \sqrt[3]{(2^3a^3b^6)(2^2a^2b)}$$

          ↑ Exponents divisible by the index, 3

$$= \sqrt[3]{2^3a^3b^6}\sqrt[3]{2^2a^2b}$$
$$= 2ab^2\sqrt[3]{4a^2b} \quad \blacktriangleleft$$

2. There are no radicals in the denominator. If the denominator contains a radical, we remove that radical by using the Fundamental Principle of Fractions. Multiply the numerator and the denominator by a factor that will make the denominator be of the form $\sqrt[n]{a^n} = a$. This process is called **rationalizing the denominator**.

Example 2

Write each radical expression in simplest radical form:

A. $\dfrac{3}{\sqrt{2}}$

We multiply the numerator and denominator by $\sqrt{2}$.

$$\frac{3}{\sqrt{2}} = \frac{3}{\sqrt{2}} \cdot \frac{\sqrt{2}}{\sqrt{2}}$$
$$= \frac{3\sqrt{2}}{\sqrt{2^2}}$$
$$= \frac{3\sqrt{2}}{2}$$

B. $\dfrac{5x}{\sqrt{x}}$

Multiply numerator and denominator by $\sqrt{x}$.

$$\dfrac{5x}{\sqrt{x}} = \dfrac{5x\sqrt{x}}{\sqrt{x}\sqrt{x}}$$

$$= \dfrac{5x\sqrt{x}}{\sqrt{x^2}}$$

$$= \dfrac{5\cancel{x}\sqrt{x}}{\cancel{x}} \qquad \text{Cancel}$$

$$= 5\sqrt{x}$$

C. $\dfrac{3}{\sqrt[3]{x}}$

Here, to make the denominator equal to $\sqrt[3]{x^3}$, we multiply numerator and denominator by $\sqrt[3]{x^2}$.

$$\dfrac{3}{\sqrt[3]{x}} = \dfrac{3\sqrt[3]{x^2}}{\sqrt[3]{x}\sqrt[3]{x^2}} = \dfrac{3\sqrt[3]{x^2}}{\sqrt[3]{x^3}}$$

$$= \dfrac{3\sqrt[3]{x^2}}{x}$$

◀

3. **There are no fractions in the radicand.** If the radicand contains fractions, we may multiply the numerator and denominator of the radicand by a factor that makes the power of the denominator evenly divisible by the index.

Example 3

Write each expression in simplest radical form:

A. $\sqrt{\dfrac{2}{3}}$

We want to make the denominator a perfect *square* (the index is 2), so we multiply the numerator and denominator by 3.

$$\sqrt{\dfrac{2}{3}} = \sqrt{\dfrac{2 \cdot 3}{3 \cdot 3}}$$

$$= \sqrt{\dfrac{6}{9}}$$

$$= \dfrac{\sqrt{6}}{\sqrt{9}} \qquad \text{Property 2}$$

$$= \dfrac{\sqrt{6}}{3}$$

7.6 Simplest Radical Form

B. $\sqrt{\dfrac{3}{5x}}$

$\sqrt{\dfrac{3}{5x}} = \sqrt{\dfrac{3 \cdot 5x}{5x \cdot 5x}}$ Multiply the numerator and denominator by $5x$ to make the denominator a perfect square

$= \sqrt{\dfrac{15x}{25x^2}}$

$= \dfrac{\sqrt{15x}}{\sqrt{25x^2}}$ Property 2

$= \dfrac{\sqrt{15x}}{5x}$

C. $\sqrt{\dfrac{5}{8}}$

If we write the denominator in prime factored form, $8 = 2^3$, we observe that we need only to multiply the numerator and denominator by 2 to make the denominator a perfect square.

$\sqrt{\dfrac{5}{8}} = \sqrt{\dfrac{5}{2^3}}$

$= \sqrt{\dfrac{5 \cdot 2}{2^3 \cdot 2}}$

$= \dfrac{\sqrt{10}}{\sqrt{2^4}}$ 2^4 is a perfect square

$= \dfrac{\sqrt{10}}{\sqrt{2^4}} = \dfrac{\sqrt{10}}{2^2}$

$= \dfrac{\sqrt{10}}{4}$

D. $\sqrt[3]{\dfrac{5}{4x}}$

Since the index is 3, we need to make the denominator a perfect cube.

$\sqrt[3]{\dfrac{5}{4x}} = \sqrt[3]{\dfrac{5}{2^2 x}}$ Prime factored form, $4 = 2^2$

$= \sqrt[3]{\dfrac{5 \cdot (2x^2)}{2^2 x(2x^2)}}$ We have two factors of 2, and we need one more to make a perfect cube. We have one factor of x, and we need two more to make a perfect cube.

$$= \sqrt[3]{\frac{10x^2}{2^3 x^3}}$$

$$= \frac{\sqrt[3]{10x^2}}{\sqrt[3]{2^3 x^3}} \qquad \text{Property 2}$$

$$= \frac{\sqrt[3]{10x^2}}{2x} \qquad \blacktriangleleft$$

4. **The index and the power of the radicand contain no common factor.** If the index and power of the radicand contain a common factor, it is removed by using Property 4.

Example 4

Write each expression in simplest radical form:

A. $\sqrt[6]{x^4}$

$$\sqrt[6]{x^4} = \sqrt[3 \cdot 2]{x^{2 \cdot 2}} \qquad \text{Common factor of 2}$$
$$= \sqrt[3]{x^2}$$

B. $\sqrt[4]{36}$

$$\sqrt[4]{36} = \sqrt[4]{6^2} \qquad \text{Common factor of 2}$$
$$= \sqrt[2 \cdot 2]{6^2}$$
$$= \sqrt{6}$$

C. $\sqrt[9]{u^6}$

$$\sqrt[9]{u^6} = \sqrt[3 \cdot 3]{u^{2 \cdot 3}} \qquad \text{Common factor of 3} \qquad \blacktriangleleft$$
$$= \sqrt[3]{u^2}$$

In all future sections, any answer to problems involving radicals should be expressed in simplest radical form.

Exercise Set 7.6

Write each of the following radical expressions in simplest radical form. Assume all variables represent positive numbers.

1. $\sqrt{20}$
2. $\sqrt{18}$
3. $\sqrt{75}$
4. $\sqrt{48}$
5. $\sqrt{45}$
6. $\sqrt{28}$
7. $\sqrt{4x^2 y^4}$
8. $\sqrt{16x^8}$

7.6 Simplest Radical Form

9. $\sqrt{2xy^5}$
10. $\sqrt{9x^3y^4}$
11. $\sqrt{\dfrac{2}{9}}$
12. $\sqrt{\dfrac{3}{4}}$

13. $\sqrt{\dfrac{5}{2}}$
14. $\sqrt{\dfrac{4}{3}}$
15. $\sqrt{\dfrac{3m}{2n}}$
16. $\sqrt{\dfrac{5x}{2y}}$

17. $\dfrac{3a}{\sqrt{2ab}}$
18. $\sqrt{\dfrac{5x}{y^3}}$
19. $\sqrt{\dfrac{8x}{2}}$
20. $\dfrac{\sqrt{9m}}{\sqrt{m}}$

21. $\sqrt{\dfrac{25}{6x}}$
22. $\dfrac{3}{\sqrt{7}}$
23. $\dfrac{5}{\sqrt{7}}$
24. $\dfrac{\sqrt{3}}{\sqrt{2}}$

25. $\dfrac{5}{\sqrt{27}}$
26. $\dfrac{7}{\sqrt{98}}$
27. $\sqrt[3]{9xy^4}$
28. $\sqrt[3]{2a^5b^2}$

29. $\sqrt[3]{24x^5y^6}$
30. $\sqrt[3]{128a^8b^3}$
31. $\dfrac{2}{\sqrt[3]{x}}$
32. $\dfrac{5}{\sqrt[3]{y}}$

33. $\dfrac{3x}{\sqrt[3]{9x}}$
34. $\dfrac{2x^2}{\sqrt[3]{16xy^2}}$
35. $\sqrt[3]{\dfrac{5}{4}}$
36. $\sqrt[3]{\dfrac{5}{9}}$

37. $\sqrt[3]{\dfrac{5}{2x^2}}$
38. $\sqrt[3]{\dfrac{7}{2a}}$
39. $\sqrt[3]{\dfrac{3a}{4b^2}}$
40. $\sqrt[3]{\dfrac{4}{25xy^2}}$

41. $\sqrt[4]{x^2}$
42. $\sqrt[6]{16}$
43. $\sqrt[9]{x^6}$
44. $\sqrt[8]{y^6}$

45. Consider the number $\dfrac{1}{\sqrt{3}}$. Find the simplest radical form for this number and explain why this form really is simpler to evaluate if we must calculate by hand (using a square root table to find $\sqrt{3}$).

Review

46. Complete the indicated addition and subtraction:

    a. $2a - 3b + 5a - b$
    b. $6p - 4 + 3p + 2$
    c. $3y + 5x - 3y - x$
    d. $5c + 2c - 3d + 7d$
    e. $\dfrac{x}{2} + 3x$
    f. $\dfrac{3y}{4} - y$
    g. $\dfrac{2x}{5} + \dfrac{x}{3}$

47. Draw the graph of the equations in two variables:

    a. $4x - y = 8$
    b. $y = -\dfrac{2}{3}x + 1$
    c. $x = 3$

7.7 Addition and Subtraction of Radical Expressions

Since radical expressions are numbers or represent numbers, we can perform arithmetic operations on them. The operations of addition, subtraction, multiplication, and division on radical expressions require only the application of previously developed principles.

Addition and Subtraction

Recall that the distributive property is used to simplify expressions by combining like terms.

$$2x + 3y - 5x + 6y = (2x - 5x) + (3y + 6y) \quad \text{Commutative and associative properties}$$
$$= (2 - 5)x + (3 + 6)y \quad \text{Distributive property}$$
$$= -3x + 9y$$

The distributive property

$$ac + bc = (a + b)c$$

is the basis for the preceding simplification. However, a, b, and c may represent *any* real numbers, including irrational numbers or radical expressions. In a similar manner, we write

$$2\sqrt{7} + 3\sqrt{5} - 5\sqrt{7} + 6\sqrt{5} = -3\sqrt{7} + 9\sqrt{5}$$

Terms of a radical expression that have identical radical parts are called **like radicals**. Two or more such terms are combined by adding their coefficients.

Example 1

Simplify each of the following expressions by combining like radicals:

A. $5\sqrt{6} - 2\sqrt{7} + 3\sqrt{6}$

$$5\sqrt{6} - 2\sqrt{7} + 3\sqrt{6} = 8\sqrt{6} - 2\sqrt{7}$$

B. $3\sqrt[3]{x} + 5\sqrt{y} - 2\sqrt[3]{x}$

$$3\sqrt[3]{x} + 5\sqrt{y} - 2\sqrt[3]{x} = \sqrt[3]{x} + 5\sqrt{y}$$

C. $6\sqrt{y} - 4\sqrt[3]{y} + \sqrt{y} - \sqrt[3]{y}$

$$6\sqrt{y} - 4\sqrt[3]{y} + \sqrt{y} - \sqrt[3]{y} = 7\sqrt{y} - 5\sqrt[3]{y}$$

Like radicals must have both identical radicands *and* identical indexes. In some expressions, radicals may not at first appear to be like radicals. Be sure to write each term of a radical expression in simplest radical form to determine which terms contain like radicals.

7.7 Addition and Subtraction of Radical Expressions

Example 2 Simplify each expression:

A. $\sqrt{2} + \sqrt{8} - \sqrt{27}$

$$\sqrt{2} + \sqrt{8} - \sqrt{27} = \sqrt{2} + \sqrt{4 \cdot 2} - \sqrt{9 \cdot 3}$$
$$= \sqrt{2} + 2\sqrt{2} - 3\sqrt{3}$$
$$= 3\sqrt{2} - 3\sqrt{3}$$

B. $\sqrt{18} + 5\sqrt{8} + 3\sqrt{27} - \sqrt{80}$

$$\sqrt{18} + 5\sqrt{8} + 3\sqrt{27} - \sqrt{80} = \sqrt{9 \cdot 2} + 5\sqrt{4 \cdot 2} + 3\sqrt{9 \cdot 3} - \sqrt{16 \cdot 5}$$
$$= 3\sqrt{2} + 5 \cdot 2\sqrt{2} + 3 \cdot 3\sqrt{3} - 4\sqrt{5}$$
$$= 3\sqrt{2} + 10\sqrt{2} + 9\sqrt{3} - 4\sqrt{5}$$
$$= 13\sqrt{2} + 9\sqrt{3} - 4\sqrt{5}$$

C. $\sqrt{\dfrac{1}{2}} + \dfrac{5}{\sqrt{8}}$

$$\sqrt{\dfrac{1}{2}} + \dfrac{5}{\sqrt{8}} = \dfrac{1}{\sqrt{2}} + \dfrac{5}{\sqrt{4 \cdot 2}}$$
$$= \dfrac{1 \cdot \sqrt{2}}{\sqrt{2}\sqrt{2}} + \dfrac{5 \cdot \sqrt{2}}{2\sqrt{2} \cdot \sqrt{2}}$$
$$= \dfrac{\sqrt{2}}{2} + \dfrac{5\sqrt{2}}{4}$$
$$= \dfrac{\sqrt{2} \cdot 2}{2 \cdot 2} + \dfrac{5\sqrt{2}}{4} \quad \text{LCD}$$
$$= \dfrac{2\sqrt{2}}{4} + \dfrac{5\sqrt{2}}{4}$$
$$= \dfrac{7\sqrt{2}}{4} \quad \blacktriangleleft$$

Exercise Set 7.7

In each expression, add or subtract as indicated by combining like terms. Assume that all variables represent positive numbers.

1. $3\sqrt{2} + 4\sqrt{2} - \sqrt{2}$
2. $\sqrt{5} + 4\sqrt{5} - 2\sqrt{5}$
3. $2\sqrt{x} + 3\sqrt{x} - 4\sqrt{x}$
4. $\sqrt{y} - 8\sqrt{y} + 3\sqrt{y}$
5. $\sqrt{45} - \sqrt{20}$
6. $\sqrt{18} - \sqrt{8}$

7. $\sqrt{12} + 3\sqrt{27}$
8. $\sqrt{48} - \sqrt{12}$
9. $\sqrt{45} + 2\sqrt{8} - \sqrt{20}$
10. $3\sqrt{27} + \sqrt{8} - 2\sqrt{12}$
11. $3\sqrt{28} - 5\sqrt{7}$
12. $6\sqrt{20} - 3\sqrt{5}$
13. $\sqrt{72} - 8\sqrt{2} + \sqrt{32}$
14. $4\sqrt{24} + 5\sqrt{6} - \sqrt{54}$
15. $\sqrt{\dfrac{1}{2}} + 2\sqrt{8}$
16. $\sqrt{\dfrac{3}{4}} - \sqrt{27}$
17. $\sqrt{\dfrac{2}{3}} + 3\sqrt{24}$
18. $\sqrt{\dfrac{5}{8}} - 2\sqrt{32}$
19. $3\sqrt{x} + 5\sqrt{4x}$
20. $\sqrt{9y} - \sqrt{4y} + \sqrt{y}$
21. $\sqrt{54} + 2\sqrt{40} - \sqrt{24}$
22. $\sqrt{8} - \sqrt{20} + 4\sqrt{2}$
23. $\sqrt{\dfrac{1}{3}} + \sqrt{12}$
24. $\sqrt{\dfrac{3}{8}} + \sqrt{\dfrac{1}{2}}$
25. $\sqrt{\dfrac{x}{2}} + 8\sqrt{x} + \sqrt{8x}$
26. $\sqrt{\dfrac{3}{2}} - \sqrt{\dfrac{2}{3}}$
27. $\sqrt[3]{24} + 2\sqrt[3]{3}$
28. $\sqrt[3]{40} + 3\sqrt[3]{5}$
29. $3\sqrt[3]{24} - \sqrt[3]{81}$
30. $5\sqrt[3]{40} - 2\sqrt[3]{5}$
31. $5x\sqrt{6x^3} - x^2\sqrt{6x}$
32. $a\sqrt[3]{8y} + 2a\sqrt[3]{27y}$
33. $m\sqrt[3]{54t} + 3m\sqrt[3]{16t}$
34. $6\sqrt{x+y} - 8\sqrt{x+y}$
35. $\sqrt[3]{32xy^4} + 2y\sqrt[3]{4xy}$
36. $\sqrt[4]{16x^3} + 3\sqrt[4]{x^3}$

37. Can terms of an expression contain like radicals but still not be like terms? Explain.

38. Why must radicals be expressed in simplest radical form before we attempt to combine them?

Review

39. Multiply:
    a. $3m(m-4)$
    b. $(2y+3)(4y-1)$
    c. $(p+6)(p-6)$
    d. $(2x+3)(2x-3)$
    e. $(x+5)^2$
    f. $(2u-7)^2$

40. Divide as indicated:
    a. $\dfrac{6w^2 - 5w}{w}$
    b. $\dfrac{8c^2 - 12c + 4}{4}$
    c. $\dfrac{-9x^3 + 6x^2 - 12x}{-3x}$

41. Write the following numbers in scientific notation:
    a. $-37{,}000$
    b. 0.000000067

7.8 Multiplication and Division of Radical Expressions

Multiplication

To multiply radical expressions, we use the familiar properties of multiplication and the properties of radicals. Multiplication is illustrated in the following example. Check to determine if the product can be simplified (simplest radical form).

Example 1

Multiply as indicated:

A. $5(6 - 2\sqrt{3})$

$$5(6 - 2\sqrt{3}) = 5 \cdot 6 - 5 \cdot 2\sqrt{3} \quad \text{Distributive property}$$
$$= 30 - 10\sqrt{3}$$

B. $6\sqrt{2}(3\sqrt{6} - 4)$

$$6\sqrt{2}(3\sqrt{6} - 4) = 6\sqrt{2} \cdot 3\sqrt{6} - 6\sqrt{2} \cdot 4 \quad \text{Distributive property}$$
$$\quad \sqrt[n]{a}\,\sqrt[n]{b} = \sqrt[n]{ab}$$
$$= 18\sqrt{12} - 24\sqrt{2}$$
$$= 18\sqrt{4 \cdot 3} - 24\sqrt{2}$$
$$= 18 \cdot 2\sqrt{3} - 24\sqrt{2} \quad \text{Simplest radical form}$$
$$= 36\sqrt{3} - 24\sqrt{2}$$

C. $(\sqrt{5} + 8)(\sqrt{5} - 2)$

$$(\sqrt{5} + 8)(\sqrt{5} - 2) = \sqrt{5}^2 + 6\sqrt{5} - 16 \quad \text{FOIL}$$
$$= 5 + 6\sqrt{5} - 16 \quad \text{Simplify}$$
$$= -11 + 6\sqrt{5} \quad \text{Combine like terms}$$

D. $(2\sqrt{7} - \sqrt{3})(\sqrt{7} + 2\sqrt{3})$

$$(2\sqrt{7} - \sqrt{3})(\sqrt{7} + 2\sqrt{3}) = 2\sqrt{7}^2 + 3\sqrt{21} - 2\sqrt{3}^2 \quad \text{FOIL}$$
$$= 2 \cdot 7 + 3\sqrt{21} - 2 \cdot 3 \quad \text{Simplify}$$
$$= 14 + 3\sqrt{21} - 6$$
$$= 8 + 3\sqrt{21}$$

E. $(\sqrt{3} + 5)^2$

$$(\sqrt{3} + 5)^2 = \sqrt{3}^2 + 10\sqrt{3} + 5^2 \quad \text{Square of a binomial}$$
$$= 3 + 10\sqrt{3} + 25$$
$$= 28 + 10\sqrt{3}$$

F. $(4\sqrt{5} - \sqrt{2})^2$

$\quad (4\sqrt{5} - \sqrt{2})^2 = 4^2\sqrt{5}^2 - 8\sqrt{10} + \sqrt{2}^2$ Square of a binomial
$\quad\quad\quad\quad\quad\quad\quad = 16 \cdot 5 - 8\sqrt{10} + 2$
$\quad\quad\quad\quad\quad\quad\quad = 80 - 8\sqrt{10} + 2$
$\quad\quad\quad\quad\quad\quad\quad = 82 - 8\sqrt{10}$

G. $(\sqrt{8} + 3)(\sqrt{2} - 4)$

$\quad (\sqrt{8} + 3)(\sqrt{2} - 4) = (2\sqrt{2} + 3)(\sqrt{2} - 4)$ Simplest radical form
$\quad\quad\quad\quad\quad\quad\quad\quad = 2\sqrt{2}^2 - 5\sqrt{2} - 12$ FOIL
$\quad\quad\quad\quad\quad\quad\quad\quad = 4 - 5\sqrt{2} - 12$
$\quad\quad\quad\quad\quad\quad\quad\quad = -8 - 5\sqrt{2}$

H. $(\sqrt{4x} + 3)(3\sqrt{x} - 1)$

$\quad (\sqrt{4x} + 3)(3\sqrt{x} - 1) = (2\sqrt{x} + 3)(3\sqrt{x} - 1)$
$\quad\quad\quad\quad\quad\quad\quad\quad\quad = 6\sqrt{x^2} + 7\sqrt{x} - 3$ FOIL
$\quad\quad\quad\quad\quad\quad\quad\quad\quad = 6x + 7\sqrt{x} - 3$

I. $(\sqrt{11} + 2)(\sqrt{11} - 2)$

$\quad (\sqrt{11} + 2)(\sqrt{11} - 2) = \sqrt{11}^2 - 2^2$ $(a - b)(a + b) = a^2 - b^2$
$\quad\quad\quad\quad\quad\quad\quad\quad = 11 - 4$
$\quad\quad\quad\quad\quad\quad\quad\quad = 7$

J. $(2\sqrt{3} - \sqrt{2})(2\sqrt{3} + \sqrt{2})$

$\quad (2\sqrt{3} - \sqrt{2})(2\sqrt{3} + \sqrt{2}) = 4\sqrt{3}^2 - \sqrt{2}^2$ Sum times a difference
$\quad\quad\quad\quad\quad\quad\quad\quad\quad\quad = 4 \cdot 3 - 2$
$\quad\quad\quad\quad\quad\quad\quad\quad\quad\quad = 12 - 2$
$\quad\quad\quad\quad\quad\quad\quad\quad\quad\quad = 10$

K. $(\sqrt{3x} - \sqrt{5})(\sqrt{3x} + \sqrt{5})$

$\quad (\sqrt{3x} - \sqrt{5})(\sqrt{3x} + \sqrt{5}) = \sqrt{3x}^2 - \sqrt{5}^2$
$\quad\quad\quad\quad\quad\quad\quad\quad\quad\quad = 3x - 5$ ◀

Examples **1I**, **1J**, and **1K** illustrate that the product of a sum times a difference in which one or both terms contains a square root radical yields an expression without radicals.

7.8 Multiplication and Division of Radical Expressions

Division

To divide a radical expression containing one or more terms by a rational number (in our examples, an integer), divide each term by that rational number.

Example 2 Divide as indicated:

A. $\dfrac{8\sqrt{11}}{2}$

$$\dfrac{8\sqrt{11}}{2} = \dfrac{\overset{4}{\cancel{8}}\sqrt{11}}{\cancel{2}} = 4\sqrt{11} \qquad \text{Fundamental Principle of Fractions}$$

B. $\dfrac{8\sqrt{5} - 12}{2}$

$$\dfrac{8\sqrt{5} - 12}{2} = \dfrac{8\sqrt{5}}{2} - \dfrac{12}{2} \qquad \text{Divide each term by 2}$$

$$= 4\sqrt{5} - 6 \qquad \blacktriangleleft$$

To divide a radical expression of one or more terms by a radical expression of one term, we rationalize the denominator as in Section 7.6. This then makes the denominator a rational number, so we may complete the division as in Example 2.

Example 3 Divide as indicated:

A. $\dfrac{5 + 3\sqrt{2}}{\sqrt{2}}$

$$\dfrac{5 + 3\sqrt{2}}{\sqrt{2}} = \dfrac{(5 + 3\sqrt{2})\sqrt{2}}{\sqrt{2}\cdot\sqrt{2}} \qquad \text{Use Fundamental Principle of Fractions to rationalize the denominator}$$

$$= \dfrac{5\sqrt{2} + 3\cdot 2}{2} \qquad \text{Distributive property}$$

$$= \dfrac{5\sqrt{2} + 6}{2} \qquad \text{Simplify}$$

or

$$\dfrac{5\sqrt{2}}{2} + 3 \qquad \text{Divide both terms of the numerator by 2}$$

B. $\dfrac{2\sqrt{6} + 3\sqrt{5}}{\sqrt{2}}$

$$\dfrac{2\sqrt{6} + 3\sqrt{5}}{\sqrt{2}} = \dfrac{(2\sqrt{6} + 3\sqrt{5})\sqrt{2}}{\sqrt{2} \cdot \sqrt{2}}$$

$$= \dfrac{2\sqrt{12} + 3\sqrt{10}}{2}$$

$$= \dfrac{2\sqrt{4 \cdot 3} + 3\sqrt{10}}{2}$$

$$= \dfrac{4\sqrt{3} + 3\sqrt{10}}{2}$$

$$= 2\sqrt{3} + \dfrac{3\sqrt{10}}{2}$$

C. $\dfrac{10 - 3\sqrt{x}}{\sqrt{x}}$

$$\dfrac{10 - 3\sqrt{x}}{\sqrt{x}} = \dfrac{(10 - 3\sqrt{x})\sqrt{x}}{\sqrt{x}\sqrt{x}}$$

$$= \dfrac{10\sqrt{x} - 3x}{x}$$

or $\dfrac{10\sqrt{x}}{x} - 3$ ◀

To divide by a radical expression of two terms in which one or both terms contains a square root radical, we multiply the numerator and denominator by the conjugate of the denominator. (Recall that the expressions $a + b$ and $a - b$ are called conjugates, and that $(a + b)(a - b) = a^2 - b^2$.) This process is also referred to as *rationalizing the denominator* since it removes radicals from the denominator.

Example 4   Divide as indicated (rationalize each denominator):

A. $\dfrac{2}{\sqrt{3} + 1}$

$\dfrac{2}{\sqrt{3} + 1} = \dfrac{2(\sqrt{3} - 1)}{(\sqrt{3} + 1)(\sqrt{3} - 1)}$ Multiply numerator and denominator by the conjugate of the denominator

$= \dfrac{2\sqrt{3} - 2}{3 - 1}$

7.8 Multiplication and Division of Radical Expressions

$$= \frac{2\sqrt{3} - 2}{2}$$ Simplify

$$= \sqrt{3} - 1$$ Divide both terms of the numerator by 2

B. $\dfrac{4 + \sqrt{2}}{4 - \sqrt{2}}$

$$\frac{4 + \sqrt{2}}{4 - \sqrt{2}} = \frac{(4 + \sqrt{2})(4 + \sqrt{2})}{(4 - \sqrt{2})(4 + \sqrt{2})}$$ Multiply the numerator and denominator by the conjugate of the denominator

$$= \frac{16 + 8\sqrt{2} + 2}{16 - 2}$$

$$= \frac{18 + 8\sqrt{2}}{14}$$

$$= \frac{9 + 4\sqrt{2}}{7}$$ Divide numerator and denominator by 2

C. $\dfrac{\sqrt{5} + 2\sqrt{3}}{2\sqrt{5} - \sqrt{3}}$

$$\frac{\sqrt{5} + 2\sqrt{3}}{2\sqrt{5} - \sqrt{3}} = \frac{(\sqrt{5} + 2\sqrt{3})(2\sqrt{5} + \sqrt{3})}{(2\sqrt{5} - \sqrt{3})(2\sqrt{5} + \sqrt{3})}$$

$$= \frac{2 \cdot 5 + 5\sqrt{15} + 2 \cdot 3}{4 \cdot 5 - 3}$$

$$= \frac{10 + 5\sqrt{15} + 6}{20 - 3}$$

$$= \frac{16 + 5\sqrt{15}}{17}$$

D. $\dfrac{\sqrt{3y}}{\sqrt{y} - 3}$

$$\frac{\sqrt{3y}}{\sqrt{y} - 3} = \frac{\sqrt{3y} \cdot (\sqrt{y} + 3)}{(\sqrt{y} - 3)(\sqrt{y} + 3)}$$ Multiply the numerator and denominator by the conjugate of the denominator

$$= \frac{\sqrt{3y^2} + 3\sqrt{3y}}{y - 9}$$

$$= \frac{y\sqrt{3} + 3\sqrt{3y}}{y - 9}$$

◀

Exercise Set 7.8

Multiply as indicated and simplify:

1. $3(5 - 2\sqrt{3})$
2. $4(2 + 3\sqrt{5})$
3. $\sqrt{5}(2 + \sqrt{5})$
4. $\sqrt{6}(\sqrt{2} + 3)$
5. $(\sqrt{2} + 1)(2\sqrt{2} - 1)$
6. $(\sqrt{3} + 2)(\sqrt{3} - 1)$
7. $(3\sqrt{3} + 1)(3\sqrt{3} - 1)$
8. $(\sqrt{6} - 2)(\sqrt{6} + 2)$
9. $(2\sqrt{5} + 1)(\sqrt{5} - 4)$
10. $(3\sqrt{3} + 2)(\sqrt{3} - 2)$
11. $(\sqrt{6} - 2)^2$
12. $(2\sqrt{2} + 3)^2$
13. $(2\sqrt{5} + 5)(\sqrt{5} + 3)$
14. $(4\sqrt{7} - 3)(\sqrt{7} + 2)$
15. $(\sqrt{3} + 1)(\sqrt{6} - 2)$
16. $(\sqrt{5} + 2)(\sqrt{10} - 3)$
17. $(\sqrt{x} + 2)(2\sqrt{x} - 5)$
18. $(\sqrt{a} - 3)(3\sqrt{a} + 2)$
19. $(5\sqrt{m} + 2)(3\sqrt{m} + 4)$
20. $(2\sqrt{b} + 5)(3\sqrt{b} - 1)$
21. $(\sqrt{7} + \sqrt{3})(\sqrt{7} - \sqrt{3})$
22. $(\sqrt{10} + 2)(\sqrt{10} - 2)$
23. $(2\sqrt{5} + 1)(2\sqrt{5} - 1)$
24. $(3\sqrt{7} - 2)(3\sqrt{7} + 2)$
25. $(\sqrt{x} - 4)(\sqrt{x} + 4)$
26. $(\sqrt{m} - 2)(\sqrt{m} + 2)$
27. $(2\sqrt{a} + \sqrt{b})(2\sqrt{a} - \sqrt{b})$
28. $(3\sqrt{a} - \sqrt{b})(3\sqrt{a} + \sqrt{b})$

Divide as indicated. Simplify answers.

29. $\dfrac{6\sqrt{5}}{2}$
30. $\dfrac{8\sqrt{3}}{4}$
31. $\dfrac{6\sqrt{5} - 3}{3}$
32. $\dfrac{8\sqrt{3} - 12}{4}$
33. $\dfrac{5\sqrt{3}}{\sqrt{6}}$
34. $\dfrac{3\sqrt{6}}{\sqrt{3}}$
35. $\dfrac{8\sqrt{7} - 2}{\sqrt{7}}$
36. $\dfrac{4\sqrt{5} + 3}{\sqrt{5}}$
37. $\dfrac{3\sqrt{x} + 8}{\sqrt{x}}$
38. $\dfrac{\sqrt{y} - 8}{\sqrt{y}}$

Divide (rationalize the denominator) and simplify: *(eliminate radicals)*

39. $\dfrac{2}{\sqrt{3} + 1}$
40. $\dfrac{3}{\sqrt{3} + 1}$
41. $\dfrac{6}{\sqrt{5} + 2}$
42. $\dfrac{12}{\sqrt{7} - 2}$
43. $\dfrac{\sqrt{3} + 1}{\sqrt{3} - 1}$
44. $\dfrac{\sqrt{5} - 2}{\sqrt{5} + 2}$
45. $\dfrac{\sqrt{3}}{5 - \sqrt{3}}$
46. $\dfrac{2}{7 - \sqrt{2}}$
47. $\dfrac{\sqrt{5} - \sqrt{3}}{\sqrt{5} + 2\sqrt{3}}$
48. $\dfrac{\sqrt{5} + \sqrt{2}}{\sqrt{5} - 2\sqrt{2}}$
49. $\dfrac{3\sqrt{6}}{\sqrt{7} - 2}$
50. $\dfrac{4\sqrt{2}}{\sqrt{11} - 3}$

51. $\dfrac{x}{\sqrt{x}-2}$ 52. $\dfrac{a}{\sqrt{a}+3}$ 53. $\dfrac{\sqrt{n}}{\sqrt{n}+3}$ 54. $\dfrac{\sqrt{y}}{\sqrt{y}-4}$

55. Summarize the steps necessary to divide expressions involving radicals. Consider all types of divisors discussed in this section.

Review

56. Solve each of the equations:
    a. $m - 3 = 0$
    b. $m + 6 = 0$
    c. $2m - 5 = 0$
    d. $5m + 13 = 0$

57. Rewrite each of the following equations so that the right member is 0:
    a. $3x^2 + x - 4 = 2x^2 + 3x + 4$
    b. $y(2y - 1) + 3 = -5 + 2y(3y - 1)$

58. Find the slope and draw the graph of $3x - 4y = 6$.

59. Draw the graph of the inequality $3x - 4y > 6$.

7.9 Review and Chapter Test

(7.1) Rules of exponents

For real numbers a and b, natural numbers m and n:

1. $a^m \cdot a^n = a^{m+n}$ Product rule
2. $(ab)^m = a^m b^m$ Power of a product rule
3. $\left(\dfrac{a}{b}\right)^m = \dfrac{a^m}{b^m}, b \neq 0$
4. $(a^m)^n = a^{mn}$ Power to a power rule
5. $\dfrac{a^m}{a^n} = \begin{cases} a^{m-n} & \text{if } m > n \\ 1 & \text{if } m = n \\ \dfrac{1}{a^{n-m}} & \text{if } n > m, a \neq 0 \end{cases}$ Quotient rule

The restrictions on the relative size of m and n are removed when integer exponents are defined. Rather than memorize this rule, it is recommended that you apply the Fundamental Principle of Fractions and cancel.

SEVEN Exponents and Radicals

Example 1

Each of the following expressions is simplified using one of the rules of exponents:

A. $x^2 \cdot x^6 = x^8$

B. $(3m)^4 = 3^4 m^4 = 81 m^4$

C. $\left(\dfrac{4}{u}\right)^3 = \dfrac{4^3}{u^3} = \dfrac{64}{u^3}$

D. $(x^3)^2 = x^6$

E. $\dfrac{x^{8-2}}{x^2} = x^{8-2} = x^6$ Cancel two factors of x

F. $\dfrac{y^3}{y^7} = \dfrac{1}{y^{7-3}} = \dfrac{1}{y^4}$ Cancel three factors of y

G. $\dfrac{b^4}{b^4} = 1$ Cancel b^4 from numerator and denominator

(7.2) Integer exponents

The rules of exponents and the properties of real numbers are used to define:

$$a^0 = 1 \quad \text{and} \quad a^{-n} = \dfrac{1}{a^n}; \quad a^n = \dfrac{1}{a^{-n}}, \quad a \neq 0 \quad \text{reciprocal}$$

With negative integer and zero exponents defined, Rule 5 of exponents may be written:

$$\dfrac{a^m}{a^n} = a^{m-n} = \dfrac{1}{a^{n-m}} \quad \text{without restrictions on } m \text{ and } n$$

Example 2

Each of the following expressions is simplified using the rules of exponents:

A. $x^{-5} \cdot x^2 = x^{(-5)+2} = x^{-3} = \dfrac{1}{x^3}$

B. $(3u)^{-3} = 3^{-3} u^{-3} = \dfrac{1}{3^3} \cdot \dfrac{1}{u^3} = \dfrac{1}{27 u^3}$

C. $\left(\dfrac{y}{5}\right)^{-2} = \dfrac{y^{-2}}{5^{-2}} = \dfrac{5^2}{y^2} = \dfrac{25}{y^2}$

D. $(v^{-3})^{-4} = v^{(-3)(-4)} = v^{12}$

E. $\dfrac{n^5}{n^{-2}} = n^{5-(-2)} = n^{5+2} = n^7$

or

$\dfrac{n^5}{n^{-2}} = \dfrac{1}{n^{-2-5}} = \dfrac{1}{n^{-7}} = n^7$

(7.3) Numbers may be written in scientific notation. A number in scientific notation has the form

$$s \cdot 10^p \text{ where } 1 \leq s < 10 \text{ and } p \in I$$

Example 3

Each of the following numbers is written in scientific notation:

A. $7820 = 7.82 \times 10^3$

B. $0.000095 = 9.5 \times 10^{-5}$

C. $4.73 = 4.73 \times 10^0$ ◀

(7.4) Definition: a is an nth root of b if $a^n = b$.

Example 4

A. -3 is a square root of 9 since $(-3)^2 = 9$.

B. 4 is a cube root of 64 since $(4)^3 = 64$. ◀

The *principal* nth root of b, written $b^{1/n}$ is defined:

$b^{1/n}$ is positive for b positive, n even or odd

$b^{1/n}$ is negative for b negative, n odd

$b^{1/n}$ is not a real number when b is negative, n is even

Example 5

Evaluate each exponential expression:

A. $16^{1/2} = 4$

B. $8^{1/3} = 2$

C. $(-8)^{1/3} = -2$

D. $(-16)^{1/4}$ is not a real number ◀

$$b^{m/n} = (b^{1/n})^m = (b^m)^{1/n}$$

$b^{m/n}$ is not a real number when $\dfrac{m}{n}$ is in lowest terms and

b is negative and n is even

The five rules of exponents remain valid for rational number exponents. They are repeated here:

For $m, n \in Q$; $a, b \in R$, $a \neq 0$, $b \neq 0$

$$a^0 = 1, \quad a^{-m} = \frac{1}{a^m}, \quad a^m = \frac{1}{a^{-m}}$$

1. $a^m \cdot a^n = a^{m+n}$ Product rule
2. $(ab)^m = a^m b^m$ Power of a product rule
3. $\left(\dfrac{a}{b}\right)^m = \dfrac{a^m}{b^m}$ Power of a quotient rule
4. $(a^m)^n = a^{mn}$ Power to a power rule
5. $\dfrac{a^m}{a^n} = a^{m-n} = \dfrac{1}{a^{n-m}},\ a \neq 0$

Example 6

Each of the following expressions is simplified using the rules of exponents. Assume each variable represents a positive real number:

A. $x^{2/3} x^{1/2} = x^{2/3 + 1/2} = x^{4/6 + 3/6} = x^{7/6}$

B. $(8m)^{-2/3} = 8^{-2/3} m^{-2/3}$
$= \dfrac{1}{8^{2/3}} \cdot \dfrac{1}{m^{2/3}}$
$= \dfrac{1}{(8^{1/3})^2 m^{2/3}}$
$= \dfrac{1}{2^2 m^{2/3}}$
$= \dfrac{1}{4 m^{2/3}}$

C. $\left(\dfrac{25}{t}\right)^{3/2} = \dfrac{25^{3/2}}{t^{3/2}} = \dfrac{(25^{1/2})^3}{t^{3/2}} = \dfrac{(5)^3}{t^{3/2}} = \dfrac{125}{t^{3/2}}$

D. $(a^{-3/2})^{-1/4} = a^{(-3/2)(-1/4)} = a^{3/8}$

E. $\dfrac{x^{-1/3}}{x^{2/3}} = x^{-1/3 - 2/3}$
$= x^{-3/3}$
$= x^{-1}$
$= \dfrac{1}{x^1}$
$= \dfrac{1}{x}$ ◀

An nth root of b may also be written in radical form: $\sqrt[n]{b}$, where b is the radicand, n is the index (generally unwritten when $n = 2$), and $\sqrt{\ }$ is the radical sign.

$\sqrt[n]{b} = b^{1/n}$. The same restrictions apply to $\sqrt[n]{b}$ as to $b^{1/n}$. That is, $\sqrt[n]{b}$ represents the principal nth root of b, and $\sqrt[n]{b}$ is not a real number when b is negative and n is even. We may change form between exponential and radical

form using
$$b^{m/n} = \sqrt[n]{b^m}$$

Example 7

Rewrite each expression:

A. $8^{2/3} = \sqrt[3]{8^2} = 2^2 = 4$

B. $x^{-5/2} = \dfrac{1}{x^{5/2}} = \dfrac{1}{\sqrt{x^5}}$

C. $\sqrt[5]{u^2} = u^{2/5}$ ◀

Any expression containing one or more radicals is called a *radical expression*. The form of a radical expression may be changed using the properties of radicals.

(7.5) Properties of radicals

Let $a, b \in R$ where $a > 0, b > 0$:

1. $\sqrt[n]{ab} = \sqrt[n]{a}\,\sqrt[n]{b}$
2. $\sqrt[n]{\dfrac{a}{b}} = \dfrac{\sqrt[n]{a}}{\sqrt[n]{b}}, \quad b \neq 0$
3. $\sqrt[n]{a^n} = (\sqrt[n]{a})^n = a$
4. $\sqrt[kn]{a^{km}} = \sqrt[n]{a^m}$

(7.6) An expression is in *simplest radical form* when all of the following conditions are satisfied:

1. The radicand contains no factor whose power is greater than or equal to the index.
2. There are no radicals in the denominator.
3. There are no fractions in the radicand.
4. The index and power of the radicand contain no common factor.

Example 8

Each of the following expressions has been written in simplest radical form. Assume all variables represent positive numbers.

A. $\sqrt{50x^3} = \sqrt{2 \cdot 5^2 \cdot x^3}$
$= \sqrt{5^2 x^2}\,\sqrt{2x}$
$= 5x\sqrt{2x}$

B. $\dfrac{3}{\sqrt{8}} = \dfrac{3}{\sqrt{4}\sqrt{2}}$

$= \dfrac{3\cdot\sqrt{2}}{2\sqrt{2}\cdot\sqrt{2}}$

$= \dfrac{3\sqrt{2}}{2\cdot 2}$

$= \dfrac{3\sqrt{2}}{4}$

C. $\sqrt{\dfrac{5}{6}} = \sqrt{\dfrac{5\cdot 6}{6\cdot 6}}$

$= \sqrt{\dfrac{30}{36}}$

$= \dfrac{\sqrt{30}}{\sqrt{36}}$

$= \dfrac{\sqrt{30}}{6}$

D. $\sqrt[3]{56a^5b} = \sqrt[3]{2^3\cdot 7a^5b}$

$= \sqrt[3]{2^3 a^3}\,\sqrt[3]{7a^2b}$

$= 2a\sqrt[3]{7a^2b}$

(7.7) To add or subtract radicals, write each in simplest radical form and then combine like radicals. Like radicals are those with identical radical parts. (Both the radicand and the index must be the same.)

Example 9

Add and subtract:

A. $5\sqrt{20} + 3\sqrt{45} = 5\sqrt{4\cdot 5} + 3\sqrt{9\cdot 5}$

$= 5\sqrt{4}\sqrt{5} + 3\sqrt{9}\sqrt{5}$

$= 5\cdot 2\sqrt{5} + 3\cdot 3\sqrt{5}$

$= 10\sqrt{5} + 9\sqrt{5}$

$= 19\sqrt{5}$

B. $6\sqrt[3]{2} - 2\sqrt[3]{16} = 6\sqrt[3]{2} - 2\sqrt[3]{8\cdot 2}$

$= 6\sqrt[3]{2} - 2\sqrt[3]{8}\sqrt[3]{2}$

$= 6\sqrt[3]{2} - 2\cdot 2\sqrt[3]{2}$

$= 6\sqrt[3]{2} - 4\sqrt[3]{2}$

$= 2\sqrt[3]{2}$

(7.8) To multiply radicals, we use properties introduced for multiplying polynomials, including the distributive property, FOIL, and special products. The resulting product should be written in simplest radical form.

Example 10

Multiply and simplify. Assume all variables represent positive numbers.

A. $\sqrt{2x}(x - \sqrt{6x}) = x\sqrt{2x} - \sqrt{12x^2}$ Distributive property
$= x\sqrt{2x} - \sqrt{4 \cdot 3 \cdot x^2}$
$= x\sqrt{2x} - 2x\sqrt{3}$

B. $(\sqrt{y} - 3)(\sqrt{y} + 3) = \sqrt{y}^2 - 3^2$ Sum times a difference
$= y - 9$

C. $(2\sqrt{m} + 1)(\sqrt{m} - 4) = 2\sqrt{m}^2 - 7\sqrt{m} - 4$ FOIL
$= 2m - 7\sqrt{m} - 4$

D. $(\sqrt{u} + 2)^2 = \sqrt{u}^2 + 4\sqrt{u} + 4$ Square of a binomial
$= u + 4\sqrt{u} + 4$

To divide by a radical, rationalize the denominator by multiplying the numerator and denominator by the appropriate radical.

Example 11

Divide:

A. $\dfrac{\sqrt{8} - 6}{\sqrt{2}} = \dfrac{(\sqrt{8} - 6)\sqrt{2}}{\sqrt{2}\sqrt{2}}$
$= \dfrac{\sqrt{16} - 6\sqrt{2}}{\sqrt{4}}$
$= \dfrac{4 - 6\sqrt{2}}{2}$
$= 2 - 3\sqrt{2}$

B. $\dfrac{\sqrt{x} + 5}{\sqrt{x} - 2} = \dfrac{(\sqrt{x} + 5)(\sqrt{x} + 2)}{(\sqrt{x} - 2)(\sqrt{x} + 2)}$
$= \dfrac{\sqrt{x}^2 + 7\sqrt{x} + 10}{\sqrt{x}^2 - 2^2}$
$= \dfrac{x + 7\sqrt{x} + 10}{x - 4}$

Exercise Set 7.9 Review

Use the rules of exponents to simplify each expression:

1. $(3x)(5x^2)$
2. $(-2x)(3x^{-2})$
3. $(8x^2)^{1/3}$
4. $u^{2/3} u^{-4/3}$
5. $\dfrac{6x}{4x^{1/3}}$
6. $\dfrac{2x^2 y^{-1}}{6x^{-2} y^2}$
7. $\dfrac{(4x^4 y)^{1/2}}{(8x^6 y)^{1/3}}$
8. $(-8ab^{-6})^{-1/3}$

Write each of the following numbers in scientific notation:

9. $-78{,}900{,}000$
10. 0.000106

Write each of the following numbers in standard form:

11. 4.38×10^{-5}
12. -6.22×10^6

Evaluate each of the following expressions where possible:

13. $8^{1/3}$
14. $16^{5/4}$
15. $-25^{1/2}$
16. $(-25)^{1/2}$
17. 13^0
18. $27^{-2/3}$

Write each of the following expressions in radical form:

19. $2x^{3/4}$
20. $(2x)^{3/4}$
21. $(x^2 + y^2)^{1/2}$
22. $(3x - 5)^{1/3}$

Write each of the following radical expressions in simplest radical form. Assume variables represent positive numbers and denominators are not zero.

23. $\sqrt{40xy^4}$
24. $\sqrt{50x^4 y^3}$
25. $\sqrt[3]{32a^2 b^8}$
26. $\dfrac{5}{\sqrt{2x}}$
27. $\dfrac{4x}{\sqrt{2x}}$
28. $\sqrt{\dfrac{x}{y}}$
29. $\sqrt{\dfrac{5}{8x}}$
30. $\sqrt[3]{\dfrac{1}{2x^2}}$

Complete each of the following operations. Write answers in simplest radical form.

31. $2\sqrt{3} + \sqrt{5} + 4\sqrt{3} - 2\sqrt{5}$
32. $\sqrt{20} + 2\sqrt{80}$
33. $2\sqrt{8} + 5\sqrt{18}$
34. $6\sqrt{2x} - \sqrt{50x}$
35. $\sqrt{2}(5 - 3\sqrt{2})$
36. $(\sqrt{x} - 4)(\sqrt{x} + 3)$
37. $(\sqrt{2x} + \sqrt{3y})^2$
38. $(\sqrt{11} + 3)(\sqrt{11} - 3)$

39. $\dfrac{8+\sqrt{28}}{\sqrt{7}}$ **40.** $\dfrac{8}{\sqrt{6}-2}$ **41.** $\dfrac{12}{\sqrt{10}+2}$

42. $\dfrac{\sqrt{5}-1}{\sqrt{5}+3}$ **43.** $\dfrac{\sqrt{x}-\sqrt{y}}{\sqrt{x}+\sqrt{y}}$

Chapter 7 Review Test

1. Use the rules of exponents to simplify each expression:

    a. $(27x^{-3}y^6)^{-1/3}$ b. $x^{-3/5}x^{1/4}$ c. $\dfrac{6x^{-2}y^3}{2xy^{-2}}$

2. Write in scientific notation:

    a. 27,100   b. 0.00051

3. Write in standard notation:

    a. $2.56 \cdot 10^{-5}$ b. $7.55 \cdot 10^7$

4. Evaluate:

    a. $8^{1/3}$ b. $4^{1/2}$ c. $-8^{1/3}$

    d. $-4^{1/2}$ e. $(-8)^{1/3}$ f. $(-4)^{1/2}$

    g. $8^{4/3}$ h. $4^{3/2}$

5. Write in radical form:

    a. $4x^{2/3}$ b. $(4x)^{2/3}$ c. $(x^2+9)^{1/2}$

6. Write each of the following expressions in simplest radical form. Assume variables represent positive numbers and denominators are not zero.

    a. $\sqrt{27a^3b^4}$ b. $\dfrac{7}{\sqrt{3x}}$ c. $\sqrt{\dfrac{a^3}{b}}$

    d. $\sqrt[3]{27a^3b^4}$ e. $\sqrt[3]{\dfrac{5}{4x^2}}$

7. Complete each of the following operations. Write answers in simplest radical form.

    a. $3\sqrt{5}-7\sqrt{3}+\sqrt{5}-2\sqrt{3}$ b. $3\sqrt{2}(\sqrt{3}-\sqrt{6})$

    c. $(\sqrt{2x}+2)(\sqrt{2x}-2)$ d. $\dfrac{5}{\sqrt{2}+\sqrt{5}}$

    e. $\sqrt{50x}-\sqrt{32x}$

CHAPTER EIGHT

QUADRATIC EQUATIONS

In Chapter 2, we discussed how to recognize and solve any equations that can be written in the form $Ax + B = 0$. Such an equation is called a *first degree* or *linear equation in one variable*, x. Note that the highest power of the variable is 1. In this chapter, we direct our attention to solving equations that can be written in the form $ax^2 + bx + c = 0$, $a \neq 0$, where the highest power of the variable is 2.

DEFINITION

An equation that can be written in the form

$$ax^2 + bx + c = 0, a \neq 0$$

where, a, b, and c are constants and x is a variable, is called a second degree or quadratic equation in one variable. The form $ax^2 + bx + c = 0$ is called the standard form for a quadratic equation.

8.1 Solving Quadratic Equations by Factoring

To solve a quadratic equation, we use the following theorem, stated without proof:

8.1 Solving Quadratic Equations by Factoring

> **ZERO FACTOR THEOREM**
>
> The product of two factors is zero if and only if one of the factors is zero. In symbols:
>
> $ab = 0$ if and only if $a = 0$ or $b = 0$

To use the zero factor theorem to solve a quadratic equation:

1. Write the equation in standard form, $ax^2 + bx + c = 0, a \neq 0$.
2. Factor the left side.
3. Set each factor equal to zero.
4. Solve the resulting linear equations.
5. Check both solutions in the original quadratic equation (optional but recommended).

Example 1

Solve and check each equation:

A. $x^2 - x - 6 = 0$

$\quad x^2 - x - 6 = 0 \qquad$ In standard form
$\quad (x - 3)(x + 2) = 0 \qquad$ Factor
$\quad x - 3 = 0 \quad$ or $\quad x + 2 = 0 \qquad$ Set each factor equal to 0
$\quad x = 3 \qquad\qquad x = -2 \qquad$ Solve both equations

Check:

$x = 3$:
$x^2 - x - 6 = 0$
$(3)^2 - (3) - 6 \stackrel{?}{=} 0$
$9 - 3 - 6 = 0$
$0 = 0 \checkmark$

$x = -2$:
$x^2 - x - 6 = 0$
$(-2)^2 - (-2) - 6 \stackrel{?}{=} 0$
$4 + 2 - 6 = 0$
$0 = 0 \checkmark$

B. $2u^2 - 5u + 3 = 0$

$\quad 2u^2 - 5u + 3 = 0$
$\quad (2u - 3)(u - 1) = 0 \qquad$ Factor the left side
$\quad 2u - 3 = 0 \quad u - 1 = 0 \qquad$ Set each factor equal to 0
$\quad 2u = 3 \qquad\quad u = 1 \qquad$ Solve
$\quad u = \dfrac{3}{2}$

Check:

$u = \dfrac{3}{2}$:

$2u^2 - 5u + 3 = 0$

$2\left(\dfrac{3}{2}\right)^2 - 5\left(\dfrac{3}{2}\right) + 3 \stackrel{?}{=} 0$

$2\left(\dfrac{9}{4}\right) - 5\left(\dfrac{3}{2}\right) + 3 = 0$

$\dfrac{9}{2} - \dfrac{15}{2} + \dfrac{6}{2} = 0$

$0 = 0$ ✓

$u = 1$:

$2u^2 - 5u + 3 = 0$

$2(1)^2 - 5(1) + 3 \stackrel{?}{=} 0$

$2(1) - 5(1) + 3 = 0$

$2 - 5 + 3 = 0$

$0 = 0$ ✓

In the remaining examples, the check will be left to the student.

C. $m^2 = 9$

$m^2 = 9$	Not in standard form
$m^2 - 9 = 0$	Add -9 to both sides
$(m - 3)(m + 3) = 0$	Factor
$m - 3 = 0 \quad m + 3 = 0$	
$m = 3 \qquad m = -3$	

D. $a^2 - 4a = \dfrac{4 - a}{3}$

$a^2 - 4a = \dfrac{4 - a}{3}$	Not in standard form. We need to clear fractions and put all non-zero terms on the same side
$3a^2 - 12a = 4 - a$	Multiply both sides by 3
$3a^2 - 11a - 4 = 0$	Add $-4 + a$ to both sides
$(3a + 1)(a - 4) = 0$	Factor the left side
$3a + 1 = 0 \qquad a - 4 = 0$	Set each factor equal to 0
$3a = -1 \qquad\qquad a = 4$	
$a = -\dfrac{1}{3}$	Solve each

E. $y^2 = \dfrac{2}{3}y$

$y^2 = \dfrac{2}{3}y$	Not in standard form

8.1 Solving Quadratic Equations by Factoring

$$3y^2 = 2y$$ Multiply both sides by 3
$$3y^2 - 2y = 0$$ Add $-2y$ to both sides
$$y(3y - 2) = 0$$ Factor. Remove common factor of y
$$y = 0 \quad 3y - 2 = 0$$
$$3y = 2$$
$$y = \frac{2}{3}$$ ◀

Occasionally, an equation has two identical solutions. In such a case, we list the solution twice and call this number a **double root**.

Example 2

A. $a^2 - 4a + 6 = 2a - 3$

$$a^2 - 4a + 6 = 2a - 3$$
$$a^2 - 6a + 9 = 0$$ Write in standard form
$$(a - 3)(a - 3) = 0$$ Factor
$$a - 3 = 0 \quad \text{or} \quad a - 3 = 0$$ Set each factor equal to 0
$$a = 3 \quad \text{or} \quad a = 3$$

We write $a = 3$ and say 3 is a double root.

B. $\dfrac{3x^2}{2} - 2x = -\dfrac{2}{3}$

$$\frac{3x^2}{2} - 2x = -\frac{2}{3}$$
$$6 \cdot \left(\frac{3x^2}{2} - 2x\right) = 6 \cdot \left(-\frac{2}{3}\right)$$ Multiply by LCD = 6
$$9x^2 - 12x = -4$$
$$9x^2 - 12x + 4 = 0$$ Standard form
$$(3x - 2)(3x - 2) = 0$$ Factor
$$3x - 2 = 0 \quad \text{or} \quad 3x - 2 = 0$$ Set each factor equal to 0
$$3x = 2 \qquad\qquad 3x = 2$$
$$x = \frac{2}{3} \qquad\qquad x = \frac{2}{3}$$

$x = \frac{2}{3}$, a double root. ◀

Exercise Set 8.1

Solve each equation by factoring. List both solutions.

1. $y^2 + y - 12 = 0$
2. $a^2 + 5a - 6 = 0$
3. $u^2 - 5u = 0$
4. $t^2 + 3t = 0$
5. $x^2 - 7x + 12 = 0$
6. $x^2 + 6x + 8 = 0$
7. $m^2 + 3m - 10 = 0$
8. $r^2 - 2r - 15 = 0$
9. $x^2 + 4x + 3 = 0$
10. $x^2 - 7x + 10 = 0$
11. $c^2 - 25 = 0$
12. $w^2 - 16 = 0$
13. $2y^2 - 3y + 1 = 0$
14. $3y^2 - 4y + 1 = 0$
15. $2a^2 + a - 15 = 0$
16. $6a^2 - 11a + 3 = 0$
17. $8m^2 = 6m + 9$
18. $6r^2 = 12r$
19. $2y^2 = 6y$
20. $x(x + 1) = -\dfrac{2}{9}$
21. $w(4w - 7) = 2$
22. $2u(u + 3) = 15 - u$
23. $x^2 + 6x - 4 = 2x + 1$
24. $x^2 - 20 = 12 - 4x$
25. $a^2 + 5(a - 1) = 2 - a^2$
26. $x(2x - 1) - 10 = 5 - 2x$
27. $\dfrac{v^2}{2} + 3v = \dfrac{v}{2} - 3$
28. $\dfrac{n^2}{3} + \dfrac{4n}{3} - 1 = \dfrac{n - 1}{2}$
29. $z^2 + 7z - 10 = \dfrac{z + 4}{2}$
30. $\dfrac{d^2 - 6d - 1}{3} = \dfrac{1 - d}{2}$

31. If $a \cdot b = 1$, does it follow that $a = 1$ or $b = 1$? Compare this statement to the zero factor theorem.

32. In the definition of a quadratic equation
$$ax^2 + bx + c = 0, \, a \neq 0$$
why is the restriction $a \neq 0$ important?

For each rectangle in the following problems, an expression for both the length and width is provided. Find the value(s) of the variable that would cause the rectangle to have the given area. Discard any value that would give a negative width or length.

33. Length $x + 1$, width $x - 2$, Area = 4

34. Length $y + 2$, width $y - 1$, Area = 4

35.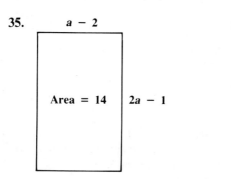

36.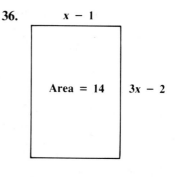

Review

37. Evaluate each of the following radicals:

   a. $\sqrt{4}$ b. $\sqrt{100}$ c. $\sqrt{49}$

38. Simplify each of the following radicals:

   a. $\sqrt{28}$ b. $\sqrt{18}$ c. $\sqrt{150}$

39. Solve the following equation *for r*:

   $2r - 5 = 3t + 7$

40. Write the equation of the line with a slope of -2 and that passes through $(0, 3)$.

8.2 Solving Quadratic Equations by the Square Root Method

A quadratic equation in one variable is an equation that can be written in the form $ax^2 + bx + c = 0$. If $b = 0$, the equation takes the simpler form

$$ax^2 + 0 \cdot x + c = 0$$
$$ax^2 = -c$$
$$x^2 = -\frac{c}{a}$$

Or, changing notation

$$x^2 = k, \text{ where } k \text{ is a constant}$$

This equation is solved using the definition of a square (Chapter 7). If k is a nonnegative constant, the solutions to

$$x^2 = k$$

are, by the definition of a square,

$$x = \sqrt{k} \quad \text{or} \quad x = -\sqrt{k}$$

which we write as

$$x = \pm\sqrt{k}$$

read "the positive or negative square root of k." We refer to this process as solving a quadratic equation by the **square root method**. For the sake of simplicity, we say we take the square root of both sides.

Example 1 Solve each equation by writing it in the form $x^2 = k$, and then taking the square root of both sides:

A. $x^2 - 9 = 0$

$\quad x^2 - 9 = 0$

$\quad\quad x^2 = 9$ Add 9 to both sides

$\quad\quad x = \pm\sqrt{9}$ Take the square root of both sides

$\quad\quad x = \pm 3$ Simplify

B. $2m^2 - 5 = 0$

$\quad 2m^2 - 5 = 0$

$\quad\quad 2m^2 = 5$ Add 5 to both sides

$\quad\quad m^2 = \dfrac{5}{2}$ Divide both sides by 2

$\quad\quad m = \pm\sqrt{\dfrac{5}{2}}$ Take the square root of both sides

$\quad\quad m = \pm\sqrt{\dfrac{5 \cdot 2}{2 \cdot 2}}$ Make the denominator a perfect square

$\quad\quad m = \pm\dfrac{\sqrt{10}}{2}$ Simplest radical form

C. $y^2 + 4 = 0$

$\quad y^2 + 4 = 0$

$\quad\quad y^2 = -4$ Add -4 to both sides

$\quad\quad y^2 = \pm\sqrt{-4}$

No real solutions exist for this equation since there is no real number square root of a negative number. ◀

8.2 Solving Quadratic Equations by the Square Root Method

If the left side of an equation can be expressed as the square of any quantity, the square root method may be used. We will use Q to represent any quantity that includes a variable and generalize the method:

TO SOLVE BY THE SQUARE ROOT METHOD

If $Q^2 = k$, $\quad k \geq 0$

Then $Q = \pm\sqrt{k}$

Example 2

Solve each equation by the square root method:

A. $(x - 3)^2 = 16$

$\quad (x - 3)^2 = 16 \qquad$ In form $Q^2 = k$

$\quad x - 3 = \pm\sqrt{16} \qquad$ Take the square root of both sides

$\quad x - 3 = \pm 4 \qquad$ Simplify

$\quad x = 3 \pm 4 \qquad$ Add 3 to both sides

$\quad x = 3 + 4 \quad$ or $\quad 3 - 4$

$\quad x = 7, -1$

B. $(y + 2)^2 = 9$

$\quad (y + 2)^2 = 9 \qquad$ In form $Q^2 = k$

$\quad y + 2 = \pm 3 \qquad$ Take the square root of both sides

$\quad y = -2 \pm 3 \qquad$ Add -2 to both sides

$\quad y = -2 + 3 \quad$ or $\quad -2 - 3$

$\quad y = 1, -5$

C. $(2u - 1)^2 = 25$

$\quad (2u - 1)^2 = 25 \qquad$ In form $Q^2 = k$

$\quad 2u - 1 = \pm 5 \qquad$ Take the square root of both sides

$\quad 2u = 1 \pm 5 \qquad$ Add 1 to both sides

$\quad u = \dfrac{1 \pm 5}{2} \qquad$ Divide both sides by 2

$\quad u = \dfrac{1 + 5}{2} \quad$ or $\quad \dfrac{1 - 5}{2}$

$\quad u = 3, -2 \qquad$ Simplify

EIGHT Quadratic Equations

D. $(t + 2)^2 = 7$

$\qquad (t + 2)^2 = 7 \qquad$ In form $Q^2 = k$

$\qquad t + 2 = \pm\sqrt{7} \qquad$ Take the square root of both sides

$\qquad t = -2 \pm \sqrt{7} \qquad$ Add -2 to both sides

The answers are generally left in this form. If numerical values are needed, a table of square roots or a calculator may be used to express the answers:

$$t = -2 + \sqrt{7} = -2 + 2.646 = 0.646$$

or $\quad t = -2 - \sqrt{7} = -2 - 2.646 = -4.646$

E. $(3a - 7)^2 = -12$

$\qquad (3a - 7)^2 = -12 \qquad$ In form $Q^2 = k$

There is no real solution, since no real number has a square that is negative.

F. $(2m - 5)^2 = 0$

$\qquad (2m - 5)^2 = 0$

$\qquad 2m - 5 = \pm\sqrt{0}$

$\qquad 2m - 5 = 0$

$\qquad 2m = 5 \qquad$ Add 5 to both sides

$\qquad m = \dfrac{5}{2} \qquad$ Divide both sides by 2

This is a double root, since the factor $2m - 5$ appears twice, and we write

$$m = \frac{5}{2}, \frac{5}{2}$$ ◀

Exercise Set 8.2

Use the square root method to solve each of the following equations, where possible:

1. $x^2 = 25$
2. $x^2 = 1$
3. $y^2 = 36$
4. $y^2 = 49$
5. $u^2 = \dfrac{25}{4}$
6. $m^2 = \dfrac{16}{25}$
7. $x^2 = 7$
8. $x^2 = 11$
9. $y^2 = -4$
10. $y^2 = -9$
11. $x^2 - 9 = 0$
12. $x^2 - 121 = 0$

13. $x^2 - 5 = 0$
14. $x^2 - 17 = 0$
15. $x^2 + 2 = 0$
16. $x^2 + 10 = 0$
17. $(x - 2)^2 = 9$
18. $(x - 1)^2 = 9$
19. $(y + 3)^2 = 1$
20. $(y + 4)^2 = 4$
21. $(a + 2)^2 = 25$
22. $(m + 4)^2 = 1$
23. $(u - 6)^2 = 25$
24. $(u - 3)^2 = 16$
25. $(2x + 1)^2 = 25$
26. $(2x - 1)^2 = 25$
27. $(3x + 1)^2 = 0$
28. $(2x + 3)^2 = 0$
29. $(x - 2)^2 = -4$
30. $(y + 1)^2 = -3$
31. $(a + 2)^2 = 5$
32. $(b + 3)^2 = 10$
33. $(y - 2)^2 = 8$
34. $(x - 3)^2 = 12$

For each square in the following problems, an expression is provided for the length of a side. Find the value(s) of the variable that would cause the square to have the given area:

35. Area = 16, $x - 2$

36. Area = 9, $y + 3$

37. Area = 11, $2x + 3$

38. Area = 15, $3y - 2$

Review

39. Square each of the following numbers, as indicated:

    a. $(6)^2$
    b. $(-6)^2$
    c. $(\frac{1}{3})^2$
    d. $(-\frac{1}{3})^2$
    e. $(\frac{3}{4})^2$
    f. $(-\frac{3}{4})^2$

40. Complete the following multiplications:

    a. $(m - 3)^2$
    b. $(m + 3)^2$
    c. $(u + 4)^2$
    d. $(u - 4)^2$
    e. $(x - \frac{2}{3})^2$
    f. $(x + \frac{2}{3})^2$

41. Solve the following system of equations by *graphing*:

    $x - 3y = -6$
    $2x + 3y = -3$

42. Factor each of the trinomials:

   a. $t^2 + 6t + 9$ b. $t^2 - 6t + 9$

   c. $y^2 + 10y + 25$ d. $y^2 - 10y + 25$

   e. $a^2 + 20a + 100$ f. $a^2 - 20a + 100$

8.3 Solving Quadratic Equations by Completing the Square

A quadratic equation written in standard form $ax^2 + bx + c = 0$, $a \neq 0$ can be solved by factoring if $ax^2 + bx + c$ is factorable (see Section 8.1). But what if it is not? We show in this section that such an equation may be rewritten in the form $Q^2 = k$ and then solved by the square root method (see Section 8.2). To rewrite a quadratic equation in the form $Q^2 = k$, we use a process called **completing the square**.

Completing the Square

Consider the perfect square expression $(x + a)^2$. Squaring this binomial, we obtain:

FIGURE 8.3-1A

$$(x + a)^2 = x^2 + 2ax + a^2$$

Observe that the third term of this expression is the square of one-half of the coefficient of x in the middle term.

$$\left[\frac{1}{2} \cdot (2a)\right]^2 = [a]^2 = a^2$$

So, if we wish to change an expression of the form $x^2 + mx$ (m a constant) into a perfect square, we take half of the coefficient of the x term, square it, and then add it to the expression $x^2 + mx$.

This can be seen geometrically as follows. Consider the area $x^2 + mx$ as in Figure 8.3-1A. To form a square, divide the area mx in two, and reposition it as in

FIGURE 8.3-1B

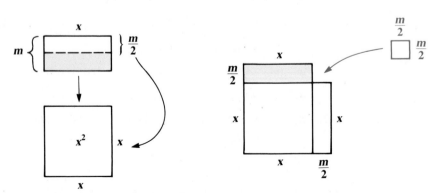

8.3 Solving Quadratic Equations by Completing the Square

Figure 8.3-1B. Now, to literally "complete the square," we add the missing corner whose dimensions are $(\frac{m}{2})$ by $(\frac{m}{2})$ and whose area is $(\frac{m}{2})^2$. Having added $(\frac{m}{2})^2$, we have now completed the square whose area is then found by "squaring" the length of a side, $x + \frac{m}{2}$. Its area is therefore $(x + \frac{m}{2})^2$ as shown in Figure 8.3-1C.

FIGURE 8.3-1C

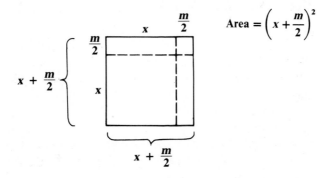

Example 1

Complete the square on each of the following expressions and then factor:

A. $x^2 + 6x$

$$\frac{1}{2} \cdot (6) = 3; \; 3^2 = 9 \qquad \text{Add}$$

$$x^2 + 6x \underline{+ 9} = (x + 3)^2 \qquad \text{Perfect square}$$

B. $y^2 - 8y$

$$\frac{1}{2} \cdot (-8) = -4; \; (-4)^2 = 16 \qquad \text{Add}$$

$$y^2 - 8y \underline{+ 16} = (y - 4)^2 \qquad \text{Perfect square}$$

———Note the agreement of the signs

C. $x^2 + 5x$

$$\frac{1}{2} \cdot (5) = \frac{5}{2}; \; \left(\frac{5}{2}\right)^2 = \frac{25}{4} \qquad \text{Add}$$

$$x^2 + 5x + \underline{\frac{25}{4}} = \left(x + \frac{5}{2}\right)^2$$

D. $w^2 - \frac{3}{4}w$

$$\frac{1}{2} \cdot \left(-\frac{3}{4}\right) = -\frac{3}{8}; \; \left(-\frac{3}{8}\right)^2 = \frac{9}{64} \qquad \text{Add}$$

$$w^2 - w + \underline{\frac{9}{64}} = \left(w - \frac{3}{8}\right)^2$$

Solving a Quadratic Equation by Completing the Square

We are now ready to rewrite a quadratic equation

$$ax^2 + bx + c = 0, a \neq 0$$

into the form

$$Q^2 = k$$

We will summarize the process and then show its application with several examples.

TO SOLVE A QUADRATIC EQUATION BY COMPLETING THE SQUARE

(Assume the variable is x)

1. Write the equation with the x^2 and x terms on the left and the constant term on the right.
2. Make the coefficient of x^2 equal to 1 by dividing both sides of the equation by the coefficient of x^2 if it is not 1.
3. Complete the square:
   a. Take one-half of the numerical coefficient of the x term.
   b. Square this number.
   c. Add the result *to both sides of the equation*!
4. Rewrite the equation by factoring the left side. It should now be in the form $Q^2 = k$.
5. Solve by the square root method.

Example 2

Solve each of the following equations by completing the square:

A. $x^2 + 6x - 7 = 0$

$$x^2 + 6x - 7 = 0$$
$$x^2 + 6x = 7 \qquad \text{Add 7 to both sides}$$

Note that the coefficient of x^2 is 1.
Complete the square:

$$\frac{1}{2} \cdot (6) = 3; \; 3^2 = 9 \qquad \text{Add 9 to both sides}$$

8.3 Solving Quadratic Equations by Completing the Square

$$x^2 + 6x + 9 = 7 + 9$$
$$(x+3)^2 = 16 \qquad \text{Rewrite}$$
$$x + 3 = \pm 4 \qquad \text{Solve by the square root method}$$
$$x = -3 \pm 4$$
$$x = 1, -7$$

You may check (strongly recommended).

B. $2x^2 + 5x - 3 = 0$

$$2x^2 + 5x - 3 = 0$$
$$2x^2 + 5x = 3 \qquad \text{Add 3 to both sides}$$
$$x^2 + \frac{5}{2}x = \frac{3}{2} \qquad \text{Divide by 2, the coefficient of the } x^2 \text{ term}$$

Complete the square:

$$\frac{1}{2} \cdot \left(\frac{5}{2}\right) = \frac{5}{4}; \ \left(\frac{5}{4}\right)^2 = \frac{25}{16} \qquad \text{Add } \frac{25}{16} \text{ to both sides}$$

$$x^2 + \frac{5}{2}x + \frac{25}{16} = \frac{3}{2} + \frac{25}{16}$$

$$\left(x + \frac{5}{4}\right)^2 = \frac{3 \cdot 8}{2 \cdot 8} + \frac{25}{16} \qquad \text{LCD} = 16$$

$$\left(x + \frac{5}{4}\right)^2 = \frac{24 + 25}{16}$$

$$\left(x + \frac{5}{4}\right)^2 = \frac{49}{16} \qquad \text{Solve by the square root method}$$

$$x + \frac{5}{4} = \pm \frac{7}{4}$$

$$x = -\frac{5}{4} \pm \frac{7}{4} \qquad \text{Subtract } \frac{5}{4} \text{ from both sides}$$

$$x = \frac{-5 \pm 7}{4}$$

$$x = \frac{1}{2}, -3$$

C. $x^2 - 4x + 1 = 0$

$$x^2 - 4x + 1 = 0$$
$$x^2 - 4x = -1 \qquad \text{Add } -1 \text{ to both sides}$$

Complete the square:

$$\frac{1}{2} \cdot (-4) = -2, (-2)^2 = 4 \qquad \text{Add 4 to both sides}$$

$$x^2 - 4x + 4 = -1 + 4$$

$$(x - 2)^2 = 3 \qquad \text{In form } Q^2 = k$$

$$x - 2 = \pm\sqrt{3}$$

$$x = 2 \pm \sqrt{3} \qquad \text{Simplest form of the answer}$$

D. $6x^2 - x - 2 = 0$

$$6x^2 - x - 2 = 0$$

$$6x^2 - x = 2 \qquad \text{Add 2 to both sides}$$

$$x^2 - \frac{1}{6}x = \frac{1}{3} \qquad \text{Divide both sides by 6}$$

$$x^2 - \frac{1}{6}x + \left(\frac{1}{12}\right)^2 = \frac{1}{3} + \frac{1}{144} \qquad \text{Add } \left(\frac{1}{2} \cdot \frac{1}{6}\right)^2 = \left(\frac{1}{12}\right)^2 = \frac{1}{144}$$
$$\text{to both sides}$$

$$\left(x - \frac{1}{12}\right)^2 = \frac{48 + 1}{144} \qquad \text{Rewrite}$$

$$\left(x - \frac{1}{12}\right)^2 = \frac{49}{144} \qquad \text{Solve by the square root method}$$

$$x - \frac{1}{12} = \pm\sqrt{\frac{49}{144}} = \pm\frac{7}{12}$$

$$x = \frac{1}{12} \pm \frac{7}{12}$$

$$x = \frac{1}{12} + \frac{7}{12} = \frac{8}{12} = \frac{2}{3}$$

or $\quad x = \frac{1}{12} - \frac{7}{12} = -\frac{6}{12} = -\frac{1}{2}$ ◀

Exercise Set 8.3

Complete the square, and then write the resulting expression as a binomial squared:

1. $x^2 + 2x$
2. $x^2 + 8x$
3. $m^2 - 6m$
4. $m^2 - 12m$
5. $u^2 - 3u$
6. $u^2 - 5u$
7. $y^2 + 7y$
8. $v^2 + 3v$

9. $x^2 + \dfrac{3}{2}x$ 10. $n^2 - \dfrac{5}{2}n$

Solve each equation by completing the square:

11. $x^2 + 2x - 8 = 0$ 12. $x^2 + 4x - 12 = 0$ 13. $m^2 - 4m + 3 = 0$
14. $y^2 - 6y + 5 = 0$ 15. $x^2 - x - 12 = 0$ 16. $x^2 + 3x - 10 = 0$
17. $a^2 + 6a + 4 = 0$ 18. $b^2 - 4b - 3 = 0$ 19. $2m^2 + m - 3 = 0$
20. $2u^2 - 7u - 4 = 0$ 21. $4y^2 - 2y - 3 = 0$ 22. $3x^2 - 10x - 8 = 0$

23. We have three techniques for solving quadratic equations: factoring, square root method, and completing the square. Discuss the claim that the technique of completing the square is the most powerful technique; that is, it can be used to solve more equations than the other technique.

Review

24. Simplify each expression:

    a. $\dfrac{8 - \sqrt{16}}{2}$

    b. $\dfrac{8 + \sqrt{16}}{2}$

    c. $\dfrac{-4 + \sqrt{20}}{2}$

    d. $\dfrac{-4 - \sqrt{20}}{2}$

25. Find the intercepts and graph the line whose equation is $3x + 5y = -30$.

26. Multiply:

    a. $(x + 3)(x^2 - 3x + 9)$ b. $(y + 4)(y - 4)$

    c. $(2m + 7)(2m - 7)$

27. Completely factor each polynomial:

    a. $6b^2 - 19b - 7$ b. $5x^3 + 40$

    c. $a^3 + a^2 + 5a + 5$

8.4 The Quadratic Formula

In the preceding section we were able to solve a quadratic equation by completing the square, but this method is somewhat lengthy and difficult. We now generalize this process by solving the equation $ax^2 + bx + c = 0$, $a \neq 0$ by completing the square and thereby develop a formula for solving *any* quadratic equation.

$$ax^2 + bx + c = 0, \; a \neq 0$$

$$ax^2 + bx = -c \qquad \text{Add } -c \text{ to both sides}$$

$$x^2 + \frac{b}{a}x = -\frac{c}{a}$$ 　　Divide both sides by a ($a \neq 0$)

$$x^2 + \frac{b}{a}x + \frac{b^2}{4a^2} = -\frac{c}{a} + \frac{b^2}{4a^2}$$ 　　Complete the square by adding $\left(\frac{1}{2} \cdot \frac{b}{a}\right)^2$ to both sides

$$\left(x + \frac{b}{2a}\right)^2 = \frac{-c}{a} \cdot \frac{4a}{4a} + \frac{b^2}{4a^2}$$ 　　Rewrite left side, and form LCD on right side

$$\left(x + \frac{b}{2a}\right)^2 = \frac{b^2 - 4ac}{4a^2}$$ 　　Simplify

$$x + \frac{b}{2a} = \pm\sqrt{\frac{b^2 - 4ac}{4a^2}}$$ 　　Take the square root of both sides

$$x = -\frac{b}{2a} \pm \frac{\sqrt{b^2 - 4ac}}{2a}$$ 　　Add $-\frac{b}{2a}$ to both sides and simplify radical

$$x = \frac{-b \pm \sqrt{b^2 - 4ac}}{2a}$$ 　　Combine numerators over common denominator

This result is called the **quadratic formula**. It represents the solution of a quadratic equation in standard form, $ax^2 + bx + c = 0$, $a \neq 0$.

The solutions to $ax^2 + bx + c = 0$, $a \neq 0$ are given by the *quadratic formula*:

$$x = \frac{-b \pm \sqrt{b^2 - 4ac}}{2a}$$

 If the equation we are to solve contains fractions, we first remove the fractions by multiplying both sides of the equation by the LCD.

Example 1　　Write each equation in standard form, free of fractions, and identify a, b, and c:

A.　$x^2 - 5 = -x$

$$x^2 - 5 = -x$$
$$x^2 + x - 5 = 0 \qquad \text{Standard form}$$
$$a = 1, b = 1, c = -5$$

8.4 The Quadratic Formula

B. $2x^2 = 5x$

$$2x^2 = 5x$$
$$2x^2 - 5x = 0 \qquad \text{Add } -5x \text{ to both sides}$$
$$a = 2, b = -5, c = 0$$

C. $\dfrac{x^2}{2} - \dfrac{2x}{3} + \dfrac{3}{4} = 0$

$$\dfrac{x^2}{2} - \dfrac{2x}{3} + \dfrac{3}{4} = 0$$

$$12 \cdot \left(\dfrac{x^2}{2} - \dfrac{2x}{3} + \dfrac{3}{4}\right) = 12 \cdot 0 \qquad \text{Multiply both sides by LCD} = 12$$

$$\overset{6}{\cancel{12}} \cdot \dfrac{x^2}{\cancel{2}} - \overset{4}{\cancel{12}} \cdot \dfrac{2x}{\cancel{3}} + \overset{3}{\cancel{12}} \cdot \dfrac{3}{\cancel{4}} = 0 \qquad \text{Distributive property}$$

$$6x^2 - 8x + 9 = 0 \qquad \text{Standard form}$$

$$a = 6, b = -8, c = 9$$

D. $n^2 = 9$ $n^2 - 9 = 0$

$$n^2 = 9$$
$$n^2 - 9 = 0 \qquad \text{Add } -9 \text{ to both sides}$$
$$a = 1, b = 0, c = -9$$

Example 2 Solve each of the following equations using the quadratic formula:

A. $p^2 + 4p - 12 = 0$

$$p^2 + 4p - 12 = 0, \quad a = 1, b = 4, c = -12$$

$$p = \dfrac{-b \pm \sqrt{b^2 - 4ac}}{2a} \qquad \text{Quadratic formula}$$

$$= \dfrac{-4 \pm \sqrt{(4)^2 - 4(1)(-12)}}{2(1)} \qquad \text{Substitute values}$$

$$= \dfrac{-4 \pm \sqrt{16 + 48}}{2} \qquad \text{Simplify}$$

$$= \dfrac{-4 \pm \sqrt{64}}{2}$$

$$= \dfrac{-4 \pm 8}{2}$$

$$p = 2, -6$$

B. $2s(s - 2) = s + 3$

$$2s(s - 2) = s + 3$$
$$2s^2 - 4s = s + 3 \qquad \text{Clear parentheses}$$
$$2s^2 - 5s - 3 = 0 \qquad \text{Add } -s - 3 \text{ to both sides}$$
$$a = 2, b = -5, c = -3$$
$$s = \frac{-b \pm \sqrt{b^2 - 4ac}}{2a}$$
$$= \frac{-(-5) \pm \sqrt{(-5)^2 - 4(2)(-3)}}{2 \cdot 2} \qquad \text{Substitute values}$$
$$= \frac{5 \pm \sqrt{25 + 24}}{4}$$
$$= \frac{5 \pm \sqrt{49}}{4}$$
$$= \frac{5 \pm 7}{4}$$
$$s = 3, -\frac{1}{2}$$

C. $x^2 = 4x$

$$x^2 = 4x$$
$$x^2 - 4x = 0 \qquad \text{Add } -4x \text{ to both sides}$$
$$a = 1, b = -4, c = 0$$
$$x = \frac{-b \pm \sqrt{b^2 - 4ac}}{2a} \qquad \text{Quadratic formula}$$
$$= \frac{-(-4) \pm \sqrt{(-4)^2 - 4(1)(0)}}{2 \cdot 1} \qquad \text{Substitute values}$$
$$= \frac{4 \pm \sqrt{16}}{2}$$
$$= \frac{4 \pm 4}{2}$$
$$x = 4, 0$$

D. $9y^2 - 6y + 1 = 0$

$$9y^2 - 6y + 1 = 0 \qquad \text{Standard form}$$
$$y = \frac{-b \pm \sqrt{b^2 - 4ac}}{2a} \qquad \text{Quadratic formula}$$

8.4 The Quadratic Formula

$$= \frac{-(-6) \pm \sqrt{(-6)^2 - 4(9)(1)}}{2 \cdot 9} \qquad \text{Substitute values}$$

$$= \frac{6 \pm \sqrt{36 - 36}}{18}$$

$$= \frac{6 \pm 0}{18}$$

$$y = \frac{1}{3}, \frac{1}{3}, \text{ a double root}$$

E. $x^2 = 2(x + 1)$

$$x^2 = 2(x + 1)$$
$$x^2 = 2x + 2 \qquad \text{Clear parentheses}$$
$$x^2 - 2x - 2 = 0 \qquad \text{Standard form}$$
$$a = 1, b = -2, c = -2$$
$$x = \frac{-b \pm \sqrt{b^2 - 4ac}}{2a}$$

$$= \frac{-(-2) \pm \sqrt{(-2)^2 - 4(1)(-2)}}{2 \cdot 1} \qquad \text{Substitute values}$$

$$= \frac{2 \pm \sqrt{4 + 8}}{2}$$

$$= \frac{2 \pm \sqrt{12}}{2}$$

$$= \frac{2 \pm 2\sqrt{3}}{2} \qquad \text{Simplify radical}$$

$$= \frac{2(1 \pm \sqrt{3})}{2} \qquad \text{Factor}$$

$$x = 1 \pm \sqrt{3} \qquad \text{Cancel 2s}$$

F. $m^2 + 4m + 6 = 0$

$$m^2 + 4m + 6 = 0 \qquad \text{Standard form}$$

$$m = \frac{-b \pm \sqrt{b^2 - 4ac}}{2a}$$

$$= \frac{-4 \pm \sqrt{4^2 - 4(1)(6)}}{2 \cdot 1}$$

$$= \frac{-4 \pm \sqrt{16 - 24}}{2}$$

$$= \frac{-4 \pm \sqrt{-8}}{2}$$

No real solution. $\sqrt{-8}$ is undefined in the real numbers.

Exercise Set 8.4

Write each equation in standard form $ax^2 + bx + c = 0$ free of fractions and identify a, b, and c. Do not solve.

1. $x^2 - 3x = 0$
2. $x^2 = 5x$
3. $x^2 = 9$
4. $x^2 = 3$
5. $2x^2 - 3x + 8 = 0$
6. $3x^2 - 5x = 7$
7. $\dfrac{x^2}{2} + 3x = \dfrac{1}{2}$
8. $\dfrac{x^2}{2} - \dfrac{1}{4} = \dfrac{3x}{2}$

Solve each of the following equations using the quadratic formula:

$$x = \frac{-b \pm \sqrt{b^2 - 4ac}}{2a}$$

9. $x^2 - 16 = 0$
10. $x^2 - 1 = 0$
11. $t^2 + 4 = 0$
12. $y^2 + 9 = 0$
13. $y^2 - 4y + 4 = 0$
14. $y^2 - 6y = -9$
15. $m^2 + 2m - 15 = 0$
16. $m^2 - 3m - 10 = 0$
17. $2x^2 + 1 = 3x$
18. $3x^2 + 5x = 2$
19. $x^2 = x + 20$
20. $x^2 = 5x + 14$
21. $2x^2 + 9x + 4 = 0$
22. $3u^2 + 6u + 4 = 0$
23. $x^2 + 4 = 6x$
24. $x^2 - 5x = 3 - x$
25. $n^2 - n - \frac{1}{2} = 0$
26. $x^2 + 3x = -\frac{7}{4}$
27. $x^2 = 3x$
28. $t^2 = -2t$
29. $w^2 - 4w + 8 = 0$
30. $x^2 - 6x + 6 = 0$
31. $x^2 + 4 = 2x$
32. $t^2 + 7 = -4t$

33. Describe the steps in solving a quadratic equation using the quadratic formula.

Review

34. Evaluate the expression $b^2 - 4ac$ for the following values:

   a. $a = 4$, $b = -5$, $c = 2$
   b. $a = 2$, $b = 6$, $c = 2$
   c. $a = 1$, $b = -6$, $c = 9$

35. Write each radical in simplest radical form:

    a. $\sqrt{54}$
    b. $\sqrt{180}$
    c. $\sqrt{\dfrac{2}{3}}$
    d. $\dfrac{5}{\sqrt{8}}$
    e. $\sqrt{75a^3b^2}$
    f. $\sqrt[3]{16a^5}$

36. Add and simplify:

    $$\dfrac{3}{a-2} + \dfrac{1}{a+2}$$

37. Simplify the complex fraction:

    $$\dfrac{2 - \dfrac{1}{x}}{\dfrac{3}{x^2} + 2}$$

8.5 Further Topics

The Discriminant

The two solutions of the quadratic equation $ax^2 + bx + c = 0$ are indicated by the quadratic formula $x = \dfrac{-b \pm \sqrt{b^2 - 4ac}}{2a}$. In this formula, the radicand $b^2 - 4ac$ is called the **discriminant**, and its value determines the nature, or type, of the roots of the equation, as indicated in the following chart.

$b^2 - 4ac$	Roots	Comments
a. Positive	Two distinct real roots; $\dfrac{-b + \sqrt{b^2 - 4ac}}{2a}$ $\dfrac{-b - \sqrt{b^2 - 4ac}}{2a}$	If $b^2 - 4ac$ is a perfect square, the roots are *rational*. If $b^2 - 4ac$ is not a perfect square, the roots are *irrational*.
b. Zero	One real root; $-\dfrac{b}{2a}$	Called a double root.
c. Negative	No real roots	There is no real number that is the square root of a negative number.

EIGHT   Quadratic Equations

Example 1

Write each equation in standard form, evaluate the discriminant, and state the nature of the roots:

A. $2(x^2 + 1) = 5x$

The equation must be written in standard form.

$2(x^2 + 1) = 5x$

$2x^2 + 2 = 5x$ Clear parentheses

$2x^2 - 5x + 2 = 0$ Add $-5x$ to both sides

$a = 2, b = -5, c = 2$

$b^2 - 4ac = (-5)^2 - 4(2)(2)$ Substitute values

$= 25 - 16$ Evaluate

$= 9$

This discriminant is positive and is a perfect square. So, the equation has *two distinct rational roots*.

B. $m^2 - 6m + 9 = 0$

$m^2 - 6m + 9 = 0$ In standard form

$a = 1, b = -6, c = 9$

$b^2 - 4ac = (-6)^2 - 4(1)(9)$ Substitute values

$= 36 - 36$ Evaluate

$= 0$

This discriminant equals zero, so the equation has *a double real root*.

C. $x^2 - 6x + 4 = 0$

$x^2 - 6x + 4 = 0$ In standard form

$a = 1, b = -6, c = 4$

$b^2 - 4ac = (-6)^2 - 4(1)(4)$ Substitute values

$= 36 - 16$ Evaluate

$= 20$

The discriminant is positive, but not a perfect square, so the equation has *two irrational roots*.

D. $4(x^2 - x + 2) = 1$

$4(x^2 - x + 2) = 1$ Not in standard form

$4x^2 - 4x + 8 = 1$ Clear parentheses

$4x^2 - 4x + 7 = 0$ Add -1 to both sides

8.5 Further Topics

$$a = 4, b = -4, c = 7$$

$b^2 - 4ac = (-4)^2 - 4(4)(7)$ Substitute values

$\qquad\qquad = 16 - 112$ Evaluate

$\qquad\qquad = -96$

This discriminant is negative, so the equation has *no real roots*. ◀

Example 2

A. Find the value of k for which the equation $x^2 + 6x + k = 0$ will have a double root.

To have a double root, the discriminant must equal zero.

$$a = 1, b = 6, c = k$$

$$b^2 - 4ac = 0$$

$6^2 - 4(1)(k) = 0$ Substitute values

$$36 - 4k = 0$$

$$36 = 4k$$

$k = 9$ Solve for k

B. Find the value for k for which the equation $2x^2 + x + k$ will have real roots.

To have real roots (double or distinct), the discriminant must be equal to or greater than zero. That is, for $a = 2, b = 1, c = k$,

$$b^2 - 4ac \geq 0$$

$1^2 - 4(2)k \geq 0$ Substitute values

$$1 - 8k \geq 0$$

$-8k \geq -1$ Add -1 to both members

$k \leq \dfrac{1}{8}$ Divide by -8. Reverse the sense of the inequality sign. ◀

Which Method?

In this chapter, we have learned to solve quadratic equations by:

1. Factoring (Section 8.1).
2. The square root method (Section 8.2).
3. Completing the square (Section 8.3).
4. The quadratic formula (Section 8.4).

In practice, the method of solution is not usually specified. You must decide which method should be used. We offer the following suggestions:

---

1. If the equation appears in the form $Q^2 = k$, use the *square root method*.

2. Otherwise, clear fractions (if any) and write the equation in standard form: $ax^2 + bx + c = 0$.

   2a. If $ax^2 + bx + c$ is factorable, solve by *factoring*.

   2b. If $ax^2 + bx + c$ is not factorable, solve by *using the quadratic formula*.

You may notice that the summary does not suggest using completing the square. Equations that cannot be solved by factoring are generally solved by use of the quadratic formula. This is somewhat easier than solving by completing the square.

Exercise Set 8.5

Use the discriminant to determine the nature of the roots of the following equations. Do not solve.

1. $x^2 + 3x - 4 = 0$
2. $x^2 + 5x + 7 = 0$
3. $2x^2 - 4x - 3 = 0$
4. $x^2 - 4x + 3 = 0$
5. $x^2 + 5 = 4x$
6. $4x^2 + 1 = 4x$
7. $t^2 + \frac{5}{2}t + 2 = 0$
8. $t^2 = 3t - 4$
9. $4y(y + 1) + 1 = 0$
10. $3m^2 + 2 = 5m$

Find the value(s) of k for which each of the following equations will have a double root:

11. $u^2 + 4u + k = 0$
12. $t^2 - 2t + k = 0$
13. $4y^2 - 12y + k = 0$
14. $9y^2 - 6y + k = 0$
15. $4v^2 + kv + 9 = 0$
16. $9v^2 + kv + 4 = 0$

Find the values of k for which the following equations will have real roots:

17. $2m^2 - 3m + k = 0$
18. $3x^2 - 3x + k = 0$
19. $3m^2 + 4m + k = 0$
20. $4y^2 + y + k = 0$

8.6 Applications

Solve each of the following equations, where possible, by the appropriate method:

21. $(u - 4)^2 = 100$
22. $(m + 2)^2 = 36$
23. $a^2 - 3a - 10 = 0$
24. $n^2 + 3n - 18 = 0$
25. $2(y^2 + 6) = 11y$
26. $3(y^2 + 4y + 3) = -(5y + 1)$
27. $(t - \frac{3}{2})^2 = \frac{7}{4}$
28. $(w + \frac{1}{3})^2 = \frac{7}{9}$
29. $p(p + 4) = 3$
30. $p(p + 6) = -7$
31. $2y(y - 3) = -7$
32. $m(m - 4) = m$

Review

33. Jane is 24 years older than her son, John. The sum of their ages is 88. Find their ages.

34. Jim has a savings account that pays 6 percent interest. Last year he earned $156 interest. What was his balance at the beginning of the year?

35. Divide:

    a. $\dfrac{5m^3 - 15m^2 + 5m}{5m}$
    b. $\dfrac{2a^3 - 5a^2 - 11a - 4}{2a + 1}$
    c. $\dfrac{c^2 - 4}{c + 3} \div \dfrac{5c + 10}{3c + 9}$

36. Multiply:

    a. $\sqrt{7}(\sqrt{14} + 3)$
    b. $(\sqrt{6} + x)(\sqrt{6} - x)$
    c. $(\sqrt{2x} - \sqrt{y})^2$

8.6 Applications

We begin this section with a review of some guidelines for solving an application problem:

1. Read the problem carefully. Determine what pertinent information is given and what needs to be found.

2. Jot down the important data. Organize the data in a table, or draw and label a sketch, if appropriate.

3. Represent one unknown with a variable. Express the other unknowns using the same variable if possible.

4. Write an equation representing the problem.

5. Solve the equation. Evaluate all requested unknowns.

6. Check the solution(s) in the statement of the problem.

EIGHT   Quadratic Equations

We consider here some applications that are represented by quadratic equations. It is important to note that although a quadratic equation generally has two solutions, both might *not* represent solutions of the problem. One (or both) solutions may be meaningless.

Example 1

The length of a rectangle is 3 feet longer than its width. Its area is 40 ft² (square feet). Find its dimensions.

Solution:

Let x = width.

$x + 3$ = length.

Draw and label a sketch.

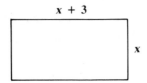

From the area formula for a rectangle, A = length · width, substitute:

$(x + 3)x = 40$

$x^2 + 3x = 40$ Clear parentheses

$x^2 + 3x - 40 = 0$ Write in standard form

$(x + 8)(x - 5) = 0$ Factor

$x + 8 = 0$ or $x - 5 = 0$ Set each factor equal to 0

$x = -8$ or $x = 5$ Solve each

In this problem, x represents the width of a rectangle, so the negative value is meaningless, and the only solution is

$x = 5$ ft, width

$x + 3 = 8$ ft, length   ◀

Example 2

The sum of a number and twice its reciprocal is $\frac{19}{3}$. Find the number.

Solution:

Let x = the number.

Then, $\dfrac{1}{x}$ = its reciprocal

and $2\left(\dfrac{1}{x}\right) = \dfrac{2}{x}$ = *twice its reciprocal.*

8.6 Applications

Translate:

The *sum* of the number and *twice its reciprocal* is $\dfrac{19}{3}$

$$x + \dfrac{2}{x} = \dfrac{19}{3}$$

Clear fractions by multiplying both sides of the equation by the LCD = $3x$.

$$3x\left(x + \dfrac{2}{x}\right) = 3x\left(\dfrac{19}{3}\right)$$

$$3x^2 + 6 = 19x$$

$3x^2 - 19x + 6 = 0$	Write in standard form
$(3x - 1)(x - 6) = 0$	Factor
$3x - 1 = 0$ or $x - 6 = 0$	Set each factor equal to 0
$3x = 1$	
$x = \dfrac{1}{3}$ or $x = 6$	Solve each

Check $x = \tfrac{1}{3}$:
  If $x = \tfrac{1}{3}$, its reciprocal is 3 and twice its reciprocal is $2 \cdot 3 = 6$. Then

$$\dfrac{1}{3} + 6 = \dfrac{1}{3} + \dfrac{18}{3} = \dfrac{19}{3} \qquad \text{Checks}$$

Check $x = 6$:
  If $x = 6$, then its reciprocal is $\tfrac{1}{6}$ and twice its reciprocal is

$$2\left(\dfrac{1}{6}\right) = \dfrac{2}{6} = \dfrac{1}{3}$$

Then

$$6 + \dfrac{1}{3} = \dfrac{18}{3} + \dfrac{1}{3} = \dfrac{19}{3} \qquad \text{Checks}$$

So, both $\tfrac{1}{3}$ and 6 are valid solutions. ◀

Example 3

It takes a boy 1 hour longer than his father to do a certain task. If it takes $1\tfrac{1}{5}$ hours ($\tfrac{6}{5}$ hours) for them to do the job together, how long would it take each to do the job alone?

Solution:
Here we use the relation $Q = R \cdot T$, where Q is the amount of work done, R is the rate of working, and T is the amount of time spent. Note that if it takes 5 hours to

do a certain task, then the *rate* of doing that task is $\frac{1}{5}$ of the task per hour, or in general, *if it takes x hours to do a task, then the rate is $\frac{1}{x}$*.

We set up a table:

Let $x =$ number of hours for the father to do the task.

$x + 1 =$ number of hours for the son to do the task.

	Hours Needed	R(Rate)
Father	x	$\dfrac{1}{x}$
Son	$x + 1$	$\dfrac{1}{x+1}$
Together	$\dfrac{6}{5}$	$\dfrac{1}{\frac{6}{5}} = \dfrac{5}{6}$

The combined rate is the sum of their individual rates and we write

$$\frac{1}{x} + \frac{1}{x+1} = \frac{1}{\frac{6}{5}}$$

$$\frac{1}{x} + \frac{1}{x+1} = \frac{5}{6}$$

Multiply both sides by the LCD $= 6x(x + 1)$.

$$6x(x+1) \cdot \left(\frac{1}{x} + \frac{1}{x+1}\right) = 6x(x+1) \cdot \frac{5}{6}$$

$6(x + 1) + 6x = 5x(x + 1)$	Multiply
$6x + 6 + 6x = 5x^2 + 5x$	Clear parentheses
$12x + 6 = 5x^2 + 5x$	Combine like terms
$0 = 5x^2 - 7x - 6$	Write in standard form
$0 = (5x + 3)(x - 2)$	Factor
$5x + 3 = 0$ or $x - 2 = 0$	Set both factors equal to 0
$5x = -3$	

8.6 Applications

$$x = -\frac{3}{5} \qquad x = 2$$

Meaningless.

Solution: $x = 2$ hours for the father.

$x + 1 = 3$ hours for the son. ◀

Example 4 If an object is thrown vertically upward at an initial velocity of v_0 ft per second (fps), then its position above ground level (s) after t seconds is given by the formula

$$s = v_0 t - 16t^2$$

Assume an object is thrown upward with an initial velocity of $v_0 = 176$ fps.

A. Find its position after 2 sec:

B. Find when it will strike the ground:

A. Substitute $t = 2$ and $v_0 = 176$

$$s = v_0 t - 16t^2$$
$$= 176(2) - 16(2)^2$$
$$= 352 - 16(4)$$
$$= 352 - 64$$
$$s = 288 \text{ ft above ground level}$$

B. When it strikes the ground, its position is $s = 0$. Substitute $s = 0$ and $v_0 = 176$.

$$s = v_0 t - 16t^2$$
$$0 = 176t - 16t^2 \qquad \text{Substitute values}$$

We now solve for t:

$$\frac{0}{16} = \frac{176t}{16} - \frac{16t^2}{16} \qquad \text{Divide by 16}$$

$$0 = 11t - t^2 \qquad \text{Standard form}$$

$$0 = t(11 - t) \qquad \text{Factor}$$

$t = 0 \quad \text{or} \quad 11 - t = 0 \qquad$ Set each factor equal to 0

$t = 0 \quad \text{or} \quad t = 11$ sec.

The first answer indicates that the object is 0 ft off the ground after 0 sec (the instant of release). The second answer, $t = 11$, is the solution to our problem. ◀

The Pythagorean Theorem

In a right triangle, the two sides that form the right angle are called the *legs* and the side opposite the right angle is the *hypotenuse*. The Pythagorean Theorem, proved in geometry, shows that the sum of the squares of the legs of a right triangle equals the square of the hypotenuse. In symbols,

$$a^2 + b^2 = c^2$$

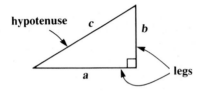

Example 5

Solve for x in the right triangle shown:

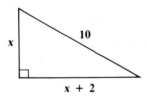

From the Pythagorean Theorem:

$$a^2 + b^2 = c^2$$

Substituting:

$x^2 + (x + 2)^2 = 10^2$	
$x^2 + x^2 + 4x + 4 = 100$	Squaring
$2x^2 + 4x - 96 = 0$	Simplify; subtract 100
$x^2 + 2x - 48 = 0$	Divide both sides by 2
$(x - 6)(x + 8) = 0$	Factor
$x - 6 = 0$ or $x + 8 = 0$	Set each factor equal to 0
$x = 6$ or $x = -8$	

↑ Meaningless

The value $x = -8$ is meaningless since we do not measure lengths with negative quantities.

So, $x = 6$
$x + 2 = 8$

Check:

$$x^2 + (x+2)^2 = 10^2$$
$$6^2 + 8^2 \stackrel{?}{=} 10^2$$
$$36 + 64 = 100$$
$$100 = 100 \checkmark$$

Exercise Set 8.6

1. Find two consecutive even natural numbers whose product is 24.
2. Find two consecutive integers whose product is 42.
3. The sum of the reciprocals of two consecutive natural numbers is $\frac{7}{12}$. Find the numbers.
4. The difference of the reciprocals of two consecutive natural numbers is $\frac{1}{20}$. Find the numbers.
5. If the length of each side of a square is increased by 1 ft, its area is increased by 13 ft². Find the length of each side of the original square.
6. If the length of each side of a square is increased by 2 ft, its area is increased by 24 ft². Find the length of each side of the original square.
7. The three sides of a right triangle have lengths that are three consecutive even integers. Find the three lengths.
8. The three sides of a right triangle have lengths that are three consecutive integers. Find the three lengths.
9. The long leg of a right triangle is 2 in. more than twice the length of the other leg. The hypotenuse has a length of 13 in. Find the length of each leg.
10. The hypotenuse of a certain right triangle is 3 cm longer than double the short leg. The other leg has a length of 12 cm. Find the length of the short leg and the hypotenuse.
11. Dale is 3 years older than Jackie. The product of their ages is 180. Find their ages.
12. Tom is 2 years younger than Tony. The product of their ages is 143. Find their ages.
13. Jan is 3 years older than Brett. The sum of the squares of their ages is 149. Find their ages.

14. Larry is 2 years younger than John. The sum of the squares of their ages is 202. Find their ages.

15. It takes John 2 more hours to paint a room than it takes Jason. Working together, they can paint a room in $2\frac{2}{5}$ hours. How long does it take each of them to paint a room?

16. It takes Al 9 minutes more to do a job than it takes Bob. Working together, they can do the job in 20 minutes. How long will it take each working alone?

17. It takes a new secretary 6 more hours to file a stack of papers than an experienced secretary. If it takes them 4 hours, working together, how long would it take each secretary working alone to file the papers?

18. It takes one inlet pipe twice as long to fill a vat than does a second inlet pipe. When both pipes are open, the vat can be filled in 2 hours. How long does it require each of the pipes to fill the vat working alone?

19. Jan takes twice as long as John to cut their lawn. If it takes $1\frac{1}{3}$ hours to cut the lawn when they work together, how long does it take each of them alone?

20. It takes Bill 3 more hours to do a task than Florence. Working together, they can do it in $3\frac{3}{5}$ hours. How long does each need to do that task alone?

21. Solve for x in the right triangle shown.

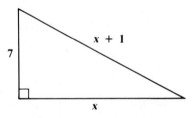

22. Solve for x in the right triangle shown.

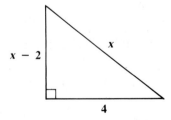

8.6 Applications

23. Find the length of the dashed diagonal line of the rectangle.

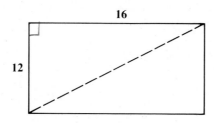

24. Find the length of the dashed diagonal line of the rectangle.

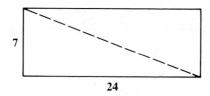

25. Find the length and width of the rectangle. (Hint: Find x first.)

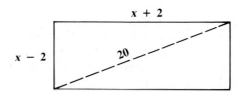

26. Find the length and width of the rectangle. (Hint: Find x first.)

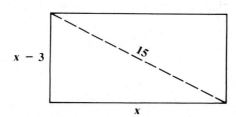

Review

27. Simplify each expression. Assume variables represent positive quantities.

   a. $\sqrt{x^2}$
   b. $\sqrt{2a}^2$
   c. $(2\sqrt{a})^2$
   d. $(\sqrt{x} + 2)^2$
   e. $(6\sqrt{y})^2$
   f. $(\sqrt{6y})^2$
   g. $(6 + \sqrt{y})^2$
   h. $(2\sqrt{x} + 3)^2$

28. Solve each of the following equations:

   a. $2u - 3 = u + 9$ b. $x(x - 1) = x + 3$

   c. $x(x + 1) = 4 + x$ d. $6y^2 - y - 1 = 0$

29. Completely factor each polynomial:

   a. $6x^2 - 17x - 14$ b. $y^3 - 125$

   c. $2ab - 5a + 4b - 10$ d. $m^2 - 8m + 15$

8.7 Solving Equations Containing Square Root Radicals

The simple equation $x = 4$ obviously has only one solution, 4. However, if we square both sides of this equation, it becomes $x^2 = 16$. This new equation is a quadratic equation whose solutions are 4 (the solution of the original equation) and -4, which is called an *extraneous solution*, meaning that it is not a solution to the original equation. In our earlier work with equations, recall that those operations on an equation that yield an *equivalent equation* are called the *elementary operations*. Squaring both sides of an equation is *not* one of the elementary operations and, as we see in the preceding example, can produce a new equation that is not equivalent to the original equation. We state without proof the following theorem:

Theorem: When both sides of an equation are squared:

a. Every solution of the original equation is also a solution of the new equation, and

b. The new equation may have solutions that are not solutions of the original equation. Each such solution is called an *extraneous solution* or *extraneous root*.

Since we will need to square both sides of certain equations in order to solve them, we review some important properties of squaring that are needed.

Properties of squaring:

$\sqrt{a^2} = a, \ a \geq 0$ Definition of a square root

$(ab)^2 = a^2 b^2$ When you square a product, each factor is squared

$(a + b)^2 = a^2 + 2ab + b^2$ Square of a binomial

The outline of the procedure for solving an equation containing radicals will first be given, followed by several examples.

8.7 Solving Equations Containing Square Root Radicals

Procedure for solving an equation containing one or two square root radicals:

1. Isolate the radical on one side of the equation. If the equation contains two radicals, isolate the more complicated one.
2. Square both sides of the equation to remove the isolated radical and simplify.
3. If the resulting equation still contains a radical, repeat Steps 1 and 2.
4. Solve the resulting equation.
5. *Check* all solutions in the original equation to see if each is valid or extraneous.

Example 1

Solve and check each equation:

A. $\sqrt{x-1} = 3$

$\sqrt{x-1} = 3$ Radical already isolated
$(\sqrt{x-1})^2 = (3)^2$ Square both sides
$x - 1 = 9$
$x = 10$

Check:
$\sqrt{x-1} = 3$
$\sqrt{10-1} \stackrel{?}{=} 3$
$\sqrt{9} = 3$
$3 = 3$ ✓

B. $\sqrt{x-1} = -3$

$\sqrt{x-1} = -3$
$(\sqrt{x-1})^2 = (-3)^2$ Square both sides
$x - 1 = 9$
$x = 10$

Check:
$\sqrt{x-1} = -3$
$\sqrt{10-1} \stackrel{?}{=} -3$
$\sqrt{9} = -3$
$3 = -3$ False

The apparent solution, $x = 10$, is extraneous. Therefore, the equation has no real solution. Note, this problem can be solved by inspection if we observe, at the beginning, that it is impossible for a principal square root, $\sqrt{x-1}$, to be negative, (-3).

C. $\sqrt{x} = 6 - x$

$$\sqrt{x} = 6 - x$$
$$(\sqrt{x})^2 = (6 - x)^2 \qquad \text{Square both sides}$$
$$x = 36 - 12x + x^2$$
$$0 = 36 - 13x + x^2 \qquad \text{Add } -x \text{ to both sides}$$
$$x^2 - 13x + 36 = 0 \qquad \text{Exchange sides}$$
$$(x - 4)(x - 9) = 0 \qquad \text{Factor}$$
$$x - 4 = 0 \quad \text{or} \quad x - 9 = 0 \qquad \text{Set each factor equal to 0}$$
$$x = 4 \qquad x = 9 \qquad \text{Apparent solutions}$$

Check:

$x = 4$:
$\sqrt{x} = 6 - x$
$\sqrt{4} \stackrel{?}{=} 6 - 4$
$2 = 2$ True

$x = 9$:
$\sqrt{x} = 6 - x$
$\sqrt{9} \stackrel{?}{=} 6 - 9$
$3 = -3$ False

This equation has only one solution, $x = 4$. The other, $x = 9$, is extraneous.

D. $\sqrt{5c + 1} - \sqrt{c - 2} = 3$

$$\sqrt{5c + 1} - \sqrt{c - 2} = 3$$
$$\sqrt{5c + 1} = 3 + \sqrt{c - 2} \qquad \text{Add } \sqrt{c - 2} \text{ to both sides to isolate one radical}$$
$$(\sqrt{5c + 1})^2 = (3 + \sqrt{c - 2})^2 \qquad \text{Square both sides}$$
$$5c + 1 = 9 + 6\sqrt{c - 2} + \sqrt{c - 2}^2 \qquad \text{Note } (a + b)^2 = a^2 + 2ab + b^2$$
$$5c + 1 = 9 + 6\sqrt{c - 2} + c - 2$$
$$5c + 1 = 7 + 6\sqrt{c - 2} + c \qquad \text{Combine like terms}$$
$$4c - 6 = 6\sqrt{c - 2} \qquad \text{Add } -c \text{ and } -7 \text{ to both sides}$$
$$2c - 3 = 3\sqrt{c - 2} \qquad \text{Divide both sides by 2}$$
$$(2c - 3)^2 = (3\sqrt{c - 2})^2 \qquad \text{Square both sides}$$
$$4c^2 - 12c + 9 = 9(c - 2) \qquad \text{Square of a binomial and square of a product}$$
$$4c^2 - 12c + 9 = 9c - 18 \qquad \text{Clear parentheses}$$

8.7 Solving Equations Containing Square Root Radicals

$4c^2 - 21c + 27 = 0$	Quadratic—standard form
$(4c - 9)(c - 3) = 0$	Factor
$4c - 9 = 0$ or $c - 3 = 0$	Set each factor equal to 0
$4c = 9$	
$c = \dfrac{9}{4}$ $c = 3$	Solve each

You may check and verify that both solutions are valid.

Exercise Set 8.7

Solve and check each equation:

1. $\sqrt{x - 1} = 2$
2. $\sqrt{x + 2} = 4$
3. $\sqrt{y + 2} = 3$
4. $\sqrt{y - 3} = 2$
5. $\sqrt{3u - 3} = 6$
6. $\sqrt{2u - 3} = 7$
7. $\sqrt{m + 3} = -2$
8. $\sqrt{2m + 2} = -4$
9. $\sqrt{2x - 3} + 7 = 2$
10. $\sqrt{3x - 1} + 5 = 3$
11. $\sqrt{3y + 1} - 5 = 3$
12. $\sqrt{2a + 3} - 7 = 2$
13. $-5 = \sqrt{x + 3} - 9$
14. $-4 = \sqrt{y + 2} - 5$
15. $\sqrt{3 - a} + 2 = 2$
16. $\sqrt{4 - x + 3} = 3$
17. $\sqrt{a + 2} = a - 4$
18. $\sqrt{a + 3} = a - 3$
19. $\sqrt{3x} = 2x - 3$
20. $\sqrt{2x} = x - 4$
21. $\sqrt{x^2 + 2} - 2 = x$
22. $\sqrt{y^2 + 1} = y - 2$
23. $\sqrt{2a + 7} - a = 2$
24. $\sqrt{3x + 4} = x$
25. $\sqrt{u + 3} + u = 3$
26. $\sqrt{y + 1} = y - 5$
27. $\sqrt{y + 1} = 1 + \sqrt{y - 2}$
28. $\sqrt{b - 2} = 5 - \sqrt{b + 3}$
29. $\sqrt{x + 3} = \sqrt{2x + 4} - 1$
30. $\sqrt{2p + 1} = \sqrt{p - 3} + 2$
31. $\sqrt{x + 1} = 1 - \sqrt{2x + 3}$
32. $\sqrt{x - 2} + 1 = \sqrt{x + 3}$

33. In an equation such as Problem 25, why is it necessary to isolate the radical before squaring both sides of the equation?

Review

34. Evaluate the polynomial $c^2 - 3c + 4$ for the given values of c:

   a. 0   b. 2   c. 4

d. 6   e. -2 f. -4

35. Evaluate $x^2 - 6x + 8$ for the given values of x:

     a. 0   b. 1   c. 2

     d. 3   e. 4   f. 5

36. Graph each equation:

     a. $2x - y = 6$ b. $x - 2y = 6$ c. $y = \tfrac{1}{2}x - 6$

37. Write each expression in simplest radical form:

     a. $\sqrt{12a^3}$ b. $\dfrac{5}{\sqrt{6}}$

     c. $\dfrac{2}{\sqrt{5} - 1}$ d. $\sqrt{5}(2\sqrt{5} + 1)$

8.8 The Graph of $y = ax^2 + bx + c$ (Optional)

A linear equation in two variables has as its graph a straight line (see Chapter 3). The equation $y = ax^2 + bx + c$ is not linear, since it contains a variable whose exponent is 2. Consequently, we would expect that the graph of such an equation would be something other than a straight line. We show in Example 1 the graph of one such equation.

Example 1

Draw the graph of the equation $y = x^2 - 2x - 8$.

To draw the graph, we select several integer values of x and, for each value, compute the corresponding value of y.

For $x = 0$:

$y = 0^2 - 2 \cdot 0 - 8$

$ = -8$

For $x = 1$:

$y = 1^2 - 2 \cdot 1 - 8$

$ = 1 - 2 - 8$

$ = -9$

8.8 The Graph of $y = ax^2 + bx + c$ (Optional)

For $x = -1$:
$$y = (-1)^2 - 2(-1) - 8$$
$$= 1 + 2 - 8$$
$$= -5$$

Continuing in this manner, we obtain the following chart of values:

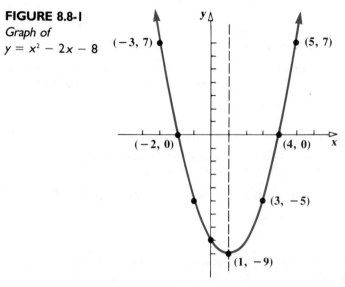

FIGURE 8.8-1
Graph of
$y = x^2 - 2x - 8$

x	y
-3	7
-2	0
-1	-5
0	-8
1	-9
2	-8
3	-5
4	0
5	7

These points are then plotted in the x-y plane and joined by a smooth curve, as shown in Figure 8.8-1. Such a curve is called a **parabola**. Note in the figure that points are mirror images or reflections of each other across the vertical dashed line shown, called the **axis of symmetry**, or more simply the **axis** of the parabola. The point where the axis intercepts the graph is the **vertex** of the parabola.

To aid in drawing the graph, the points where the graph crosses the axes can be found. Recall every point on the y-axis has an x-coordinate of zero. So, to find the y-intercept, we set $x = 0$ and compute $y = 0^2 - 2 \cdot 0 - 8 = -8$. Similarly, to find the x-intercept(s), if any exist, set $y = 0$ yielding, in this example,

$$y = x^2 - 2x - 8$$
$$0 = x^2 - 2x - 8$$

Solve this quadratic equation by factoring:

$$0 = (x - 4)(x + 2)$$
$$x - 4 = 0 \quad \text{or} \quad x + 2 = 0$$
$$x = 4 \qquad\qquad x = -2$$

The x-intercepts in this example are (4, 0) and (-2, 0), as can be seen in the graph.

EIGHT   Quadratic Equations

It can be shown that the axis has as its equation, $x = -\dfrac{b}{2a}$. This can be easily remembered by noting that this is the first term of the quadratic formula:

$$x = \dfrac{-b \pm \sqrt{b^2 - 4ac}}{2a}$$

The value $x = -\dfrac{b}{2a}$ is also the x-coordinate of the vertex. ◀

The General Case

Rather than plotting points to draw the graph of $y = ax^2 + bx + c$, we prefer to graph by plotting only the special points, as indicated in the following generalization.

TO DRAW THE GRAPH OF THE PARABOLA $y = ax^2 + bx + c$

1. Sketch the axis of symmetry, a vertical line whose equation is

$$x = \dfrac{-b}{2a}$$

2. Substitute $x = -\dfrac{b}{2a}$ into the equation to find the y-coordinate of the vertex and then plot the vertex.

3. Find the intercepts and plot.

    a. Set $x = 0$, yielding $(0, c)$

    b. Set $y = 0$, then solve

    $$ax^2 + bx + c = 0$$

    to find the x-intercepts (if one or two exist)

4. The parabola opens up (∪) if a is positive and opens down (∩) if a is negative.

5. If necessary, select one or two values of x and compute the corresponding y-values and plot them.

Example 2

Graph each of the following parabolas:

A. $y = x^2 - 2x + 3$

$y = x^2 - 2x + 3$; $a = 1, b = -2, c = 3$

8.8 The Graph of $y = ax^2 + bx + c$ (Optional)

1. Axis of symmetry:
$$x = \frac{-b}{2a} = \frac{-(-2)}{2(1)} = \frac{2}{2} = 1; \; x = 1$$

2. When $x = 1$, $y = 1^2 - 2(1) + 3$
$$= 1 - 2 + 3$$
$$= 2 \quad \text{Vertex: } (1, 2)$$

3. Intercepts:
   a. Set $x = 0$, then $y = 3$; $(0, 3)$ y-intercept
   b. Set $y = 0$ and solve $x^2 - 2x + 3 = 0$

   Not factorable

   $$x = \frac{2 \pm \sqrt{4 - 4(1)(3)}}{2} \quad \text{Substitute into quadratic formula}$$
   $$= \frac{2 \pm \sqrt{4 - 12}}{2}$$
   $$= \frac{2 \pm \sqrt{-8}}{2}$$

   No real solutions, therefore no x-intercepts.

4. The parabola opens up ($a = 1 > 0$).

5. Select another value of x, such as $x = 3$, giving
$$y = 3^2 - 2(3) + 3$$
$$= 9 - 6 + 3 = 6$$
$(3, 6)$ a point on the graph

Sketch:

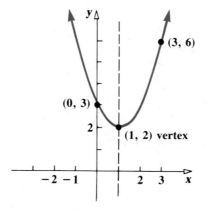

B. $y = 6 + x - x^2$

Note: $a = -1, b = 1, c = 6$

1. Axis of symmetry:

$$x = \frac{-b}{2a} = \frac{-1}{2(-1)} = \frac{-1}{-2} = \frac{1}{2}; x = \frac{1}{2}$$

2. When $x = \frac{1}{2}$

$$y = 6 + x - x^2$$
$$= 6 + \frac{1}{2} - \left(\frac{1}{2}\right)^2 = 6 + \frac{1}{2} - \frac{1}{4}$$
$$= \frac{24 + 2 - 1}{4}$$
$$y = \frac{25}{4} \quad \text{Vertex:} \left(\frac{1}{2}, \frac{25}{4}\right)$$

3. Intercepts:

   a. When $x = 0, y = 6;$ $(0, 6)$ y-intercept

   b. When $y = 0$

   $$0 = 6 + x - x^2$$
   $$x^2 - x + 6 = 0$$
   $$(x - 3)(x + 2) = 0$$
   $$x = 3 \quad \text{or} \quad x = -2 \quad (3, 0), (-2, 0) \text{ } x\text{-intercepts}$$

4. Parabola opens *down* $(a = -1 < 0)$.

5. No further points should be needed.

Sketch:

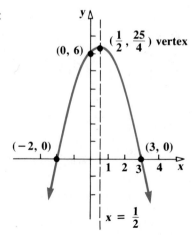

Exercise Set 8.8

Sketch the graph of each of the following parabolas:

1. $y = x^2 - 4x + 3$
2. $y = x^2 - 6x + 8$
3. $y = x^2 - 4x - 5$
4. $y = x^2 + 4x - 5$
5. $y = x^2 - 4x + 4$
6. $y = x^2 - 6x + 9$
7. $y = 2x - x^2$
8. $y = 3x - x^2$
9. $y = x^2 - 9$
10. $y = x^2 - 4$
11. $y = 2x^2 - 5x + 2$
12. $y = 2x^2 - 5x - 3$
13. $y = -x^2 + 2x - 4$
14. $y = -x^2 - 4x - 5$
15. $y = x^2 + 3$
16. $y = -1 - x^2$
17. $y = (x + 2)^2$
18. $y = (x - 1)^2$
19. $y = (x + 2)^2 - 3$
20. $y = (x - 1)^2 + 2$

21. Compare the equations and the graphs in exercises 17 and 19. How are they related to each other?

22. Compare the equations and the graphs in exercises 18 and 20. How are they related to each other?

Review

23. Solve each equation:
    a. $x - 4 = 0$
    b. $(x - 4)(x - 2) = 0$
    c. $(x - 4)(x - 2) = 3$
    d. $\sqrt{x - 4} = 3$
    e. $\sqrt{x - 5} = \sqrt{x - 1}$

24. Solve the following equation for y:
    $$2x - 3y = 12$$

25. Find the slope of the graph of $2x - 3y = 12$.

26. Draw each of the graphs:
    a. $2x - 3y = 12$
    b. $2x - 3y > 12$

8.9 Review and Chapter Test

(8.1) A quadratic equation is an equation that can be written in the form $ax^2 + bx + c = 0$, $a \neq 0$, called *standard form*. A quadratic equation in standard form may be solved by factoring if the left side is factorable and setting each linear factor equal to zero.

Example 1

Solve the equation $2x^2 - x + 3 = 9$.

$$2x^2 - x + 3 = 9$$
$$2x^2 - x - 6 = 0 \qquad \text{Subtract 9 from both sides}$$
$$(2x + 3)(x - 2) = 0 \qquad \text{Factor the left side}$$
$$(2x + 3) = 0 \quad \text{or} \quad x - 2 = 0 \qquad \text{Set each factor to equal 0}$$
$$2x = -3$$
$$x = -\frac{3}{2} \quad \text{or} \quad x = 2 \qquad \text{Solve both equations} \quad \blacktriangleleft$$

(8.2) A quadratic equation that is written, or may easily be written, in the form $Q^2 = k$, where Q is an expression containing the variable and k is a non-negative constant, may be solved using the definition of a square:

$$\text{If } Q^2 = k \quad \text{then} \quad Q = \pm\sqrt{k}$$

Example 2

Solve each equation:

A. $x^2 = 16$

$$x^2 = 16$$
$$x = \pm\sqrt{16} \qquad \text{Square root of both sides}$$
$$x = \pm 4$$

B. $(3a - 5)^2 = 12$

$$(3a - 5)^2 = 12$$
$$3a - 5 = \pm\sqrt{12} \qquad \text{Square root of both sides}$$
$$3a = 5 \pm 2\sqrt{3} \qquad \text{Add 5 and simplify the radical}$$
$$a = \frac{5 \pm 2\sqrt{3}}{3} \qquad \text{Divide both sides by 3}$$

Note, the answer may be left in this final form, or a calculator with a square root key may be used to evaluate.

$$a = \frac{5 + 2\sqrt{3}}{3} \approx 2.821$$

$$a = \frac{5 - 2\sqrt{3}}{3} \approx 0.512$$

(8.3) A quadratic equation may be written from the standard form $ax^2 + bx + c = 0$ into the form $Q^2 = k$ and then solved by the square root method. The process is called *solving by completing the square*. To complete the square of an expression of the type $x^2 \pm mx$, add $(\frac{1}{2} \cdot m)^2$.

Example 3

Complete the square and factor the following expressions:

A. $x^2 - 6x$

$$\text{Add } \left[\frac{1}{2} \cdot (6)\right]^2 = (3)^2 = 9$$

$$x^2 - 6x + 9 = (x - 3)^2$$

B. $y^2 + \frac{3}{4}y$

$$\text{Add } \left[\frac{1}{2} \cdot \frac{3}{4}\right]^2 = \left(\frac{3}{8}\right)^2 = \frac{9}{64}$$

$$y^2 + \frac{3}{4}y + \frac{9}{64} = \left(y + \frac{3}{8}\right)^2$$

We solve a quadratic equation by completing the square as in Example 4.

Example 4

Solve $2y^2 - 5y + 4 = 0$ by completing the square:

$$2y^2 - 5y + 2 = 0$$

$2y^2 - 5y \quad = -2 \quad$ Put the constant on the right

$y^2 - \frac{5}{2}y \quad = -1 \quad$ Divide both sides by the coefficient of the squared term (2)

Complete the square:

$y^2 - \frac{5}{2}y + \frac{25}{16} = -1 + \frac{25}{16} \quad$ Add $\left[\frac{1}{2} \cdot \frac{5}{2}\right]^2$ to both sides

$\left(y - \frac{5}{4}\right)^2 = \frac{-16 + 25}{16} \quad$ Rewrite the left side as a binomial squared, and simplify the right side

$\left(y - \frac{5}{4}\right)^2 = \frac{9}{16} \quad$ The equation is now in the form $Q^2 = k$ and can be solved by the square root method

$$y - \frac{5}{4} = \pm\sqrt{\frac{9}{16}}$$

$$y = \frac{5}{4} \pm \frac{3}{4} \qquad \text{Add } \frac{5}{4} \text{ and simplify radical}$$

$$y = \frac{5+3}{4} \quad \text{or} \quad \frac{5-3}{4}$$

$$y = 2, \frac{1}{2}$$

◀

(8.4) Solving the quadratic equation $ax^2 + bx + c = 0$ by completing the square yields the *quadratic formula*:

$$x = \frac{-b \pm \sqrt{b^2 - 4ac}}{2a}, \quad a \neq 0$$

This formula may be used to solve *any* quadratic equation.

Example 5 Use the quadratic formula to solve the equation $\frac{y^2 + 3}{4} = y + 1$.

$$\frac{y^2 + 3}{4} = y + 1 \qquad \text{Not in standard form}$$

$$y^2 + 3 = 4y + 4 \qquad \text{Multiply by } 4 = \text{LCD}$$

$$y^2 - 4y - 1 = 0 \qquad \text{Standard form}$$

$$a = 1, b = -4, c = -1$$

$$y = \frac{-b \pm \sqrt{b^2 - 4ac}}{2a} \qquad \text{Quadratic formula}$$

$$= \frac{-(-4) \pm \sqrt{(-4)^2 - 4(1)(-1)}}{2(1)} \qquad \text{Substitute } a = 1, b = -4, c = -1$$

$$= \frac{4 \pm \sqrt{16 + 4}}{2}$$

$$= \frac{4 \pm \sqrt{20}}{2}$$

$$= \frac{4 \pm 2\sqrt{5}}{2}$$

$$= \frac{\cancel{2}(2 \pm \sqrt{5})}{\cancel{2}} \qquad \text{Simplify}$$

$$y = 2 \pm \sqrt{5}$$

◀

(8.5) The nature of the roots of a quadratic equation $ax^2 + bx + c = 0, a \neq 0$, can be predicted by evaluating the *discriminant*, $b^2 - 4ac$.

$b^2 - 4ac$		Roots	Comments
a.	Positive	Two distinct real roots $$x = \frac{-b + \sqrt{b^2 - 4ac}}{2a}$$ $$x = \frac{-b - \sqrt{b^2 - 4ac}}{2a}$$	If $b^2 - 4ac$ is a perfect square, the roots are rational. If $b^2 - 4ac$ is not a perfect square, the roots are *irrational*.
b.	Zero	One real root; $-\frac{b}{2a}$	Called a double root.
c.	Negative	No real roots	There is no real number that is the square root of a negative number.

Example 6

Determine the nature of the roots of each equation:

A. $3x^2 - 2x - 8 = 0$

$3x^2 - 2x - 8 = 0$

$a = 3, b = -2, c = -8$

$b^2 - 4ac = (-2)^2 - 4(3)(-8)$

$= 4 + 96$

$= 100$ A perfect square

The equation has two rational roots.

B. $9x^2 + 12x + 4 = 0$

$9x^2 + 12x + 4 = 0$

$a = 9, b = 12, c = 4$

$b^2 - 4ac = (12)^2 - 4(9)(4)$

$= 144 - 144$

$= 0$

The equation has a double root.

C. $x^2 + 6x + 11 = 0$

$x^2 + 6x + 11 = 0$

$a = 1, b = 6, c = 11$

$$b^2 - 4ac = (6)^2 - 4(1)(11)$$
$$= 36 - 44$$
$$= -8$$

This equation has no real roots. ◀

It is suggested that the method used to solve a quadratic equation is determined as follows:

1. If the equation can be easily written in the form $Q^2 = k$, use the *square root* method.
2. Otherwise, write the equation in standard form $ax^2 + bx + c = 0$, $a \neq 0$.
   a. If the left side is factorable, solve by *factoring*.
   b. If the left side is not factorable, solve using the *quadratic formula*.

(8.6) To solve an equation containing one or two square root radicals:

1. Isolate a radical.
2. Square both sides and simplify.
3. If the equation still contains a radical, repeat Steps 1 and 2.
4. Solve the equation.
5. Check all solutions.

Example 7

Solve $\sqrt{x + 5} = \sqrt{x} + 1$.

$$\sqrt{x + 5} = \sqrt{x} + 1$$
$$(\sqrt{x + 5})^2 = (\sqrt{x} + 1)^2 \qquad \text{Square both sides}$$
$$x + 5 = x + 2\sqrt{x} + 1$$
$$4 = 2\sqrt{x} \qquad \text{Simplify}$$
$$2 = \sqrt{x} \qquad \text{Divide by 2}$$
$$(2)^2 = (\sqrt{x})^2 \qquad \text{Square both sides}$$
$$4 = x$$

Check:
$$\sqrt{4 + 5} \stackrel{?}{=} \sqrt{4} + 1$$
$$\sqrt{9} = 2 + 1$$
$$3 = 3 \checkmark$$
◀

(8.7) Applications are solved by translating to an equation and solving the equation. All solutions should be checked in the original problem.

8.9 Review and Chapter Test

Example 8

Jim can do a job in 1 more hour than Mary. Working together they can do the job in $1\frac{5}{7}$ hours. How long does it take each to complete the job alone? (Note: $1\frac{5}{7} = \frac{12}{7}$.)

	Time Required Alone	Rate
Jim	$x + 1$	$\dfrac{1}{x + 1}$
Mary	x	$\dfrac{1}{x}$
Together	$\dfrac{12}{7}$	$\dfrac{1}{\frac{12}{7}} = \dfrac{7}{12}$

$$\frac{1}{x+1} + \frac{1}{x} = \frac{7}{12}$$

LCD = $12(x)(x + 1)$

$$12x(x+1) \cdot \frac{1}{x+1} + 12x(x+1) \cdot \frac{1}{x} = 12x(x+1) \cdot \frac{7}{12}$$

$$12x + 12(x + 1) = x(x + 1)7$$

$$12x + 12x + 12 = 7x^2 + 7x$$

$$24x + 12 = 7x^2 + 7x$$

$$0 = 7x^2 - 17x - 12$$

$$0 = (7x + 4)(x - 3) \quad \text{Factor}$$

$7x + 4 = 0$ or $x - 3 = 0$ Set each equal to 0

$x = -\dfrac{4}{7}$ $x = 3$ hours, Mary's time

$x + 1 = 4$ hours, Jim's time

Negative time is meaningless here. ◀

(8.8) (Optional) To draw the graph of a parabola whose equation is of the form $y = ax^2 + bx + c$:

1. Sketch the axis of symmetry, whose equation is $x = -\frac{b}{2a}$
2. Find the vertex at $\left(-\frac{b}{2a}, \quad \right)$.
   ↳ Find the y-coordinate by substituting $x = -\frac{b}{2a}$ into the original equation.
3. Find the intercepts (when they exist):
   a. Set $x = 0$, solve for y ($y = c$).

b. Set $y = 0$, solve $ax^2 + bx + c = 0$.
If this equation has one or two real solutions, they represent the x-intercept(s).

4. The parabola opens up if $a > 0$; the parabola opens down if $a < 0$.

5. Plot additional point(s) if necessary.

Example 9

Graph $y = x^2 - 5x$.

$y = x^2 - 5x$; $a = 1, b = -5, c = 0$

1. Axis:

$$x = \frac{-b}{2a} = \frac{-(-5)}{2(1)} = \frac{5}{2}; \quad x = \frac{5}{2}$$

2. Substitute $x = \frac{5}{2}$.

$$y = x^2 - 5x$$
$$= \left(\frac{5}{2}\right)^2 - 5\left(\frac{5}{2}\right)$$
$$= \frac{25}{4} - \frac{25}{2} \cdot \frac{2}{2}$$
$$= \frac{25}{4} - \frac{50}{4}$$
$$= -\frac{25}{4} \quad \text{Vertex: } \left(\frac{5}{2}, -\frac{25}{4}\right)$$

3. Intercepts:

a. Set $x = 0$: $y = x^2 - 5x$
$= 0^2 - 5(0)$
$= 0$

y-intercept: $(0, 0)$

b. Set $y = 0$: $0 = x^2 - 5x$
$0 = x(x - 5)$
$x = 0, x = 5$

x-intercepts: $(0, 0), (5, 0)$

4. Parabola opens up ($a = 1, a > 0$).

5. No additional points needed.

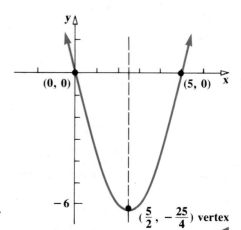

Exercise Set 8.9 Review

Solve each equation *by factoring*:

1. $4a^2 - 9a + 2 = 0$
2. $m^2 + 3m = 0$
3. $u^2 - 100 = 0$

Solve each equation *by the square root method*:

4. $y^2 = 81$
5. $(2x + 5)^2 = 16$
6. $(x - 3)^2 = 7$
7. $(3t + 4)^2 = -20$

Solve each equation *by completing the square*:

8. $2t^2 + 11t + 5 = 0$
9. $C^2 - 8C + 13 = 0$

Use the discriminant to determine the nature of the roots; then solve each equation *by using the quadratic formula*:

10. $x^2 + 8x + 15 = 0$
11. $2y^2 + y - 1 = 0$
12. $x^2 + 4x - 1 = 0$
13. $y^2 - 3y = 0$
14. $m^2 + 2m + 3 = 0$

Find the value of k so that each equation will have a double root:

15. $x^2 - 4x + k = 0$
16. $x^2 + kx + 16 = 0$

Solve each of the following equations by the appropriate method:

17. $m^2 - 2m - 2 = \dfrac{m}{3}$
18. $(t + 4)^2 = 5$
19. $y^2 - y - 4 = y + 2$

Solve each equation. Check for extraneous solutions:

20. $\sqrt{2x - 1} = 5$
21. $\sqrt{2x - 2} = x - 1$
22. $\sqrt{2x - 1} = \sqrt{x + 1}$

23. The length of a rectangle is 2 cm longer than the width, and the area is 35 cm². Find the length and width.

24. It takes Janie 2 more hours than it takes Ralph to clean the basement. If it takes them $\frac{4}{3}$ hour to clean it working together, how long does it take each, working alone?

Sketch the graph of each parabola:

25. $y = x^2 + 2x - 3$
26. $y = 9 - x^2$
27. $y = x^2 + 4x + 7$

Chapter 8 Review Test

Solve each of the following equations by the method indicated:

1. Solve by factoring:

   a. $6y^2 + y - 2 = 0$
   b. $x^2 + 6x = 0$

2. Solve by the square root method:

   a. $x^2 = 7$
   b. $(2a - 3)^2 = 16$

3. Solve by completing the square:

   a. $y^2 + 4y + 3 = 0$
   b. $3m^2 + 7m - 6 = 0$

4. Solve each equation by using the quadratic formula:

   a. $x^2 + 4x - 3 = 0$
   b. $2m^2 + 11m + 5 = 0$

5. Use the discriminant to determine the nature of the roots of each equation. Do not solve.

   a. $t^2 - t + 1 = 0$
   b. $5y^2 - 8y - 4 = 0$
   c. $4m^2 - 4m + 1 = 0$
   d. $p^2 - 2p - 6 = 0$

6. Solve each equation by any method:

   a. $2y = \dfrac{-1}{y + 3}$
   b. $(2p + 7)^2 = 4$
   c. $3t(t - 4) - 3t = 2(t - 5) - 10$

7. Solve and check:

   $\sqrt{x + 4} + 2 = x$

8. The sides of a right triangle are three consecutive even integers. Find the three lengths.

9. Sketch the graph of $y = x^2 - 3x - 4$.

CHAPTER NINE

RELATIONS AND FUNCTIONS

This chapter is a brief introduction to relations and functions. You will study these topics in more detail in subsequent mathematics course. Relations and functions and the associated notation are used extensively in the fields of business, economics, physical and natural science, mathematics, engineering, and many others.

9.1 Relations and Functions

We often use the words "relation" or "related to" in our day-to-day conversation. Examples include saying a student's grades are related to his/her time spent studying; the cost of repair of an appliance is related to the number of hours required to fix it; each student is assigned (or related to) a student number; and the price of an item is related to the demand.

A relation may be viewed as a pairing of elements in one set with elements in some other set. These pairings can be represented as ordered pairs. An element of the first set is the first component of the ordered pair, and its corresponding element from the second set is the second component of the ordered pair.

> **DEFINITION**
>
> **A relation is a set of ordered pairs. The set of all first components is the domain of the relation, and the set of all second components is the range of the relation.**

Example 1

Set R is the set showing 4 college roommates and their student numbers. Find the domain and range of R.

$$R = \{(\text{Tom}, 12345), (\text{Jerry}, 12543), (\text{Bill}, 13027), (\text{Rick}, 13692)\}$$

The domain is the set of all first components:

Domain = {Tom, Jerry, Bill, Rick}

The range is the set of all second components:

Range = {12345, 12543, 13027, 13692}

We can view these pairings by showing the sets:

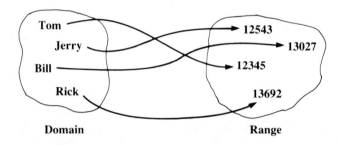

In mathematics and physical science, the members of the domain and range are often numbers, and the relation may be expressed with a formula or graph (or both).

Example 2

Write an equation and draw the graph of the relation that expresses the distance (d) traveled in time (t) of between 0 and 10 hours when the rate is 50 mph.

From the formula $d = rt$, with the rate $r = 50$, we write $d = 50t$. To show that the time is between 0 and 10 hours, we write $0 \leq t \leq 10$, or together:

$$d = 50t, \; 0 \leq t \leq 10$$

To draw the graph, plot t on the horizontal axis and d on the vertical axis.

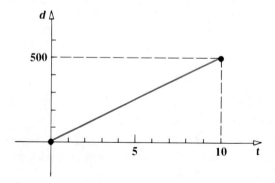

Traditionally, the variable representing the domain (in the preceding example, t) is associated with the horizontal axis, and the variable representing the range (above, d) is associated with the vertical axis. Although a relation is often expressed by an equation or formula, it is not necessary, as illustrated in the following example.

Example 3

Draw the graph of the given relation, R, and indicate the domain and the range.

$$R = \{(0, 0), (1, 1), (1, -1), (2, 4), (2, -4)\}$$

Graph:

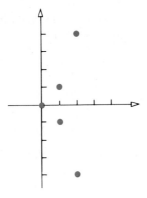

Domain: $\{0, 1, 2\}$

Range: $\{0, 1, -1, 4, -4\}$ ◀

Functions

A function is a special type of relation in which each domain element is associated with a single range element. More formally:

DEFINITION

A function is a relation in which no two distinct ordered pairs have the same first component.

Example 4

Draw the graph of the relation

$$\{(0, 0), (1, 1), (-1, 1), (2, 4), (-2, 4)\}$$

Specify its domain and range. Is it a function?

Graph: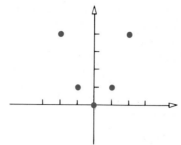

Domain: $\{-2, -1, 0, 1, 2\}$

Range: $\{0, 1, 4\}$

Yes, it is a function since, according to the definition, no two distinct ordered pairs have the same first component. That is, for each x, there is one and only one y. ◀

Example 5 Draw the graph of the relation $\{(2, 3), (2, -4)\}$. Specify its domain and range. Is it a function?

Graph: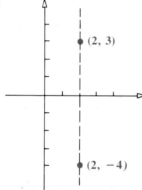

Domain: $\{2\}$

Range: $\{3, -4\}$

No, it is not a function. $(2, 3)$ and $(2, -4)$ have the same first component. ◀

The preceding example is almost trivial but is included to illustrate the graphical interpretation of a relation that *is* or *is not* a function. Observe in the graph of Example 5 that since the two points have the same first component, they lie on the same vertical line (shown dashed in the graph). This observation is generalized and called the *vertical line test*.

VERTICAL LINE TEST

A relation is a function if every vertical line intersects its graph in at most one point.

9.1 Relations and Functions

Example 6

Indicate whether each of the following is the graph of a function:

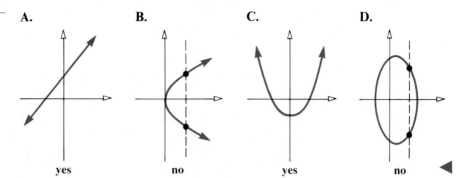

A. yes   B. no   C. yes   D. no

In parts of **B** and **D** of Example 6, the relations whose graphs are shown are not functions since we are able to find at least one vertical line that intersects the graphs at more than one point. These graphs violate the vertical line test.

Exercise Set 9.1

In Exercises 1–4, write each relation as a set of ordered pairs. Indicate whether the relation is or is not a function. List the elements of the domain and range of each.

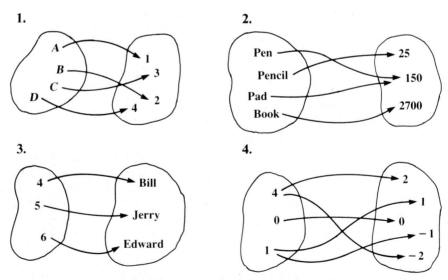

In Exercises 5–10, indicate the domain and range of each relation; draw the graph of each relation, and indicate whether the relation is or is not a function.

5. $A = \{(1, -3), (2, -2), (3, -1), (4, 0), (5, 1)\}$

6. $B = \{(-2, 4), (-1, 1), (0, 0), (1, 1), (2, 4)\}$
7. $C = \{(-4, 1), (0, 1), (4, 1)\}$
8. $D = \{(0, 0), (1, 0), (2, 0), (3, 0)\}$
9. $E = \{(-2, 2), (-2, -2), (2, 2), (2, -2)\}$
10. $F = \{(0, 0), (1, -1), (2, -1), (3, 0), (1, 1), (2, 1)\}$

In Exercises 11–22, x is a domain element and is associated with the horizontal axis, and y is a range element and is associated with the vertical axis. Draw the graph of each relation. Indicate whether the relation is or is not a function.

11. $y = 2x + 1$
12. $y = 3x - 1$
13. $y = 2x + 1, 0 \leq x \leq 5$
14. $y = 3x - 1, -2 \leq x \leq 2$
15. $y \leq 2x + 1$
16. $y > 3x - 1$
17. $2x + 3y = 6$
18. $3x - 2y = 6$
19. $y = 4$
20. $y = 2$
21. $x = -1$
22. $x = 5$

In Exercises 23–31, indicate whether the graph is or is not the graph of a function.

23.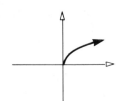
24.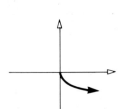
25.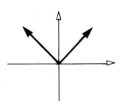

26.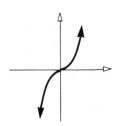
27.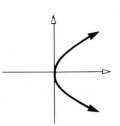
28.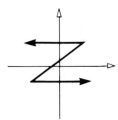

29.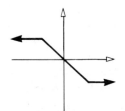
30.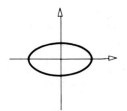
31.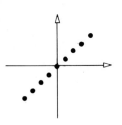

32. What is the distinction between a relation and a function?

Review

33. Evaluate $x^2 - 3x + 5$ for the given values of x:

    a. 2   b. -2 c. 4   d. -3

34. Evaluate, or simplify, the expression $2x - 3$ by replacing x with:

    a. 0   b. 3   c. -2 d. $2 + h$ e. $3u$

35. Given $y = 3x + 4$, fill in the following table:

x	y
-2	
0	
2	
4	

36. Solve the following system of equations:

    $$\frac{x}{2} - \frac{y}{2} = 5$$

    $$\frac{2x}{3} + \frac{y}{4} = 3$$

37. Solve each equation:

    a. $2y(y + 2) = y - 4(y - 1)$

    b. $a^2 - 2a - 4 = 0$

9.2 Function Notation

We frequently use letters to symbolize relations and functions. We can, for example, refer to a function f whose ordered pairs (x, y) satisfy the equation $y = 2x + 3$. We write

$$f: y = 2x + 3$$

By the definition of a function, each x determines a single value of y and we say "y is a function of x." A more compact way of writing this is

$$y = f(x), \quad \text{read} \quad \text{"}y = f \text{ of } x\text{"}$$

So, we write
$$f(x) = 2x + 3$$
Using this notation, we observe that y and $f(x)$ are interchangeable. However, the latter gives us the following notational advantage. Rather than writing:

when $x = 5$, $y = 13$

we write
$$f(5) = 13$$
Here, $f(5)$ is the *value of f at* $x = 5$.
  Similarly, for the function $f(x) = 2x + 3$:
$$f(2) = 2(2) + 3 = 4 + 3 = 7$$
$$f(0) = 2(0) + 3 = 0 + 3 = 3$$
$$f(-4) = 2(-4) + 3 = -8 + 3 = -5$$
$$f(a) = 2(a) + 3 = 2a + 3$$
$$f(2 + h) = 2(2 + h) + 3 = 4 + 2h + 3 = 7 + 2h$$

In the notation $f(x)$, we think of f as describing the operation(s) performed on x. Thus, $f(x) = 2x + 3$ means, given any number or expression (x), multiply that number or expression times 2, then add 3. This is occasionally viewed as a "function machine," where x is some element from the domain (or input) operated on by some process f, and the corresponding result (or output) is $f(x)$, a member of the range. See Figure 9.2-1.

FIGURE 9.2-1
Function Machine

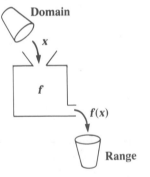

$f(x) = 2x + 3$
$f(5) = 2(5) + 3$
$f(5) = 13$
Input   Output
$(5, 13) \in f$

Any letter or combination of letters may be used to specify a function, and any letter may be used to indicate the domain variable as in the following examples.

Example 1

Let $g(u) = u^2 - 5$. Find $g(2)$, $g(5)$, $g(0)$, and $g(2 + h)$:
$$g(2) = (2)^2 - 5 = 4 - 5 = -1$$
$$g(5) = (5)^2 - 5 = 25 - 5 = 20$$

9.2 Function Notation

$$g(0) = 0^2 - 5 = 0 - 5 = -5$$
$$g(2 + h) = (2 + h)^2 - 5 = 4 + 4h + h^2 - 5 = -1 + 4h + h^2$$

Example 2

Let $Q(t) = t^2 - 2t + 6$. Find $Q(0)$, $Q(2)$, $Q(-3)$, and $Q(3 + h)$:

$$Q(0) = 0^2 - 2(0) + 6 = 0 - 0 + 6 = 6$$
$$Q(2) = (2)^2 - 2(2) + 6 = 4 - 4 + 6 = 6$$
$$Q(-3) = (-3)^2 - 2(-3) + 6 = 9 + 6 + 6 = 21$$
$$Q(3 + h) = (3 + h)^2 - 2(3 + h) + 6$$
$$= 9 + 6h + h^2 - 6 - 2h + 6$$
$$= 9 + 4h + h^2$$

Using ordered pair notation, $Q(2) = 6$ and we write $(2, 6) \in Q$. Similarly, $(0, 6) \in Q$, $(-3, 21) \in Q$, and $(4, 14) \in Q$.

In some problems, it is not uncommon to be required to add, subtract, multiply, and divide function values, as in the following example.

Example 3

Given $f(x) = x^2 + 1$ and $G(x) = 3x - 1$, find the value of each of the following expressions:

A. $f(2) + G(2)$

$$f(2) + G(2) = [(2)^2 + 1] + [3(2) - 1]$$
$$= (4 + 1) + (6 - 1)$$
$$= 5 + 5$$
$$= 10$$

B. $f(3) - G(1)$

$$f(3) - G(1) = [(3)^2 + 1] - [3(1) - 1]$$
$$= (9 + 1) - (3 - 1)$$
$$= 10 - 2$$
$$= 8$$

C. $f(-2) \cdot G(0)$

$$f(-2) \cdot G(0) = [(-2)^2 + 1] \cdot [3(0) - 1]$$
$$= [4 + 1][0 - 1]$$
$$= [5][-1]$$
$$= -5$$

D. $\dfrac{f(7)}{G(7)}$

$$\dfrac{f(7)}{G(7)} = \dfrac{[(7)^2 + 1]}{[3(7) - 1]}$$

$$= \dfrac{(49 + 1)}{(21 - 1)}$$

$$= \dfrac{50}{20}$$

$$= \dfrac{5}{2} \quad \blacktriangleleft$$

Further, a function of a function, or **composition of functions**, is indicated when the result of one operation is used for a subsequent operation. Using the function machine concept, we say the output of one machine is the input to a second machine. The notation and machine equivalents are shown in Figure 9.2-2.

FIGURE 9.2-2
Composition of Functions

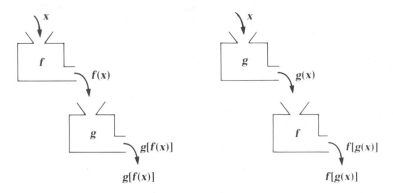

Example 4

Given $f(x) = 3 - x$ and $g(x) = 2x^2$, evaluate each of the following expressions:

A. $g[f(6)]$

$g[f(6)] = g[3 - 6]$ Evaluate "inner" function first (f)

$\quad\quad\quad\; = g(-3)$

$\quad\quad\quad\; = 2(-3)^2$ Evaluate "outer" function (g)

$\quad\quad\quad\; = 2 \cdot 9$

$\quad\quad\quad\; = 18$

B. $f[g(2)]$

$f[g(2)] = f[2(2)^2]$ Evaluate inner function first

$\quad\quad\quad\; = f(8)$

$$= 3 - 8$$
$$= -5$$
Evaluate outer function

c. $g[f(2)]$

$$g[f(2)] = g[3-2]$$
$$= g(1)$$
$$= 2(1)^2$$
$$= 2$$

Note, in general, $f[g(x)] \neq g[f(x)]$. ◀

Function notation and these function operations are used extensively in future courses.

Exercise Set 9.2

1. Given $f = \{(0, 2), (1, 6), (2, 12), (3, 20)\}$:
   a. Find $f(0)$
   b. Find $f(2)$
   c. If $f(a) = 20$, find a

2. Given $F = \{(0, 0), (1, 0), (2, 2), (3, 6)\}$:
   a. Find $F(1)$
   b. Find $F(3)$
   c. If $F(t) = 0$, find t

3. Given $f(x) = 5x - 2$, evaluate or simplify each expression:
   a. $f(0)$ b. $f(2)$ c. $f(-3)$
   d. $f(5)$ e. $f(-3+h)$ f. $f(x+h)$

4. Given $g(x) = 2x + 6$, evaluate or simplify each expression:
   a. $g(0)$ b. $g(1)$ c. $g(4)$
   d. $g(-3)$ e. $g(4+h)$ f. $g(x+h)$

5. Given $P(a) = a^2 - 3$, evaluate each expression:
   a. $P(0)$ b. $P(2)$ c. $P(-2)$
   d. $P(-4)$ e. $P(4)$ f. $P(5)$

6. Given $H(y) = 4 - y^2$, evaluate each expression:
   a. $H(1)$
   b. $H(-1)$
   c. $H(3)$
   d. $H(-3)$
   e. $H(4)$
   f. $H(-4)$

7. Given $f(x) = 2x - 1$ and $g(x) = 3x + 5$, evaluate each expression:
   a. $f(3)$
   b. $g(3)$
   c. $f(3) + g(3)$
   d. $f(2)$
   e. $g(2)$
   f. $f(2) - g(2)$
   g. $f(3) \cdot g(-2)$
   h. $\dfrac{f(8)}{g(0)}$
   i. $f[g(-1)]$
   j. $g[f(-1)]$

8. Given $a(t) = 4t - 3$ and $b(t) = 3t + 5$, evaluate each expression:
   a. $a(1)$
   b. $b(1)$
   c. $a(1) + b(1)$
   d. $a(2)$
   e. $b(2)$
   f. $a(2) - b(2)$
   g. $a(3) \cdot b(2)$
   h. $\dfrac{a(5)}{b(-2)}$
   i. $a[b(3)]$
   j. $b[a(3)]$

9. The area, A, of a square is a function of the length of a side, s, given by $A(s) = s^2$. Evaluate each expression:
   a. $A(2)$
   b. $A(3)$
   c. $A(4)$

10. The distance, s, that an object falls is a function of time, t, in seconds, written $s(t) = 16t^2$. Evaluate each expression:
    a. $s(1)$
    b. $s(2)$
    c. $s(4)$

11. Consider the function $f(x) = 2x + 3$. Explain why $f(m + 3)$ and $f(m) + 3$ do not represent equivalent statements.

Review

12. Draw the graph of each of the equations:
    a. $y = 2x - 3$
    b. $y = -\tfrac{1}{3}x + 1$
    c. $y = x$
    d. $y = 4$

13. Simplify each radical:
    a. $\sqrt{12}$
    b. $\sqrt{50x^3}$
    c. $\dfrac{3}{\sqrt{6}}$
    d. $\sqrt{\dfrac{7x}{8}}$

14. Multiply:

   a. $3x^2(2x - 5)$

   b. $3\sqrt{5}(2\sqrt{5} - 5)$

15. Divide:

   a. $\dfrac{3x^3 - 9x^2}{3x^2}$

   b. $\dfrac{5\sqrt{2} - 3}{\sqrt{2}}$

   c. $\dfrac{3}{\sqrt{5} + 2}$

9.3 Linear Functions

Functions typically can be classified in many ways. Here we present one special function, the linear function (used very frequently in applications), and also look at several forms of it.

A linear function is one that is written in the form

LINEAR FUNCTION

$f(x) = mx + b$

The linear function $f(x) = mx + b$ can be written in the equivalent form $y = mx + b$, which is the slope-intercept form of a straight line (see Section 3.4). Therefore, the graph of a linear function is a straight line whose slope is m and whose y-intercept is b. To draw its graph, the domain variable (x) is associated with the horizontal axis, and y [or $f(x)$] is associated with the vertical axis.

Example 1

Draw the graph of each linear function. Label the axes appropriately.

A. $L(x) = 4x - 3$

The variable x is associated with the horizontal axis. The line has a y-intercept of -3 and a slope of 4.

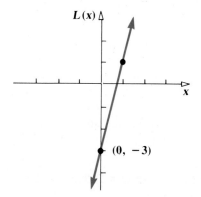

B. $F(t) = -\frac{2}{3}t + 5$

The graph of this function is a straight line with intercept 5 and slope of $-\frac{2}{3}$. Recall a line whose slope is negative falls to the right.

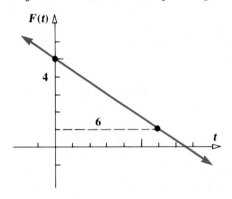

It is not uncommon that a graph is only meaningful for certain members of the domain, as in the next example.

Example 2

Draw the graph of each linear function, noting the restrictions on the domain:

A. $G(x) = 3x - 2, x \geq -2$

Here the restriction $x \geq -2$ means the graph is only to be drawn for x-values at or to the right of -2. The graph has a y-intercept of -2 and a slope of 3.

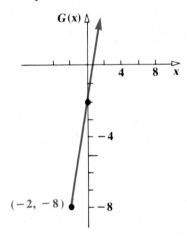

Note that the dot at $(-2, -8)$ shows the inclusion of the endpoint. The arrow on the other end of the line shows that the graph continues in that direction.

B. $D(n) = -n + 3; 1 \leq n < 5$

This graph is a line with y-intercept 3 and a slope of -1. It is drawn only for values of *n* between 1 and 5. To show 1 is included in the graph, we draw a dot at the point (1, 2); to show 5 is not included in the graph, we draw a hole at the point (5, −2).

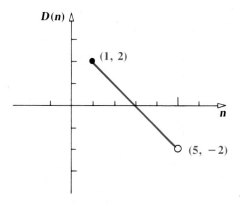

Constant Function

A constant function is a special case of a linear function with $m = 0$. That is, we can define a constant function as one that can be written in the form:

CONSTANT FUNCTION

$f(x) = b$

The graph of a constant function is a horizontal line passing through (0, *b*).

Example 3

Draw the graph of each function:

A. $C(x) = 4$

The graph is a horizontal line made up of all points whose *y*-coordinate is 4.

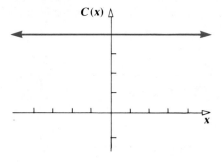

B. $H(t) = -2$
The graph is a horizontal line with y-intercept $(0, -2)$.

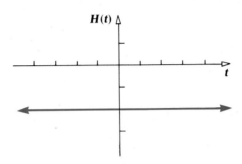

If we again picture a "function machine," a constant function is one whose output is constant regardless of the input. The function $G(x) = 3$ is a machine whose input may be any number, but the output remains constant. The domain of such a function is all real numbers, but the range contains only one element 3. Some ordered pairs belonging to G include $(2, 3)$, $(-7, 3)$, $(\frac{1}{2}, 3)$, $(\sqrt{7}, 3)$, $(-\frac{4}{5}, 3)$, etc.

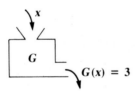

The Identity Function

The function $f(x) = x$ is equivalent to the equation $y = x$, whose graph is a straight line with a y-intercept of 0 and a slope of 1. Some ordered pairs belonging to $f(x) = x$ include $(1, 1)$, $(5, 5)$, $(7, 7)$, $(-\frac{3}{4}, -\frac{3}{4})$, $(0, 0)$, etc. Note the name *identity function* suggests that each range element is *identical* to the corresponding domain element. Its graph is shown here. Often, since this function is unique, it is given a special symbol. We will denote the identity function as:

IDENTITY FUNCTION

$Id(x) = x$

9.3 Linear Functions

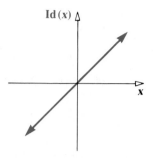

If a function machine is used to represent the identity function, it can be viewed as a hollow box through which the input passes unchanged. In your next course, the identity function will take on special importance.

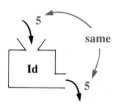

Exercise Set 9.3

Identify each of the following functions as linear, constant, or identity, whichever is most specific. Draw the graph of each function. Note any restrictions on the domain variable.

1. $f(x) = \frac{1}{2}x + 4$
2. $f(x) = \frac{1}{2}x - 2$
3. $g(x) = -1$
4. $g(x) = 3$
5. $h(x) = x$
6. $h(x) = 2x + 1$
7. $F(x) = 4 - x$
8. $F(x) = 6 - 2x$
9. $G(u) = 2u - 4, u \geq 0$
10. $G(t) = \frac{1}{2}t + 3, t \geq 0$
11. $H(r) = 3r - 5, 0 \leq r \leq 6$
12. $H(n) = 2n + 6, -3 \leq n \leq 1$
13. $T(x) = 4, 0 \leq x < 10$
14. $N(x) = -2, -5 \leq x \leq 5$
15. $S(t) = 8 - 3t$
16. $N(u) = 12 - 2u$
17. $f(x) = \frac{12 - 2x}{4}$
18. $g(x) = \frac{20 - 3x}{5}$

19. $L(w) = \frac{3}{4}w - \frac{2}{3}$
20. $G(u) = u$
21. $I(x) = x, x \geq 0$
22. $Q(x) = 2x, x \leq 0$
23. $I(x) = x, x > 0$
24. $Q(x) = 2x, x < 0$

Review

25. Solve each equation:
   a. $3a - 7 = 5 + a$
   b. $(b - 2)(2b + 3) = 9$
   c. $\sqrt{x - 3} + 2 = 5$
   d. $y^2 - 4y + 1 = 0$
26. On a number line, graph the solution of $2x - 5 = 10$.
27. In a Cartesian plane, graph the solution of $2x - 5y = 10$.

9.4 Review and Chapter Test

(9.1) A *relation* is a set of ordered pairs. The set of all first components is the *domain*, and the set of all second components is the *range* of the relation.

Example 1

For the relation
$$A = \{(2, 6), (1, 1), (0, 0), (-1, 2), (-2, 7)\}$$
The domain of $A = \{2, 1, 0, -1, -2\}$, and the range of $A = \{6, 1, 0, 2, 7\}$.

To draw the graph of a relation, the domain is associated with the horizontal axis, and the range is associated with the vertical axis. The graph of relation A is shown in the figure:

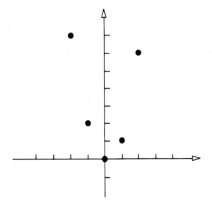

A *function* is a relation in which no two distinct ordered pairs have the same first component. Relation A of Example 1 is a function. The relation

$$B = \{(4, -2), (2, 0), (0, 2), (2, 4), (4, 5)\}$$

is not a function, since it has more than one ordered pair with the same first component; $(4, -2)$ and $(4, 5)$ are examples.

If the graph of a relation is drawn, it can be determined if it is or is not the graph of a function by the *vertical line test*: The relation is a function if each vertical line intersects the graph in at most one point. The graph of relation B is shown in the figure:

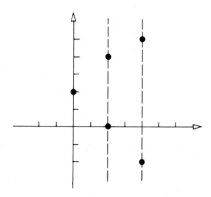

The dashed vertical lines are shown intersecting the graph in more than one point, indicating that relation B is not a function.

(9.2) If a relation is a function, we can write $y = f(x)$, read "y equals f of x." The function can be defined by an equation, such as $f(x) = x^2 - 6$. Using this notation, we can write, for example, "$f(4) = 4^2 - 6 = 10$," rather than "when $x = 4$, $y = 10$."

Example 2

For the function $f(x) = x^2 - 6$, evaluate or simplify each of the following expressions:

A. $f(3)$

$$f(3) = 3^2 - 6 = 9 - 6 = 3$$

B. $f(7)$

$$f(7) = 7^2 - 6 = 49 - 6 = 43$$

C. $f(a + 4)$

$$f(a + 4) = (a + 4)^2 - 6$$
$$= a^2 + 8a + 16 - 6$$
$$= a^2 + 8a + 10$$

◀

A function $f(x)$ can be viewed as a function machine for which x is the input (from the domain), f is a description of the operation(s) performed on x, and $f(x)$ is the output.

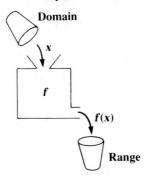

Values of functions can be added, subtracted, multiplied, divided, or used in composition.

Example 3

Let $f(x) = 3x - 1$ and $G(x) = 2 - x^2$. Evaluate each of the following expressions:

A. $f(2) + G(5)$

$$\begin{aligned}
f(2) + G(5) &= [3 \cdot 2 - 1] + [2 - 5^2] \\
&= (6 - 1) + (2 - 25) \\
&= 5 + (-23) \\
&= -18
\end{aligned}$$

B. $f(3) \cdot G(3)$

$$\begin{aligned}
f(3) \cdot G(3) &= (3 \cdot 3 - 1)(2 - 3^2) \\
&= (9 - 1)(2 - 9) \\
&= (8)(-7) \\
&= -56
\end{aligned}$$

C. $f[G(4)]$

$$\begin{aligned}
f[G(4)] &= f[2 - 4^2] \qquad &&\text{Evaluate inner function } (G) \\
&= f[2 - 16] \\
&= f(-14) \\
&= 3(-14) - 1 \qquad &&\text{Evaluate outer function } (f) \\
&= -42 - 1 \\
&= -43
\end{aligned}$$

(9.3) Some special functions include the following:

A *linear function* is one that can be written in the form $f(x) = mx + b$. Its graph is a straight line with a slope of m and a y-intercept b.

A *constant function* is one of the form $f(x) = b$, where b is a constant. Its graph is a horizontal line made up of all points whose y-coordinate is b.

The *identity function* is the function $f(x) = x$ or $\text{Id}(x) = x$ whose graph is the straight line whose slope $= 1$ and whose y-intercept is 0 (the origin).

Example 4

Identify and graph each function:

A. $F(x) = \frac{4}{3}x - 2$

This is a linear function whose graph is a straight line with slope $= \frac{4}{3}$ and y-intercept -2.

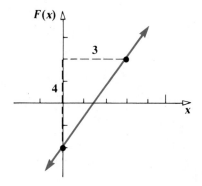

B. $G(x) = x$

This is the identity function.

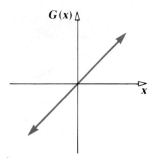

C. $H(x) = -2, \ -1 \leq x < 3$

This is a constant function whose graph is the horizontal line through $(0, -2)$, drawn only for x-values between -1 and 3.

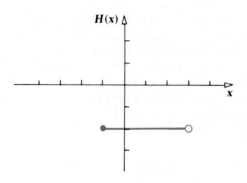

Exercise Set 9.4 Review

For each of the following relations, indicate the domain and range, indicate whether the relation is or is not a function, and sketch the graph:

1. $M = \{(-2, 5), (-1, 2), (0, 1), (1, 2), (2, 5)\}$
2. $N = \{(5, -2), (2, -1), (1, 0), (2, 1), (5, 2)\}$
3. $A = \{(-3, 3), (-2, 2), (-1, 1), (0, 0), (1, 1), (2, 2)\}$

Indicate whether each graph is or is not a function:

4.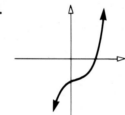

5.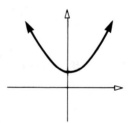

6.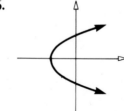

7.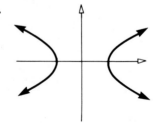

Let $F(x) = x^2 + 3$ and $G(u) = 3u - 2$. Evaluate or simplify each of the following expressions:

8. $F(5)$
9. $G(4)$
10. $F(-3)$

11. $G(-2)$
12. $F(2) + G(2)$
13. $F(4) - G(4)$
14. $F(-1) \cdot G(-1)$
15. $F(7) \cdot G(0)$
16. $\dfrac{F(5)}{G(2)}$
17. $F[G(2)]$
18. $G[F(2)]$

Classify and sketch each of the following functions:

19. $f(x) = 2x + 5$
20. $g(x) = -3$
21. $h(x) = x$
22. $F(x) = \dfrac{12 - 6x}{2}$
23. $G(x) = 2x - (x - 3)$
24. $H(x) = 4x - 1$

Chapter 9 Review Test

1. Given $F = \{(-3, -7), (-1, -3), (0, -1), (2, 3), (4, 7)\}$:
   a. Find the domain of F
   b. Find the range of F
   c. Find $F(2)$
   d. If $F(x) = 7$, find x
   e. Graph F
   f. Is F a function?

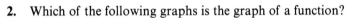

2. Which of the following graphs is the graph of a function?

   a.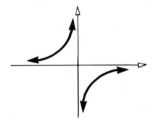
   b.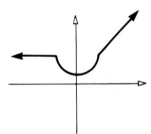
   c.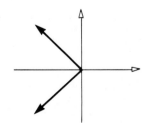
   d.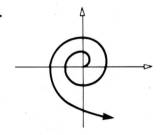

3. Let $f(x) = x^2 - 2x + 3$ and $g(x) = 3x - 1$. Evaluate or simplify each of the following expressions:

   a. $f(-2)$ b. $g(-2)$ c. $f(3) \cdot g(1)$

   d. $f[g(2)]$ e. $g[f(2)]$ f. $g(2) \cdot f(2)$

4. Draw the graph of each function. Note any restrictions on the domain variable.

   a. $f(x) = 3x - 1$ b. $g(m) = 2m + 3,\ -2 \leq m \leq 2$

   c. $h(x) = x,\ x \geq 0$ d. $k(x) = \dfrac{10 - 15x}{5}$

SELECTED ANSWERS

Exercise Set 1.1

1. True
3. True
5. False
7. True
9. False
11. True
13. True
15. False
17. True
19. False
21. $\{1, 3, 7, 9\}$
23. $\{1, 2, 4, 6, 7, 8, 10\}$
25. ϕ
27. $\{3, 4, 5, 6, 7, 8\}$
29. $\{1, 2, 3, 4, 5\}$
31. $\{4, 5, 6, 7, 8\}$
33. {January, February, March, April, September, October, November, December}
35. $\{7, 8, 9, 10, \ldots, 68, 69, 70\}$
37. $\{41, 42, 43, \ldots\}$

Exercise Set 1.2

1.
3.
5.
7.

9. (a) 3 (b) 3, 0
 (c) $3, -2, 0, \frac{2}{3}, 4.67$
 (d) $3, -2, 0$ (e) $\sqrt{3}, \pi$
11. (a) none (b) none
 (c) $\frac{1}{2}, -\frac{4}{3}, 1.3, -4.71$
 (d) none (e) $-\pi, \sqrt{5}$

13.–37. (odd exercises)

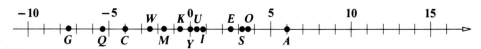

Exercise Set 1.3

1. $4 > 1$
3. $0 > -3$
5. $\pi > 0$
7. $-2 < 3$
9. $\pi > -\pi$
11. $-3 > -7$
13. $5.6 > 5$
15. $\pi > \sqrt{5}$
17. $C < A$
19. $B > C$
21. $0 > B$
23. $A > B$
25. $-4 < B$
27. $C < -\dfrac{1}{2}$
29. $-5 < B$
31. 3
33. -3
35. 4
37. 3
39. -5
41. 0
43. 8
45. 5
47. -5
49. -4
51. 15
53. -26
55. -13
57. 4
59. -4
61. Regardless of where we are on the number line, the distance from B to A is *negative* if we must move to the *left* to get to A from B.

Exercise Set 1.4

1. (a) 7 (b) -4 (c) 0 (d) 5.8 (e) -8
3. (a) 4 (b) 6 (c) 5 (d) 0 (e) 7.1
5. (a) 4 (b) -5 (c) -30 (d) 7
7. (a) -6 (b) -6 (c) -23 (d) 23 (e) 0
9. (a) $\dfrac{1}{4}$ (b) $-\dfrac{1}{6}$ (c) $\dfrac{8}{7}$ (d) -4 (e) 3

11.

Number (x)	Negative ($-x$)	Absolute Value $\lvert x \rvert$	Reciprocal $\dfrac{1}{x}$
$+3$	-3	3	$\dfrac{1}{3}$
5	-5	5	$\dfrac{1}{5}$
$\dfrac{2}{3}$	$-\dfrac{2}{3}$	$\dfrac{2}{3}$	$\dfrac{3}{2}$
$\dfrac{7}{9}$	$-\dfrac{7}{9}$	$\dfrac{7}{9}$	$\dfrac{9}{7}$
0	0	0	Undefined
-4	4	4	$-\dfrac{1}{4}$
$-\dfrac{5}{6}$	$\dfrac{5}{6}$	$\dfrac{5}{6}$	$-\dfrac{6}{5}$
$-\dfrac{1}{3}$	$\dfrac{1}{3}$	$\dfrac{1}{3}$	$-\dfrac{3}{1} = -3$
1	-1	1	1

Exercise Set 1.5

13.

Number (x)	Negative ($-x$)	Absolute Value $\|x\|$	Reciprocal $\frac{1}{x}$
$+6$	-6	6	$\frac{1}{6}$
4	-4	4	$\frac{1}{4}$
2.15	-2.15	2.15	$\frac{1}{2.15}$
$\frac{3}{4}$	$-\frac{3}{4}$	$\frac{3}{4}$	$\frac{4}{3}$
-0	0	0	Undefined
-5	5	5	$-\frac{1}{5}$
$-\frac{1}{6}$	$\frac{1}{6}$	$\frac{1}{6}$	$-\frac{6}{1} = -6$
-3.25	3.25	3.25	$-\frac{1}{3.25}$
-0.4	0.4	0.4	$-\frac{1}{0.4}$
$-\frac{7}{8}$	$\frac{7}{8}$	$\frac{7}{8}$	$-\frac{8}{7}$

15. No. Taking the negative of a number changes its sign. For example, $-(-3) = 3$, a positive number.

17. The number zero is its own negative. Any positive number, or zero, is its own absolute value. The number 1 is its own reciprocal, and -1 is its own reciprocal.

Exercise Set 1.5

1. commutative
3. commutative
5. closure for addition
7. distributive
9. additive inverse
11. identity
13. additive inverse
15. multiplicative inverse
17. identity
19. identity
21. associative
23. associative
25. (a) commutative (b) commutative
    (c) associative (d) identity
    (e) closure (f) closure
27. (a) commutative (b) commutative
    (c) closure
    (d) commutative and associative
29. (a) commutative (b) distributive
    (c) commutative (twice)
    (d) commutative
31. (a) additive inverse (b) commutative
    (c) distributive
    (d) multiplicative inverse
33. (a) commutative (b) commutative
    (c) associative (d) commutative
35. (a) additive (b) additive inverse
    (c) identity (d) multiplicative inverse

	Number	Additive Inverse	Multiplicative Inverse
37.	7	-7	$\frac{1}{7}$
39.	0	0	none
41.	-5	5	$-\frac{1}{5}$
43.	$\sqrt{2}$	$-\sqrt{2}$	$\frac{1}{\sqrt{2}}$

45. 16
47. 16
49. 26
51. 32
53. 33
55. 20
57. 55
59. 19
61. 28
63. 16
65. 2
67. 62
69. 17
71. 156
73. 13
75. 6
77. 144
79. 198
81. 195
83. $10x - 3y$

Exercise Set 1.6

1.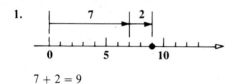
$7 + 2 = 9$

3.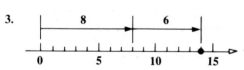
$8 + 6 = 14$

5.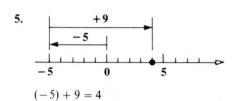
$(-5) + 9 = 4$

7.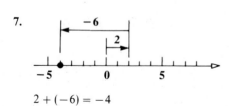
$2 + (-6) = -4$

9.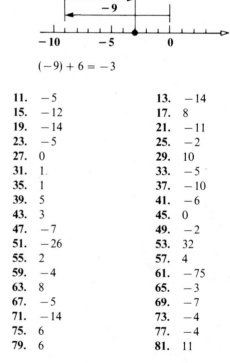
$(-9) + 6 = -3$

11. -5
13. -14
15. -12
17. 8
19. -14
21. -11
23. -5
25. -2
27. 0
29. 10
31. 1
33. -5
35. 1
37. -10
39. 5
41. -6
43. 3
45. 0
47. -7
49. -2
51. -26
53. 32
55. 2
57. 4
59. -4
61. -75
63. 8
65. -3
67. -5
69. -7
71. -14
73. -4
75. 6
77. -4
79. 6
81. 11

Exercise Set 1.7

1. -28
3. 45
5. -3
7. 4
9. -6
11. 0
13. undefined
15. 3
17. -3
19. -4
21. 24
23. 10
25. -9
27. -10
29. -6
31. 9
33. 3
35. -3
37. undefined
39. -9
41. -5
43. -2
45. $\frac{4}{1}$
47. -5
49. 6
51. 3
53. -7
55. 12
57. -6
59. 0

Exercise Set 1.8

	Constants	Variables
1.	3	x
3.	$\pi, 2$	r
5.	$\frac{1}{2}$	A, b, h
7.	$\pi, 3$	V, r, h
9.	48, 32, 2	v, t
11.	none	y, m, x, b

13. 14
15. 6
17. 2
19. 4
21. 2
23. 6
25. 18
27. 4
29. 10
31. -24
33. 1
35. 2
37. 19
39. -72
41. -4
43. 4
45. 14
47. 417
49. 7
51. 6
53. 15
55. 9
57. 36
59. 150.72
61. 4800 cubic feet

Exercise Set 1.9

1. $\{3, 8\}$ [1.1]
2. $\{1, 2, 3, 4, 5, 8\}$ [1.1]
3. ϕ [1.1]
4. True [1.1]
5. False [1.1]
6. False [1.1]
7. True [1.1]
8. True [1.1]
9. False [1.1]
10. [number line] [1.2]
11. [number line] [1.2]
12. [number line] [1.2]
13. [number line] [1.2]
14. [number line] [1.2]
15. (a) 2 (b) $2, 0$ (c) $2, 0$ (d) $2, 0, \frac{1}{2}, 2.3$ (e) $-\pi, \sqrt{2}$ [1.2]
16. $A < C$ [1.3]
17. $B > A$ [1.3]
18. $0 > B$ [1.3]
19. $0 < D$ [1.3]
20. $A < 0$ [1.3]
21. $C > B$ [1.3]
22. 4 [1.3]
23. -4 [1.3]
24. -4 [1.3]
25. -1 [1.3]
26. 0 [1.3]
27. -3 [1.3]
28. 0 [1.3]

29. [1.4]

Number (x)	Negative (−x)	Absolute Value |x|	Reciprocal $\frac{1}{x}$
+7	−7	7	$\frac{1}{7}$
2	−2	2	$\frac{1}{2}$
0	0	0	none
−4	4	4	$-\frac{1}{4}$
$\frac{3}{8}$	$-\frac{3}{8}$	$\frac{3}{8}$	$\frac{8}{3}$
$-\frac{5}{6}$	$\frac{5}{6}$	$\frac{5}{6}$	$-\frac{6}{5}$
$\frac{1}{7}$	$-\frac{1}{7}$	$\frac{1}{7}$	7
$-\frac{1}{9}$	$\frac{1}{9}$	$\frac{1}{9}$	−9

30. 2 [1.5] **31.** 19 [1.5] **48.** −4 [1.6] **49.** −3 [1.6]
32. 28 [1.5] **33.** 10 [1.5] **50.** 3 [1.7] **51.** −12 [1.7]
34. 48 [1.5] **35.** 3 [1.5] **52.** undefined [1.7] **53.** −8 [1.7]
36. commutative [1.5] **54.** 0 [1.7] **55.** −14 [1.7]
37. associative [1.5] **56.** −3 [1.7] **57.** −3 [1.7]
38. distributive [1.5] **58.** −6 [1.8] **59.** 2 [1.8]
39. additive inverse [1.5] **60.** −5 [1.8] **61.** −6 [1.8]
40. −4 [1.6] **41.** 18 [1.6] **62.** 14 [1.8] **63.** −26 [1.8]
42. −6 [1.6] **43.** 4 [1.6] **64.** 6 [1.8] **65.** −12 [1.8]
44. 12 [1.6] **45.** 5 [1.6] **66.** −4 [1.8] **67.** −4 [1.8]
46. 6 [1.6] **47.** 2 [1.6] **68.** 96 [1.8] **69.** 32 [1.8]

Chapter 1 Review Test

1. (a) true (b) true (c) false
  (d) true (e) true (f) false

2. (a) 2 (b) 0, 2 (c) −6, 0, 2
  (d) $-6, -\frac{3}{2}, 0, 2, \frac{9}{2}$ (e) $\sqrt{6}$
  (f) all members of C

3.

Exercise Set 2.2

4. (a) 9   (b) −8

5.

	Number (x)	Negative ($-x$)	Absolute Value $\lvert x \rvert$	Reciprocal $\frac{1}{x}$
(a)	$-\frac{3}{4}$	$\frac{3}{4}$	$\frac{3}{4}$	$-\frac{4}{3}$
(b)	$\frac{5}{2}$	$-\frac{5}{2}$	$\frac{5}{2}$	$\frac{2}{5}$
(c)	0	0	0	none
(d)	$\sqrt{5}$	$-\sqrt{5}$	$\sqrt{5}$	$\frac{1}{\sqrt{5}}$
(e)	-7	7	7	$-\frac{1}{7}$

6. (a) $-\frac{5}{8} < 2$ (b) $\frac{5}{8} > -2$
   (c) $-\frac{5}{8} < 0$ (d) $\frac{5}{8} > 0$

7. $A-3, B-2, C-5, D-1, E-2, F-6,$
   $G-4, H-3, I-5, J-2$

8. (a) 14   (b) 13

9. (a) -4 (b) 12   (c) -32
   (d) -2 (e) 28   (f) undefined
   (g) 0   (h) -12
   (i) 52   (j) -11

10. (a) $3^2 xy^3 z^2$ or $9xy^3 z^2$
   (b) $2 \cdot 2 \cdot 5 \cdot m \cdot m \cdot m \cdot m \cdot n \cdot n$

Exercise Set 2.1

1. $3x + 12$
3. $5x - 15$
5. $2x^2 + 6x$
7. $6y^2 + 18y$
9. $-18m + 6$
11. $5a + 3m - 1$
13. $3x + 4 - 2a + b$
15. $2x + 3 - 5y - 4z$
17. $5(x^2 + 3x - 2)$
19. $x(a + b)$
21. $x(3x + 1)$
23. $4(3x^2 + 2x + 1)$
25. $5x + y$
27. $5a + 7b$
29. $4m - 3p$
31. $3u$
33. $2x^2 + 9x - 4$
35. $3m^2 + 3m + 5$
37. $4a + 1$
39. $x^2 + x - 8$
41. $4p^2 + p + 3$
45. $m^2 - 21$
49. $9x$
53. $-x^2 + x + 8$
57. (a) $3x + 2y$ (b) $3(x + y)$
59. (a) closure   (b) distributive
   (c) commutative   (d) identity
61. (a) -6 (b) 19
   (c) undefined   (d) 0
43. $-x^2 + 9x + 6$
47. $4p - 18$
51. $x^2 - x - 33$
55. $S = 2x(x + 2y)$

Exercise Set 2.2

1. 6
3. 6
5. 18
7. 120
9. 60
11. $8x$
13. $12xy$
15. $24xy$
17. $15xy$
19. $xy(x + y)$
21. 4
23. 6
25. 15
27. 3
29. $12x - 4y + 9$
31. $3x + 6$

33. LCD = 12; $4x + 3$
35. LCD = 24; $9 - 4x$
37. LCD = $2u$; $u^2 + 4$
39. LCD = xy; $x^2 + y^2$
41. LCD = $5x$; $20 - 3x$
43. 42
45. 230
47. 406
49. $100x + 192$
51. $2406 - 1000y$
53. 10; 31
55. 100; 206
57. 1000; $50x + 7$
59. 1000; $3015x - 60$
61. 10,000; $7000b + 10004c$
63. Decimals are a special form of fractions, having denominators of 10, 100, 1000 . . . . So multiplying by 10, 100, 1000, . . . is very similar to multiplying by the LCD.
65. (a) 24 (b) 8 (c) 1
    (d) 0 (e) 7 (f) $100x - 108$

Exercise Set 2.3

1. identity
3. identity
5. conditional
7. conditional
9. $x = 0, y = 5; x = 5, y = 0; x = 2, y = 3; \ldots$
11. $x = 2, y = 0; x = 0, y = -4; x = 3, y = 2; \ldots$
13. $x = 0, y = 2; x = 5, y = 7; x = 3, y = 5; \ldots$
23. 1
25. -5
27. -10
29. (a) 20 (b) 23 (c) 45
    (d) 54
31. (a) $-y + 18$ (b) $6a + 2$
    (c) $-3b + 3$

Exercise Set 2.4

1. $x = 4$
3. $y = 8$
5. $u = 4$
7. $a = 3$
9. $x = 1$
11. 1
13. 1
15. -3
17. $\dfrac{4}{3}$
19. $-\dfrac{7}{3}$
21. no solution
23. identity
25. $\dfrac{2}{3}$
27. $\dfrac{3}{5}$
29. $\dfrac{3}{2}$
31. 1
33. 2
35. 2
37. 0
39. identity
41. -2
43. 4
45. -5
47. 1
49. $\dfrac{7}{3}$
51. $\dfrac{7}{8}$
53. 7
55. $\dfrac{40}{3}$
57. 2.4
59. $\dfrac{51}{35}$
61. $-\dfrac{11}{34}$
63. $-\dfrac{48}{7}$
65. 3
67. $-\dfrac{11}{2}$
69. $\dfrac{9}{7}$
71. 5
73. -8
75. -35
77. $\dfrac{151}{5}$
79. $\dfrac{373}{148}$
81. $2x = 5$
83. $3z + 6 = 12$
85. $2(x + 3) = 3x$
87. (a) $P = 46, A = 126$ (b) $l = 29$
    (c) $w = 11$
    (d) $P = 2l + 2w$: constants 2; variables P, l, w; $A = lw$: constants, none; variables A, l, w
89. $2a^3bc^2$

Exercise Set 2.5

1. $r = \dfrac{I}{Pt}$
3. $R = \dfrac{E}{I}$
5. $r = \dfrac{d}{t}$
7. $x = \dfrac{-by - c}{a}$
9. $M_1 = \dfrac{Fd^2}{kM_2}$
11. $C = \dfrac{5F - 160}{9}$
13. $L = \dfrac{P - 2W}{2}$
15. $h = \dfrac{3V}{\pi r^2}$
17. $L_2 = \dfrac{L_1 W_2}{W_1}$
19. $h = \dfrac{A - 2\pi r^2}{2\pi r}$
21. $x = \dfrac{-a^2 + b^2 + c^2}{2bc}$

23.
```
 -5        0        5       10
──┼──┼──┼──┼──┼──┼──┼──┼──┼──►
     C   E       B   D   A
```

25. (a) $8x + 7$ (b) $2m + 22$
    (c) $2x^2 - x + 12$

Exercise Set 2.6

1. (a) $=$ (b) $<$ (c) $<$
   (d) $<$ (e) $>$ (f) $>$

3. (a) 3 is less than x
   (b) x is less than 8
   (c) 2 is less than or equal to $3 \cdot 8$
   (d) $x + 3$ is greater than or equal to 15
   (e) $x - 2$ is less than or equal to 15
   (f) $2(3 + x)$ is not equal to $x - 3$

5. (a) $-5 < 0$ (b) $3 + 4 < 8$
   (c) $x + 2 > 9$ (d) $x \ne 5$
   (e) $x + 3 \ge -4$ (f) $-5 \le x - 9$

7. (a) false (b) true   (c) true
   (d) false (e) true   (f) true

9. (a)
```
        0         4
◄───────┼─────────◉──────►
```

(b)
```
        0              3
────────┼──────────────◉──►
```

(c)
```
       -2         0
────────◉─────────┼──────►
```

(d)
```
       -6         0
────────◉─────────┼──────►
```

(e)
```
              -½   0
◄──────────────◉───┼──────►
```

(f)
```
        0              12
────────┼──────────────●──►
```

(g)
```
       -6              0
────────●──────────────┼──►
```

(h)
```
        0              9
◄───────┼──────────────●──►
```

11. (a)
```
       -2              6
────────◉──────────────◉──►
```

(b)
```
        0              5
────────◉──────────────◉──►
```

(c)
```
              -½   0
────────────────◉──◉──────►
```

(d)
```
              4         7
──────────────●─────────◉──►
```

(e)
```
       -9              -1
────────◉──────────────◉──►
```

(f)
```
       -4              3
────────●──────────────●──►
```

13. $x < 1$

15. $x < 1$

17. $x \le 3$

19. $x \ge 7$
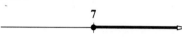

21. $x \ge -2$

23. $x > -\dfrac{1}{2}$

25. $x < -4$

27. $x > -10$

29. $-5 < x \le 2$

31. $-\dfrac{3}{2} \le x < 4$

33. $-4 < x \le 4$
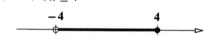

35. $4 < x \le 9$

37. 6 weeks

39. (a) 6 (b) 8
(c) $x = \dfrac{-3y + 12}{2}$

Exercise Set 2.7

1. (a) $N + 3$ (b) $N - 4$ (c) $2N$
   (d) $2N - 3$
3. (a) $T + 20$ (b) $T - 10$ (c) $3T$
   (d) $\dfrac{1}{3}T$
5. (a) 30 cents (b) 30 cents
   (c) 200 cents (d) 2845 cents
   (e) $10d$ cents (f) $5n$ cents
   (g) $100(20 - c)$ cents
7. (a) \$25 (b) \$18 (c) \$30
   (d) \$21.25 (e) $0.04x$ dollars
   (f) $P + 0.06P$ dollars
9. (a) 120 mi (b) 180 mi
   (c) 60 mi (d) $\dfrac{44}{3}$ mi
   (e) $40x$ mi (f) $\dfrac{2z}{3}$ mi

11. (a) 9 (b) 19
    (c) $9 - x$ (d) $x - 15$
    (e) $A + 2$ (f) $2A - 9$
13. (a) $d + 5$ (b) $d - 3$
    (c) $2d - 2$ (d) $\dfrac{1}{2}d + 4$
15. (a) $E + 2$ and $E + 4$ (b) $3E$
    (c) $E + 1$ (d) $E - 7$
    (e) $\dfrac{E}{2} + 5$ (f) $E + (x + 7)$
17. (a) $20 - n$ (b) $5n$
    (c) $0.10(20 - n)$
19. (a) $2S - 15$ (b) $75S$
    (c) $2(2S - 15)$
    (d) $0.75S + 2(2S - 15)$
    (e) $75S + 200(2S - 15)$
21. 16, 22

Exercise Set 2.8

23. 41 (home) to 35 (visitors)
25. 13 free throws, 27 field goals
27. text $19, calculator $24
29. width 105 ft, length 125 ft
31. $1800 at 5 percent, $1500 at 6 percent
33. boy's age = 18, his sister's age = 21
35. woman's age = 33, her daughter's age = 8
37. $1300
39. ten 15¢ stamps, forty 25¢ stamps
41. 11:40 A.M.
43. fourteen 30¢, sixteen 25¢
45. 38
47. 12, 14, 16
49. at least 3 years
51. 64 mph or faster   ANS is 64
53. any odd integer of size 23 or higher; 23
55. (a) $y = -3x + 5$ (b) $y = \dfrac{-x+5}{3}$
   (c) $y = \dfrac{-3x+9}{-2}$
57. (a) commutative   (b) commutative
   (c) associative   (d) associative

Exercise Set 2.8

1. $(2 + y)x^2$ [2.1]
2. $(6 - x)a$ [2.1]
3. $(5m + 1)m$ [2.1]
4. $(x - 1)y$ [2.1]
5. $-3x$ [2.1]
6. $x^2 - 2x$ [2.1]
7. $5x - 2$ [2.1]
8. $8y + 6$ [2.1]
9. [2.3]
10. [2.3]
11. $x = 3, y = 2; x = 5, y = 1; x = 7, y = 0$ and others [2.3]
12. $x = 0, y = -2; x = 5, y = 3; x = 2, y = 0$ and others [2.3]
13. conditional [2.3]
14. identity [2.3]
15. $\dfrac{5}{4}$ [2.4]
16. $\dfrac{11}{7}$ [2.4]
17. $\dfrac{1}{21}$ [2.4]
18. $\dfrac{9}{8}$ [2.4]
19. $-\dfrac{9}{2}$ [2.4]
20. 30 [2.4]
21. $\dfrac{15}{8}$ [2.4]
22. $\dfrac{89}{40}$ [2.4]
23. -46 [2.4]
24. 1.501 [2.4]
25. $y = \dfrac{-ax - c}{b}$ [2.5]
26. $P = \dfrac{I}{rt}$ [2.5]
27. $A = \dfrac{2S - NL}{N}$ [2.5]
28. $R_2 = \dfrac{E - IR_1}{I}$ [2.5]
29. $b = \dfrac{ad}{c}$ [2.5]
30. $x < \dfrac{13}{3}$

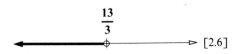 [2.6]

31. $x \leq -\dfrac{1}{2}$

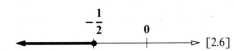 [2.6]

32. $x \leq -2$

 [2.6]

33. $x \geq -5$

[2.6]

34. $-\dfrac{2}{3} < x < \dfrac{8}{3}$

[2.6]

35. $\dfrac{1}{3} \leq x < \dfrac{7}{3}$

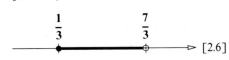 [2.6]

36. trousers $28, jacket $42 [2.7]
37. son 18; mother 43 [2.7]
38. 7 nickels, 11 quarters [2.7]
39. $3500 at 8 percent, $6500 at 11 percent
40. The smallest integer may be 19, 20, 21, ... The smallest value is 19. [2.7]

Chapter 2 Review Test

1. (a) $5(3x + 5y)$ (b) $2x(x - 4)$
   (c) $(4x + 1)x$
2. (a) $5b + 4$ (b) $-2y^2 - 5y - 3$
   (c) $8m + 24$
3. (a) 24 (b) $19 - 2x$
   (c) $270 - 401y$
4. (a) 5 (b) 4
   (c) 4 (d) -8
   (e) 450
5. (a) $y = \dfrac{-3x + 12}{-5}$ (b) $P = \dfrac{A}{1 + RT}$
   (c) $a = \dfrac{3m - 12}{2}$
6. (a)
   (b)
   (c)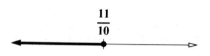
7. (a) $m \le 5$
   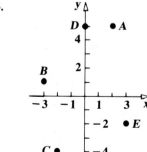
   (b) $u > 1$

   (c) $t \le \frac{11}{10}$

   (d) $-2 < x < 3$
   (e) $-13 \le x \le 2$
8. (a) $A = 1.08P$ (b) $2160
   (c) $3800
9. length 25 ft, width 9 ft
10. twenty 25¢, eight 15¢

Exercise Set 3.1

1.

3.

5. $A(2, 0)$, $B(3, 2)$, $C(0, 1)$, $D(-1, 3)$, $E(-4, 0)$, $F(-4, -2)$, $G(0, -3)$

7. (a) I;   (b) IV;   (c) III;   (d) II

9. (a) $(+, +)$;   (b) $(-, +)$;   (c) $(-, -)$;   (d) $(+, -)$

11.

13.

15. an infinite number

17. (a) $y = 5$;   (b) $y = -1$;   (c) $y = 9$

19.

Exercise Set 3.2

1.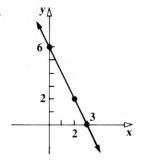

x	y
0	6
2	2
3	0

5.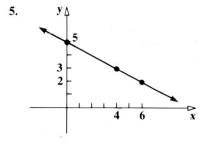

x	y
0	5
4	3
6	2

3.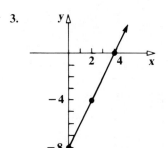

x	y
4	0
2	-4
0	-8

7.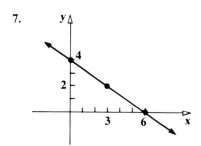

x	y
0	4
3	2
6	0

530   Selected Answers

9.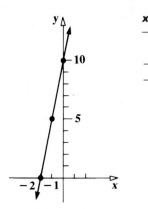

x	y
0	10
−1	5
−2	0

15.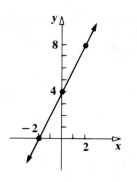

x	y
0	4
−2	0
2	8

11.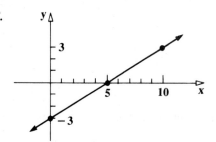

x	y
0	−3
5	0
10	3

17.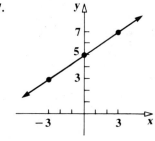

x	y
0	5
−3	3
3	7

19.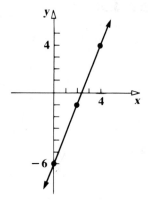

x	y
0	−6
2	−1
4	4

13.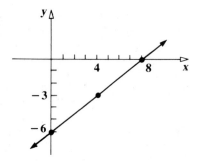

x	y
0	−6
8	0
4	−3

21.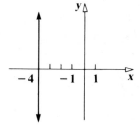

Exercise Set 3.3

23.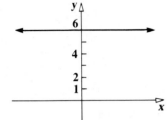

25.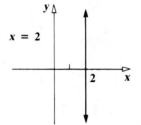

27. (a) yes (b) no
    (c) yes (d) no
29. (a) not linear (b) not linear
    (c) linear (d) linear
31. Infinite number of lines through one point. One line through two fixed points. In general, no lines can be drawn through three distinct points.
33. (a) $y = \dfrac{3}{2}x - 9$ (b) $y = -9$
    (c) $x = 6$
35. (a) $2x + 4y - 9$ (b) $-2x - 9$
    (c) $5a + 2$

Exercise Set 3.3

1.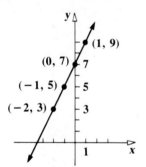

3.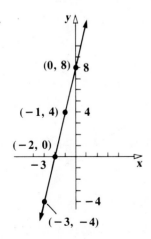

5.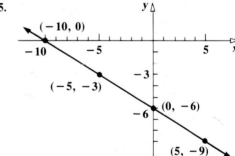

7.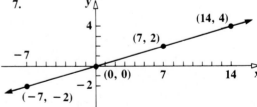

Selected Answers

9. $y = \frac{2}{3}x - 6$

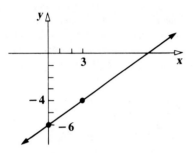

11. $y = x - 5$

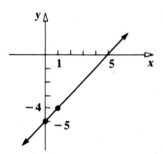

13. $y = -\frac{2}{3}x + 2$

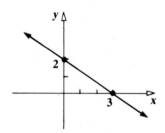

15. $y = \frac{1}{3}x - 5$

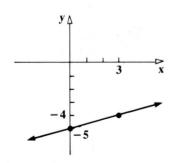

17. $y = -\frac{3}{4}x + \frac{9}{4}$

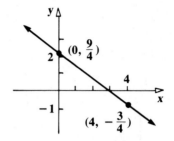

19. $y = -\frac{4}{5}x - \frac{13}{5}$

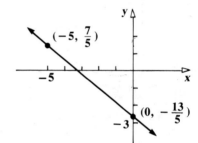

21.

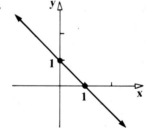

23.

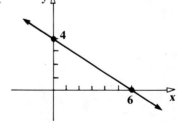

Exercise Set 3.4

25.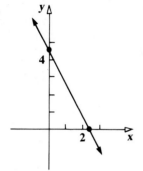

27.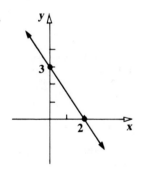

29.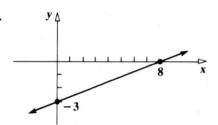

31.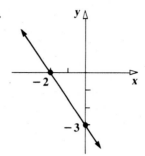

33. $2x + 3y = 6$
35. vertical line: $x = -3$
37. (a) 6; (b) -6; (c) -6;
(d) 6; (e) undefined; (f) 0
39. (a) T (b) F (c) T
(d) T (e) F (f) F

Exercise Set 3.4

1. $y = 2x - 5$
   $m = 2; b = -5$
3. $y = x + 4$
   $m = 1; b = 4$
5. $y = -2x + 6$
   $m = -2; b = 6$
7. $y = \frac{2}{3}x - \frac{8}{3}$
   $m = \frac{2}{3}; b = -\frac{8}{3}$
9. $m = 3$
11. $m = -1$
13. $m = 1$
15. $m = \frac{5}{4}$
17. $m = 0$
19. m is undefined

21.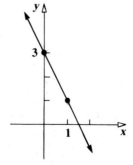

23.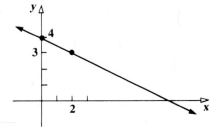

25.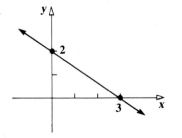

27.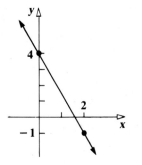

29.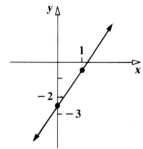

31.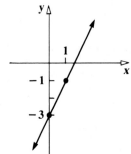

33.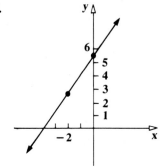

35. (a) $b = 4$ (b) $m = -1$
(c) $y = -x + 4$

37. (a) $b = 3$ (b) $m = \dfrac{2}{5}$
(c) $y = \dfrac{2}{5}x + 3$

39. (a) $b = 5$ (b) $m = 0$
(c) $y = 5$

41. (a) $x < 5$

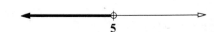

(b) $x \leq -1$

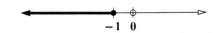

(c) $-4 < x \leq 3$

43. (a)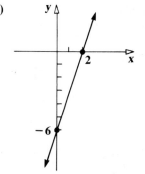

(c)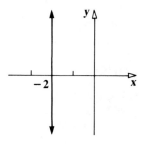

(b)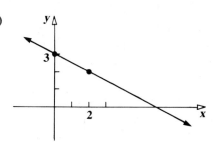

(d)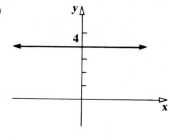

Exercise Set 3.5

1.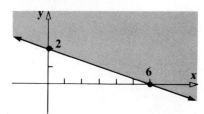

5.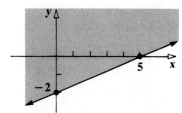

3.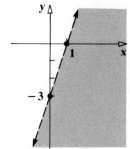

7.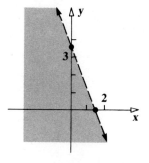

9.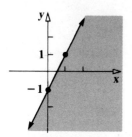

11.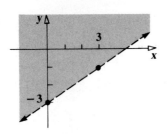

13.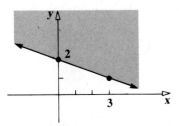

15.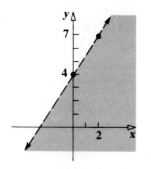

17.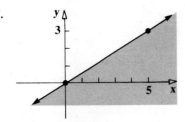

19.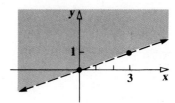

21.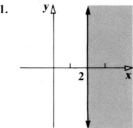

23.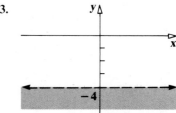

25.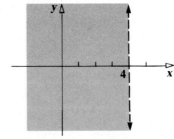

27.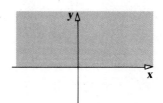

29. $x = \dfrac{b}{1+a}$

31. $m = -\dfrac{2}{3}$

Exercise Set 3.6

1. $3x - y = 10$
3. $2x + y = 10$
5. $3x - y = -16$
7. $2x - 3y = 14$
9. $x + 3y = 20$
11. $5x + 4y = 18$
13. ~~$3x + 2y = 12$~~ wrong
    $3x + 8y = 12$ right
15. $y = \frac{2}{3}x - \frac{5}{3}$
17. $y = 4x + 6$
19. $y = \frac{3}{2}x - 7$
21. $y = \frac{2}{5}x + \frac{8}{5}$
23. $2x - y = 0$
25. $2x - 3y = 7$
27. $3x - 7y = 5$
29. $3x - 2y = 6$
31. $y = \frac{3}{4}x - \frac{5}{4}$
33. $y = \frac{2}{3}x + \frac{4}{3}$
35. $y = -\frac{4}{5}x + 3$
37. $y = \frac{3}{2}x + 3$
39. $y = \frac{1}{2}x + \frac{1}{2}$
41. $y = 3$
43. $A(1, 3)$
    $B(-2, -4)$
    $C(-4, 3)$
    $D(0, 2)$
    $E(-3, 0)$
    $F(3, -5)$
45. $y \leq 3$

47. (a) -48 (b) -9 (c) -12
49.

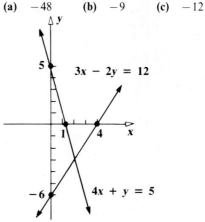

Exercise Set 3.7

1. **A.** $(3, -2)$ [3.1]
   a. abscissa: 3
   ordinate: -2
   b.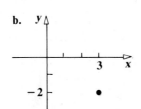
   c. QIV

   B. $(-2, -3)$ [3.1]
   a. abscissa: -2
   ordinate: -3
   b.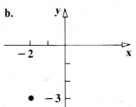
   c. QIII

   C. $(-4, 1)$ [3.1]
   a. abscissa: -4
   ordinate: 1
   b.
   c. QII

   D. $(4, 2)$ [3.1]
   a. abscissa: 4
   ordinate: 2
   b.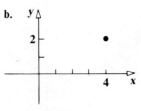
   c. QI

538 Selected Answers

2. [3.1]
A(3, −1)
B(−4, 1)
C(0, 4)
D(−2, 0)
E(−2, −4)
F(4, 1)
G(0, 0)

3. (a) [3.2]

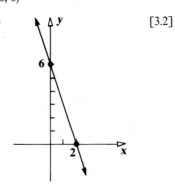

(b) [3.2]

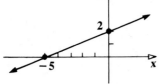

(c) [3.2]

(d) [3.2]

3. (e) [3.2]

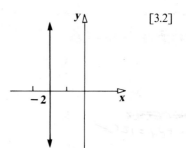

(f) [3.2]

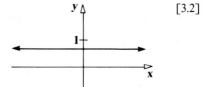

4. (a) $m = 2 > 0$ [3.4]
rises to right
(b) $m = -\dfrac{2}{5} < 0$ [3.4]
falls to right
(c) $m = $ undefined [3.4]
vertical
(d) $m = 0$ [3.4]
horizontal

5. (a) $y = 3x - 2$ [3.4]
$m = 3;\ b = -2$

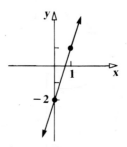

(b) $y = -\dfrac{2}{3}x + 6$ [3.4]
$m = -\dfrac{2}{3}$
$b = 6$

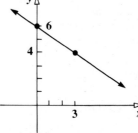

5. (c) $y = 3$
$m = 0$
$b = 3$ [3.4]
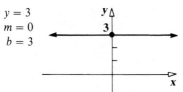

6. (a) [3.5]
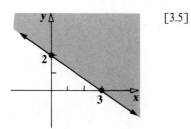

(b) [3.5]
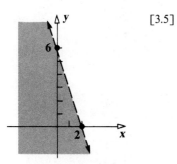

(c) [3.5]
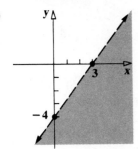

6. (d) [3.5]
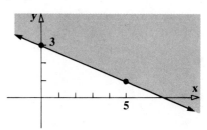

(e) [3.5]
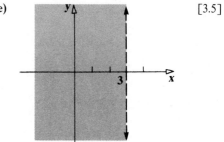

(f) [3.5]
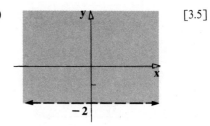

7. [3.6]
(a) $3x + 4y = -17$
(b) $7x - 3y = -25$

Chapter 3 Review Test

1.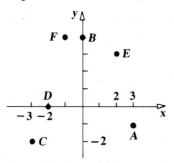

2. $P(-1, 3)$
$Q(2, 0)$
$R(3, -2)$
$S(-3, -2)$
$T(3, 2)$
$U(0, -3)$

3. (a) $m = \dfrac{3}{2}$ (b) $m = -\dfrac{1}{2}$
(c) $m = -\dfrac{3}{7}$

4. (a)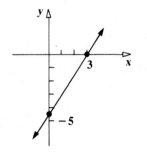

(b)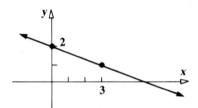

(c)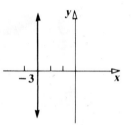

(d)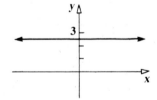

5. (a)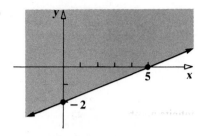

(b)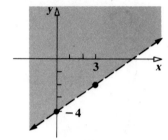

(c)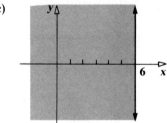

6. (a) yes (b) no
   (c) yes (d) no
7. (a) $3x - 5y = 20$ (b) $x + 2y = 1$
   (c) $4x + 5y = 11$ (d) $x = -3$

Exercise Set 4.1

1. $(5, -2)$
3. $(-4, -2)$
5. $(4, 12)$
7. $(2, 1)$
9. same line; infinite solutions
11. same line; infinite solutions
13. $(2, 1)$
15. no solution
17. $(2, -1)$
19. The two graphs can intersect in a single point, can be parallel lines, or can be colinear (the same line).
21. (a) $x - 2y$ (b) $-3x$ (c) $8x$
    (d) $-4y$
23. (a) 6 (b) 6 (c) -2
    (d) $-\dfrac{27}{5}$

Exercise Set 4.2

1. (1, 4)
3. (2, 1)
5. (−5, 9)
7. (0, 2)
9. no solution
11. (−2, −11)
13. (4, 0)
15. infinite number of solutions
17. $\left(\dfrac{29}{13}, -\dfrac{4}{13}\right)$
19. $\left(\dfrac{33}{65}, \dfrac{42}{65}\right)$
21. (10, 4)
23. (16, −2)
25. (3, −2)
27. (2, 5)
29. (−2, 4)
31. (a) $y = -3x + 6$ (b) $x = 3y + 2$
  (c) $x = \dfrac{5}{2}y + 6$ (d) $y = \dfrac{2}{5}x - \dfrac{12}{5}$
33. (a) 3 (b) 6
  (c) 8

Exercise Set 4.3

1. (−4, −2)
3. infinite number of solutions
5. (−2, −11)
7. (0, −2)
9. (4, 0)
11. no solution
13. $\left(\dfrac{1}{2}, -3\right)$
15. (4, −5)
17. (5, 2)
19. (1, 3)
21. (−1, 3)
23. (3, 4)
25. $\left(\dfrac{2}{7}, -\dfrac{3}{7}\right)$
27. $\left(\dfrac{2}{5}, -\dfrac{3}{5}\right)$
29. (8, −3)
31. (a) a system should be solved by elimination if none of the variables has a numerical coefficient of 1 or −1.
  (b) a system should be solved by substitution if one of the equations is written with one variable isolated; or if one can easily be written in that form because one of the variables has a numerical coefficient of 1 or −1.
33. (5, 2)
35. (a) $240 (b) $45.50
  (c) $0.09x$ dollars (d) $0.045y$ dollars

Exercise Set 4.4

1. 25 and 13
3. $38, trigonometry book; $17, calculator
5. 63° and 27°
7. 53 field goals, 16 free throws
9. length 11 cm., width 7 cm.
11. $7000 at 6 percent, $3000 at 5 percent
13. The man is 38 years old and his son is 14.
15. 11 dimes, 3 nickels
17. eight 15¢ stamps, sixteen 25¢ stamps
19. 30 liters of 15 percent solution, 20 liters of 40 percent solution
21. 30 pounds of 55¢ and 20 pounds of $1.00 candy
23. (a) $x \geq 8$

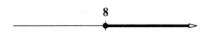

  (b) $-3 \leq x < \dfrac{13}{2}$

25. $B = \dfrac{2A - bh}{h}$

Selected Answers

Exercise Set 4.5

1.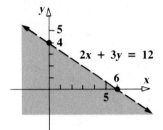

3.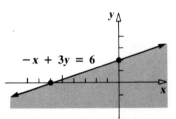

5.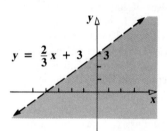

7.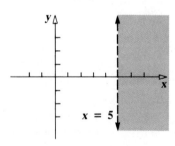

9.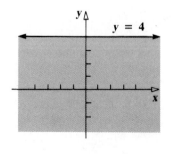

11.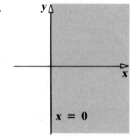

13.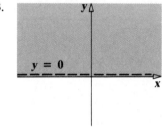

15.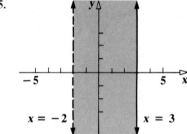

17.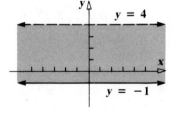

Exercise Set 4.5

19.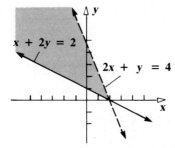

21.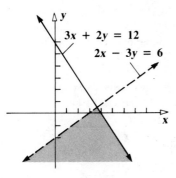

23.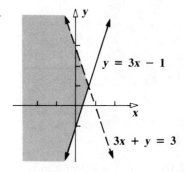

25.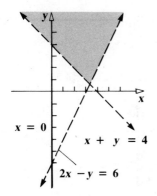

27.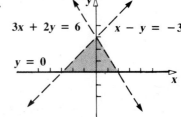

29.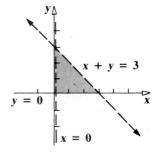

31.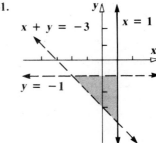

33. (a) $b \cdot b \cdot b$ (b) $6 \cdot m \cdot n \cdot n \cdot n$
    (c) $2 \cdot x \cdot x \cdot y$ (d) $5 \cdot (x + y)(x + y)$

35. (a) 9   (b) 2
    (c) 18   (d) 60

Exercise Set 4.6

1. [4.1]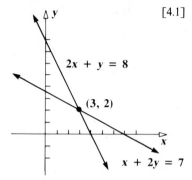

2. [4.1]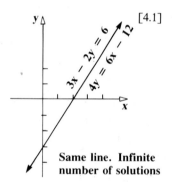
Same line. Infinite number of solutions

3. [4.1]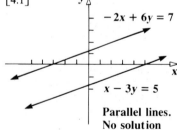
Parallel lines. No solution

4. [4.1]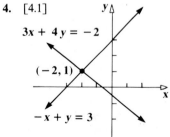

5. $\left(\dfrac{8}{3}, \dfrac{7}{3}\right)$ [4.2]
6. $\left(\dfrac{5}{3}, -\dfrac{2}{3}\right)$ [4.2]
7. (2, 1) [4.2]
8. no solution [4.2]
9. (1, −3) [4.2]
10. infinite number of solutions [4.2]
11. (−36, −25) [4.2]
12. infinite number of solutions [4.2]
13. (24, 15) [4.2]
14. (10, 0) [4.2]
15. (1, −2) [4.3]
16. (2, 3) [4.3]
17. (1, 1) [4.3]
18. no solution [4.3]
19. infinite number of solutions [4.3]
20. $\left(\dfrac{27}{11}, -\dfrac{31}{11}\right)$ [4.3]
21. suit, $325; shoes, $105 [4.4]
22. 15 nickels, 12 quarters [4.4]

23. 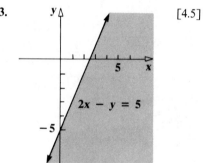 [4.5]

24. 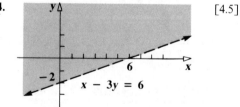 [4.5]

25. 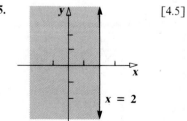 [4.5]

Exercise Set 4.6

26. 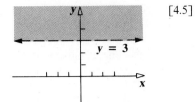 [4.5]

27. 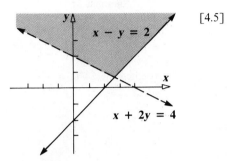 [4.5]

28. 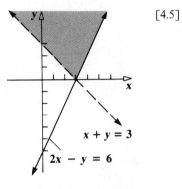 [4.5]

29. 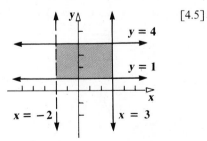 [4.5]

30. 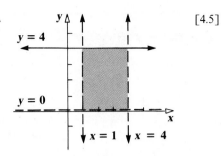 [4.5]

31. 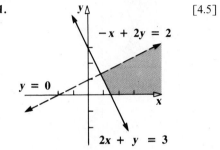 [4.5]

32. 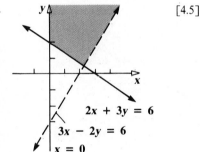 [4.5]

Chapter 4 Review Test

1. A and D are solutions
2. (a)

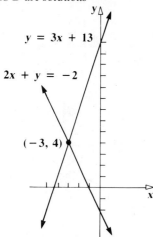

(b)

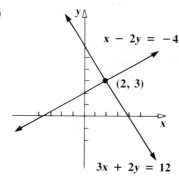

(c)

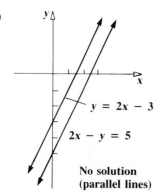

No solution (parallel lines)

3. (a) $\left(\dfrac{4}{5}, -\dfrac{22}{5}\right)$ (b) (9, 7)

4. (a) (7, 0) (b) infinite number of solutions
5. (a) Let B = Bruce's age and J = Joanne's age.
$$B = J + 3$$
$$B + J = 105$$
Bruce is 54, Joanne is 51.

(b) Let x = amount at 8 percent, and y = amount at 6 percent.
$$x + y = 7000$$
$$0.08x = 0.06y$$
$3000 at 8 percent, $4000 at 6 percent.

6. (a)

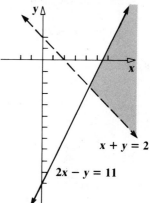

(b)

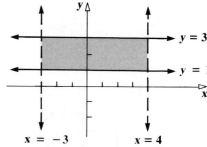

(c)

Exercise Set 5.1

1. polynomial   3. polynomial   5. not polynomial   7. not polynomial
9. not polynomial

	Polynomial	Terms	Numerical Coefficient	Literal Part	Number of Terms	Number of Variables	Degree
11.	$2x^2 - 5x + 1$	$2x^2$ $-5x$ 1	2 -5 1	x^2 x none	trinomial	1	2
13.	$x + 2y - 3$	x $2y$ -3	1 2 -3	x y none	trinomial	2	1
15.	$x^2 + xy - y^2$	x^2 xy $-y^2$	1 1 -1	x^2 xy y^2	trinomial	2	2
17.	$2xy$	$2xy$	2	xy	monomial	2	2

19. 0   21. 5
23. 18   25. 39
27. Degree of the term having highest degree
29. $\dfrac{3}{5}$

31.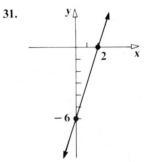

Exercise Set 5.2

1. $-3x + 4y$
3. $3x^2 - 2x$
5. $4y + 3$
7. $-2y^2 + y + 6$
9. $2xy + 3yz - 4xz$
11. $4a$
13. (a) $9x$ (b) $-2y$ (c) $-a$
    (d) $-2xy$
15. (a) $10x^2$ (b) $3m$ (c) $-5z$
    (d) 0
17. (a) $-2x$ (b) $-7x^2$ (c) $10ab$
    (d) $7m$ (e) $-6u$
19. (a) $4x^2 + 10x - 11$
    (b) $9a^2 + 10a - 2$
    (c) $6a^2 + 12a - 14$
    (d) $4x^3 + 4x^2y - 2xy^2$
21. (a) $y^2 - 10y - 2$
    (b) $3x^3 - 2x^2 - 4$
    (c) $-8m^2$
23. (a) $4x^2 + 10x - 11$
    (b) $9a^2 + 10a - 2$

(c) $6a^2 + 12a - 14$
(d) $4x^3 + 4x^2y - 2xy^2$
25. (a) $y^2 - 10y - 2$ (b) $3x^3 - 2x^2 - 4$
(c) $-8m^2$
27. $2a + 7$
29. $x^2 + 2x - 4$
31. $4y^2 - 3y - 6$
33. $4x^2 - 2x - 4$
35. $-x^2 - 5xy - y^2$
37. $x^2 - 6x + 2$
39. $-4x^2 - 4x + 10$
41. $14x + 6$
43. $9a$
45. $6l + 2$
47. (a) $c \cdot c \cdot c$ (b) $c \cdot c \cdot c \cdot c \cdot c$
(c) c^8
49. (a) -12 (b) -12 (c) -30
(d) 0

Exercise Set 5.3

1. $6x^3$
3. $-21x^3yw$
5. $-16a^3b^3$
7. $-12u^5$
9. $-8a^4b^4$
11. $2x^2 + 8x$
13. $3x^3 - 9x^2$
15. $x^4 + 2x^3 - 3x^2$
17. $-3a^3 + 15a$
19. $12x^4y - 42x^2y^3$
21. $6x^3y^2 + 9x^2y^2 - 33xy^2$
23. 184
25. 140
27. $30x^4 - 24x^3 + 18x^2 - 42x$
29. $-8a^5 + 24a^4 + 48a^3 - 16a^2$
31. $x^2 - 3x - 10$
33. $xy - 5x + 2y - 10$
35. $2x^2 + 5x - 12$
37. $x^3 - x^2 + 2$
39. $2x^3 + x^2y - 5xy^2 + 2y^3$
41. $3a^3 + 7a^2 - 9a + 2$
43. $2x^2 + x - 15$
45. $2a^3 - a^2 - 3a$
47. $6x^2 + xy - y^2$
49. $x^3 + 7x^2 + 6x - 8$
51. $x^3 + 5x^2 + 2x - 8$
53. $x^4 + 4x^3 - x - 2$
55. $x^4 - x^3 - 6x^2 + 4x + 8$
57. $x^4 + 2x^3 - 3x^2 - 4x + 4$
59. $x^6 + 6x^5 + 9x^4 - 4x^3 - 9x^2 + 6x - 1$
61. $x^4 + 10x^3 + 35x^2 + 50x + 24$
63. $A = 2x^3 - x^2 - x$
$P = 2x^2 + 2x + 2$
65. $A = 2x^2 + 2x - 4$
$P = 6x + 6$
67. Degree of product is sum of degrees of factor polynomials.
69. (a) $6p^2 + 11p - 35$
(b) $3u^2 + 20uv - 12v^2$
(c) $4m^2 - 25$
(d) $9c^2 + 30c + 25$
(e) $3x^6 + 19x^3 - 14$
71. $(x, y) = (6, -2)$

Exercise Set 5.4

1. $a^2 + 8a + 12$
3. $y^2 + 7y - 8$
5. $b^2 - 11b + 28$
7. $x^2 - 5x - 14$
9. $2m^2 + 3m - 5$
11. $6y^2 - 11y + 3$
13. $x^2 + 6x + 9$
15. $y^2 - 8y + 16$
17. $4x^2 + 20xy + 25y^2$
19. $16b^2 - 24b + 9$
21. $9x^4 - 6x^2y + y^2$
23. $x^2 - 9$
25. $4x^2 - 49$
27. $x^4 - 1$
29. $a^2 - 4b^2$
31. 399
33. 625
35. $-x + 6$
37. $6x^2 + 7x + 6$
39. $6x$
41. $3u^3 + 13u^2$
43. $-M^3 + 4M^2 + 11M - 2$
45. $2x^2 + 5x + 3$
47. $4x^2 + 12x + 9$
49. $x^2 - 9$
51. $V = 3x^2 - 3x - 6$
53. $V = 12x^3 + 36x$
55. $V = 4x^3 + 36x^2 + 33x + 8$
57. (a) $r \cdot r \cdot r \cdot r \cdot r \cdot r \cdot r \cdot r \cdot r$
(b) $r \cdot r \cdot r \cdot r$
(c) r^5
59. 329

Exercise Set 5.5

1. $2x^3$
3. $-4y^3$
5. $6y$
7. $2x^2 + 4x$
9. $4x^3 + 3x - 7$
11. $-2x^2 + \frac{5}{4}x - 1$
13. $\frac{1}{2}x - y + \frac{3}{2}$
15. $4m^4 + 2m^3 - \frac{5}{2}m^2$
17. $x + 2$
19. $x + 3$
21. $x^2 + 2x + 1$
23. $x^2 + 2x - 3$
25. $4a^2 + 2a - 1; R = 1$
27. $a - 3$
29. $2x + 5$
31. $3x^3 - 2x^2 + x - 6$
33. $y^2 + 2y + 5$
35. $3a^2 + a - 2$
37. $3x + 3; R = 5x - 1$
39. $5x^2 - 5x + 7; R = -10x + 8$
41. $W = 2y - 3$
43. $L = 2x + 5$
45. $L = x^2 + 4x - 1$
47. (a) $3x^2 - 6x + 12$
    (b) $3m^3 + 9m^2 - 6m$
    (c) $x^4 + 3x^3 - 8x^2$
    (d) $2b^2 + b - 15$
    (e) $4a^2 - 12ab + 5b^2$
49. (a) $t = 2$ (b) $y = -4$

Exercise Set 5.6

1. $8x^2 - 12x$
3. $6b^3 - 4b^2 + 2b$
5. $2x^2y + 4xy^2 - 6xy$
7. $6m^3 + 9m^2 + 12m$
9. $6x^3y - 9x^2y^2 + 12x^2y$
11. $2x(3x - 4)$
13. $3x(2x^2 + 5x + 2)$
15. $5xy^2(2x^3 - 3xy - y^2)$
17. $7a^2(a^4 - a^3 + 3a^2 - 2)$
19. $-3y(2y^2 + 3y - 8)$
21. $3x^2y^3(5x^2 + 3x - 1)$
23. $3a^2(1 + 3a - 4a^2)$
25. $3x^2y(3x^2 - 4xy + 5y^2)$
27. $7x^2y(2x^3 - 3x^2y + 5xy^2 + y^3)$
29. $8x^2y^2(2x^2 - 9xy + y^2)$
31. $4x^4(3x^4 - 2x^3 + 5x^2 - 4x + 1)$
33. $6x^2y(5x^3 - 3x^2 + 2x - 1)$
35. $w = 2x^2 - 3x + 1$
37. $w = xy^2 - 3x + 2y^2$
39. (a) $x^2 + 8x + 15$ (b) $x^2 - 2x - 15$
    (c) $x^2 + 2x - 15$ (d) $x^2 - 8x + 15$
41. (a) $1 \cdot 12$ $-1 \cdot (-12)$
       $2 \cdot 6$ $-2 \cdot (-6)$
       $3 \cdot 4$ $-3 \cdot (-4)$
    (b) $1 \cdot -15$
       $3 \cdot -5$
       $5 \cdot -3$
       $15 \cdot -1$
43. (a) $x^3 - x^2 - 7x + 3$ (b) degree $= 3$

Exercise Set 5.7

1. $(y + 5)(y - 3)$
3. $(m - 5)(m - 2)$
5. $3(x^2 + 3x - 5)$
7. $a(a + 7)(a - 3)$
9. $2x(x + 5)(x + 7)$
11. $(a + 3b)(a - 2b)$
13. $(x - 3y)(x - 4y)$
15. $(m + 4n)(m + n)$
17. $(k - 6m)(k - 5m)$
19. $4x^2(x - 3)(x + 2)$
21. $2x^2y^2(y + 4)(y + 7)$
23. $3xy(x - 4y)(x + 3y)$
25. $m^2n^3(m + 7n)(m + 4n)$
27. $x^2y(2x - 6y + 15)$
29. $3x^2y(x + 2y + 5)$
31. $(x - 8)(x - 5)$
33. $(x + 3y)(x - 2y)$
35. $2xy(x - 3)(x - 1)$
37. $-2a^2(a + 3b)(a + 6b)$
39. (a) $6a^2 + 17a + 12$
    (b) $6a^2 + a - 12$
    (c) $6a^2 - a - 12$
    (d) $6a^2 - 17a + 12$
41. (a) 4 and 3
    (b) no such numbers
    (c) 4 and -3
    (d) -12 and 1

Exercise Set 5.8

1. $(x + 4)(x + 2)$
3. $(2x - 3y)(x + y)$
5. $(a + 8)(a - 2)$
7. $(3m - n)(2m - n)$
9. $(x + 6)(x + 2)$
11. $(x - 2)(x - 3)$
13. $(x - 3)^2$
15. $(x + 6)(x - 2)$
17. $(2x - 1)(x + 6)$
19. $2(x + 3)(x + 2)$
21. $(3x - 5)(2x + 1)$
23. $(6x - 5)(x + 2)$
25. $5x(2x + 3)(x + 3)$
27. $3(x - 4)^2$
29. $(3x - 1)(2x + 9)$
31. $3x(x^2 + 2x + 2)$
33. prime
35. $(9a - 8)(a - 1)$
37. $2(m - 2)^2$
39. $(7a + 2)(a - 3)$
41. $(2x + 5)(x + 7)$
43. $4x^2(x + 3)(x - 2)$
45. $2xy^2(3x + 2y - 3)$
47. $(3m + 2n)^2$
49. $2xy^2(3x - y)(5x + 2y)$
51. $(4x - 3y)(x - 2y)$
53. $mn(6a - 1)(a + 4)$
55. (a) $9x^2 - 4$ (b) $x^4 - 1$
    (c) $x^2 - 9y^2$ (d) $4x^2 - 49$
    (e) $x^2 + 4x + 4$
    (f) $a^2 - 8a + 16$
    (g) $4a^2b^2 + 12ab + 9$
    (h) $9m^2 - 12mn + 4n^2$
57. $w = 7$ in. width
    $2w - 1 = 13$ in. length

Exercise Set 5.9

1. $(x + 6)(x + 3)$
3. $(y + 5)(y - 5)$
5. $(x + 10y)(x - 10y)$
7. $(2a - 1)(a + 4)$
9. $(2x + 5)(4x + 9)$
11. $8(x + 1)(x - 1)$
13. $x(2x - 3)(6x + 5)$
15. $(R^2 + 2)(R^2 - 2)$
17. $4x(x + 5)(x - 1)$
19. $2(x + 9)(x - 9)$
21. prime
23. $(9A - 2)(A + 6)$
25. $5x(2x - 1)(x + 2)$
27. $2(y + 1)(y - 12)$
29. $(x^2 + 9)(x + 3)(x - 3)$
31. $x^2(2x - 7)(6x - 5)$
33. $x(8x + 3)(x - 4)$
35. $5x(x^2 + 4)(x + 2)(x - 2)$
37. $k = 9$
39. $k = 24$
41. $k = 25$
43. (a) $x^3 + y^3$ (b) $x^3 + 8$
    (c) $x^3 + 27$
45. (a) $x^2 + 8x + 15$
    (b) $x^4 - 2x^2 - 15$
    (c) $x^3 + 5x^2 + 3x + 15$
    (d) $xy + 5x + 3y + 15$

Exercise Set 5.10

1. $8x^3$
3. x^3y^3
5. $27x^3$
7. $125x^3y^3$
9. $343a^6$
11. $512x^3y^6$
13. $(2x)^3$
15. $(10x^2y)^3$
17. $(4m^3n^2)^3$
19. $(a + 4)(a^2 - 4a + 16)$
21. $(2x - 3)(4x^2 + 6x + 9)$
23. $(y - 5)(y^2 + 5y + 25)$
25. $(3b - 2)(9b^2 + 6b + 4)$
27. $(x - y)(x^2 + xy + y^2)$
29. $(10m + 3n)(100m^2 - 30mn + 9n^2)$
31. $(x + 5)(x + 3)$
33. $(a - 2)(a - 5)$
35. $(a + 4)(x + y)$
37. $(n + 2)(n + 3)$
39. $(y - 2)(y + 5)$
41. $(x + 6)(x - 1)$
43. $(b + 2)(b^2 + 3)$
45. $(x - 2)(x^2 + 4)$
47. $(a + 2)(a + 3)(a - 3)$
49. $3(x - 1)(x^2 + 2)$
51. (a) $y^2 - 4y - 21$ (b) $u^2 - 36$
    (c) $2a^2 + 9ab + 10b^2$
    (d) $4v^2 + 28v + 49$
53. $m = \dfrac{3}{5}$

Exercise Set 5.11

1. $(x + 3)(x + 6)$
3. $(y + 5)(y - 5)$
5. $(2a - 1)(a + 4)$
7. $(2x + 5)(4x + 9)$
9. $8(x + 1)(x - 1)$
11. $x(6x + 5)(2x - 3)$
13. $(R^2 + 2)(R^2 - 2)$
15. $(x + 5)(x^2 - 5x + 25)$
17. $4x(x + 5)(x - 1)$
19. $2(x + 9)(x - 9)$
21. $(2a + 5)(4a^2 - 10a + 25)$
23. prime
25. $(9A - 2)(A + 6)$
27. $5x(2x - 1)(x + 2)$
29. $(x + 2)(x + 5)$
31. $2(y - 12)(y + 1)$
33. $(x^2 + 9)(x + 3)(x - 3)$
35. $(y - 4)(y + 1)$
37. $x^2(6x - 5)(2x - 7)$
39. $x(8x + 3)(x - 4)$
41. $3y(y + 2)(y^2 - 2y + 4)$
43. $5x(x^2 + 4)(x + 2)(x - 2)$
45. $3x(x - 6)(x^2 + 2)$
47. $a = 2$

49. (a)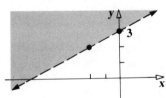

(b)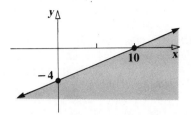

(c)

Exercise Set 5.12

1. polynomial [5.1]
2. not polynomial [5.1]
3. not polynomial [5.1]
4. not polynomial [5.1]
5. (a) binomial [5.1] (b) 1 (c) 2
6. (a) trinomial [5.1] (b) 2 (c) 3
7. (a) monomial [5.1] (b) 2 (c) 2
8. (a) monomial [5.1] (b) 0 (c) 0
9. (a) binomial [5.1] (b) 1 (c) 1
10. (a) binomial [5.1] (b) 2 (c) 4
11. 23 [5.1]
12. -3 [5.1]
13. 39 [5.1]
14. 5 [5.1]
15. 9 [5.2]
16. $4y^2 + 6y - 3$ [5.2]
17. $-4m$ [5.2]
18. $2x^2 + 3xy - 2y^2$ [5.2]
19. $2p^2 - 6p - 3$ [5.2]
20. 10 [5.2]
21. $-10x^3y^2$ [5.3]
22. $3x^4y - 9x^3y^2 + 3x^2y^3$ [5.3]
23. $x^2 + x - 20$ [5.3]
24. $x^3 - 3x^2 - 4x + 12$ [5.3]
25. $x^3 - x^2 - x + 10$ [5.3]
26. $3y^4 + 7y^3 - 7y^2 - 12y + 8$ [5.3]
27. $4x^2 - 4xy + y^2$ [5.4]
28. $3a^4 + 10a^2 - 8$ [5.4]
29. $2x^2 - 4x$ [5.5]
30. $5xy + 7x^2 - 1$ [5.5]
31. $x + 3$ [5.5]
32. $x^2 + 2x + 4$ [5.5]

33. $2m^3 + 2m^2 - m + 1; R = 1$ [5.5]
34. $2c^3 + 6c^2 + 15c + 39; R = 102c - 35$ [5.5]
35. $3x(3x + 1)(3x - 1)$ [5.6]
36. $7m^3(m^2 + 3m + 5)$ [5.6]
37. $(2x - 3y)^2$ [5.9]
38. $(5m + 7y)(5m - 7y)$ [5.9]
39. $2(4x^2 + 9y^2)$ [5.6]
40. $2y(3x^3 - 4xy + 5x^2y - 2)$ [5.6]
41. $2m^2n(m + 5n)^2$ [5.9]
42. $(x^2 + 4)(x + 2)(x - 2)$ [5.9]
43. $(x + 5)(x + 3)$ [5.7]
44. $(y + 7)(y - 5)$ [5.7]
45. $(3x + 2y)(2x - y)$ [5.8]
46. prime [5.7]
47. $2x^2(x - 4)(x - 3)$ [5.8]
48. $(5x - 2y)(x + 3y)$ [5.8]
49. $(6x - 5)(3x + 2)$ [5.8]
50. $3x^2(6x - 5)(2x - 1)$ [5.8]
51. $(9x - 4)(2x - 1)$ [5.8]
52. $(2x + 3)(6x + 5)$ [5.8]
53. $(a - 5)(a^2 + 5a + 25)$ [5.10]
54. $(x - 5)(x^2 + 3)$ [5.10]

Chapter 5 Review Test

1. (a) 4 (b) 2 (c) 3 (d) 4 (e) 17
2. $-2m^2 - 6m + 9$
3. $3x^2 - 8x + 8$
4. (a) $-6a^3b^2$ (b) $-6x^5y^3$
   (c) $2a^3b^2 - 6a^2b^3 + 8ab^4$
   (d) $y^4 + y^2 - 6$
   (e) $6x^2 - 5xy - 6y^2$
   (f) $2m^3 - 7m^2 + 14m - 12$
   (g) $m^2 + 6mq + 9q^2$
   (h) $2x^4 - 5x^3 + 4x^2 - 7x + 6$
   (i) $4x^2 - y^2$
5. (a) $-2xy + 1 - x^2y$
   (b) $2x + 1; R = 4$
   (c) $y^2 - 4y + 4$
   (d) $2x - 8; R = 19x + 7$
6. (a) $9x(2x - 1)$
   (b) $2x(2x + 1)(2x - 1)$
   (c) $(3a + 4b)(3a - 4b)$
   (d) $(x - 5)(x + 3)$
   (e) $(3x + 2y)(x - 4y)$
   (f) $3y^2(2y + 1)(y - 3)$
   (g) $(x^2 + 9)(x + 3)(x - 3)$
   (h) $(2x + 3)^2$
   (i) $(3x + 2)(2x - 1)$
   (j) $2x(2x + 3)(4x^2 - 6x + 9)$

Exercise Set 6.1

1. $\dfrac{7}{9}$
3. $\dfrac{3x^2}{2}$
5. $\dfrac{y}{3}$
7. $2x$
9. $\dfrac{2}{3}$
11. $\dfrac{2x^2}{3y}$
13. $\dfrac{2x}{7y}$
15. $\dfrac{5}{2x}$
17. $-\dfrac{1}{2}$
19. $\dfrac{x - 1}{4x}$
21. $\dfrac{2}{x + 5}$
23. $\dfrac{-(2 + x)}{x - 3}$
25. $\dfrac{x - 4}{x + 4}$
27. $\dfrac{x - 4}{x + 3}$
29. $\dfrac{-(4 + x)}{x - 4}$ or $\dfrac{4 + x}{4 - x}$
31. $\dfrac{x + y}{x + 3}$
33. $\dfrac{a + 2}{b + 1}$
35. $\dfrac{2a + b}{2a - b}$
37. $\dfrac{2y}{3x}$
39. $\dfrac{x - y}{2x + y}$
41. $15x$
43. $3xy$
45. $4bx^3y^2$
47. $4a^3b - 2a^2bm$
49. $3xy + 6y^2$
53. (a) $6x^2 - 3x$
    (b) $6x^2 + 5x - 4$
    (c) $6x^3 + 17x^2 + 6x - 8$
    (d) $b - a$

Exercise Set 6.2

1. $\dfrac{9}{10}$
3. $\dfrac{3}{5}$
5. $\dfrac{2}{3}$
7. $\dfrac{4}{3}$
9. 2
11. $\dfrac{15x}{2}$
13. $9y$
15. $\dfrac{10x^2}{y}$
17. $\dfrac{1}{3y^2}$
19. $12x^3y$
21. $\dfrac{7}{15x}$
23. $2x$
25. $(x+3)(x+1)$
27. $\dfrac{2}{(x+1)(x-1)}$
29. 2
31. $\dfrac{(2x+1)^2(x-1)}{x+4}$
33. $\dfrac{6(x-2)}{(x+3)^2}$
35. $\dfrac{(x+2)^2(x+1)}{4(2x-1)}$
37. $\dfrac{-(x+3)^2}{2(x+4)}$
39. $\dfrac{-(x+7)^2}{18(x+5)}$
41. $\dfrac{a+3}{a+2}$
43. $\dfrac{4(x+3)}{x}$
45. $\dfrac{x+3}{x+1}$
47. (a) 24 (b) 24 (c) 48 (d) $6x^2$

Exercise Set 6.3

1. 1, 2, 5, 10
3. 1, 2, 3, 4, 6, 8, 12, 24
5. 1, 11
7. 1
9. 1, 2, 3, 6, 7, 14, 21, 42
11. 3, 6, 9, 12, 15, ...
13. 7, 14, 21, 28, 35, ...
15. 2, 4, 6, 8, 10, ...
17. 11, 22, 33, 44, 55, ...
19. 15, 30, 45, 60, 75, ...
21. $3 \cdot 5$
23. 5^2
25. 11
27. $3^2 \cdot 5$
29. $2^3 \cdot 3^2$
31. $2^2 \cdot 5^2$
33. $2 \cdot 3 \cdot 5^3$
35. 15
37. 12
39. 24
41. 30
43. 105
45. 90
47. 60
49. 450
51. $24a^3$
53. $10x^2y^3$
55. $6m(m-n)$
57. $8x^2(2x-3)$
59. $9(a+4)^2$
61. $24a^3b^3c^3$
63. $12y(y+3)(y-3)$
65. $(3y+4)(3y-4)(y-2)$
67. —
69. (a) $\dfrac{1}{c} \cdot a + \dfrac{1}{c} \cdot b = \dfrac{a}{c} + \dfrac{b}{c}$
    (b) $\dfrac{1}{c} \cdot a - \dfrac{1}{c} \cdot b = \dfrac{a}{c} - \dfrac{b}{c}$
71. (a) $x^2 + 4x$ (b) $x^2 + 3x - 4$ (c) $x^3 + 3x^2 - 4x$

Exercise Set 6.4

1. $\dfrac{3x+5}{2y}$
3. $\dfrac{-x}{3y}$
5. $\dfrac{6x+y}{3y}$
7. $\dfrac{1}{x+2}$
9. 2
11. 4
13. $\dfrac{y+2}{y+1}$
15. $\dfrac{2}{y-1}$
17. $\dfrac{7+2x}{x-3}$
19. $\dfrac{-x^2+6x-7}{x-2}$
21. $\dfrac{6a^2+10a}{(a-3)(a+4)}$
23. $\dfrac{5x^2-2x-5}{(x+1)(x-1)}$
25. $\dfrac{2x+11}{(x+1)(x+3)}$
27. $\dfrac{2x+9}{(x+3)(x-1)(x+4)}$
29. $\dfrac{3x-1}{(x+3)(x-3)}$
31. $\dfrac{5x}{3}$

33. $\dfrac{2x^2 + 2x + 4}{(x+1)(x-1)}$

35. (a) $\dfrac{6}{x+2}$ (b) $\dfrac{b+1}{3}$

   (c) $\dfrac{5(c-2)}{c-1}$

37. (a) commutative for addition
    (b) associative for multiplication
    (c) identity   (d) multiplicative inverse

Exercise Set 6.5

1. $\dfrac{11}{20}$

3. $\dfrac{34}{39}$

5. x

7. $\dfrac{1}{x-3}$

9. $\dfrac{8x+2}{5x}$

11. $\dfrac{1}{x-y}$

13. $\dfrac{x+3}{2x+1}$

15. $\dfrac{(2x+3)(x-2)}{2x-6}$

17. $\dfrac{x+4}{2x+3}$

19. $\dfrac{3xy - x^2y^2 + 1 - x^2 - 2y^2}{(y-x)(1-xy)}$

21. $\dfrac{-1}{x+3}$

25. (a) $\dfrac{2}{y+3}$ (b) $\dfrac{m}{2m-7}$

   (c) $\dfrac{a^2 + 2a + 4}{b+3}$

Exercise Set 6.6

1. $\dfrac{1}{2}$

3. -4

5. 15

7. 39

9. -6

11. 4

13. 2

15. -9

17. 2

19. -2

21. $-\dfrac{3}{2}$

23. no solution (2 is extraneous)

25. $\dfrac{28}{9}$

27. Identity. Valid for all real numbers except 2.

29. 4

31. -1

33. no solution $\left(-\dfrac{1}{2}\text{ is extraneous}\right)$

37. (a) $\dfrac{2}{3}$ (b) $\dfrac{4}{5}$ (c) $\dfrac{3x}{y}$ (d) $\dfrac{2}{3}$

39. (a) 30 (b) 5

Exercise Set 6.7

1. (a) $\dfrac{8}{5}$ (b) $\dfrac{5}{8}$ (c) $\dfrac{5}{13}$

3. 120

5. 88

7. 12 years

9. 15

11. 3

13. 4.63

15. 4.49

17. 31,200 words

19. 52 min.

21. 9 in.

23. 66 in.

25. $12,000, $36,000 and $48,000

27. 30°, 60°, and 90°

29. 117 ml.

31. 1080 ml.

33. 15

35. 8.5

39. (a) $x = 0$ (b) $x = 0$ (c) $c = 3$
    (d) $u = -4$ or 3 (e) $m = 2$ or -2

41. (a) 12 hrs (b) $\dfrac{x}{40}$ hrs (c) $\dfrac{175}{x}$ hrs

Exercise Set 6.8

1. 18
3. 4, 20
5. 8
7. 5
9. 3
11. $\dfrac{7}{5}$
13. 4 hours
15. 36 hours
17. 60 hours
19. 16 mph
21. 180 mph
23. 55 mph and 65 mph
25. (a) x^9 (b) $-10x^9$ (c) $6x^3 - 15x^8$ (d) $\dfrac{x^3}{8}$
27. (a) $-\dfrac{3}{2}$ (b) $-\dfrac{3}{4}$ (c) $-\dfrac{3}{2}$

Exercise Set 6.9

1. $\dfrac{-(2+x)}{x+3}$
2. $\dfrac{3(2x+y)^2}{8a}$
3. $\dfrac{n}{m}$
4. $\dfrac{3a+5b}{2a}$
5. $15x^2y$
6. $2x^3 - 4xy$
7. $x^2 - 6x + 8$
8. $m^2 + 4m + 4$
9. $\dfrac{4m^3}{9n}$
10. $\dfrac{9x^2y^3}{49mn^3}$
11. $\dfrac{x+1}{2y}$
12. $\dfrac{x-2}{x+2}$
13. $\dfrac{by^3}{3a^2x^2}$
14. 3^3
15. $2^2 \cdot 3^2$
16. 23 is prime
17. $2^2 \cdot 5^3$
18. $20(x+1)(x-1)$
19. $72x^3y^2z^3$
20. $(x+3)(x-3)(x-2)$
21. $4m^2(2m+5)(m-3)$
22. $\dfrac{31}{36}$
23. $\dfrac{1}{70}$
24. $\dfrac{8x+7}{(x+2)(x-1)}$
25. $\dfrac{20ac - 8ab^2 + 5b^2c^2}{10a^2b^2c^2}$
26. $\dfrac{4x-8}{(x+4)(x-4)}$
27. $\dfrac{x^2-3x+4}{(x-2)(x-1)}$
28. $\dfrac{6x}{(2x+3)(x-2)(x+1)}$
29. $\dfrac{28}{15x}$
30. $\dfrac{22}{29}$
31. $\dfrac{1}{a-2}$
32. $\dfrac{x+2}{x-3}$
33. -1
34. 5
35. $-\dfrac{21}{29}$
36. $-\dfrac{9}{2}$
37. $-\dfrac{3}{4}$
38. 1.76
39. 1.05
40. 3 hr. 45 min.
41. 50°, 60°, 70°
42. $6\dfrac{2}{3}$ days
43. 15 mph

Chapter 6 Review Test

1. $\dfrac{x-y}{x+2y}$
2. (a) $20x^3y$ (b) $6y^2 + 13y + 6$
3. (a) $\dfrac{2acx}{3bm}$ (b) $\dfrac{x+5}{x-1}$ (c) $\dfrac{(y+2)(y-4)}{(y+3)(y+1)}$ (d) $\dfrac{-m+23}{(m-3)(m+2)}$ (e) $\dfrac{9y - 8xy + 12x^2}{12x^2y^2}$ (f) $\dfrac{5x-27}{(x-5)(x+2)(x-3)}$
4. $\dfrac{x-3}{x+4}$
5. (a) 2 (b) $\dfrac{6}{7}$
6. 4 hr. 10 min.
7. $7\dfrac{1}{2}$ days

Exercise Set 7.1

1. x^4
3. $4m^2$
5. a^5b^5
7. $\dfrac{8}{b^3}$
9. c^6
11. 4
13. $\dfrac{1}{2a}$
15. $\dfrac{9u^2}{25t^2}$
17. $2a^2b^2$
19. $\dfrac{2x^3}{3}$
21. $\dfrac{p^6}{4}$
23. $27x^3y^6$
25. $\dfrac{a^2b^2}{4}$
27. $\dfrac{m^3}{p}$
29. $\dfrac{-3w}{4uv^3}$
31. -1
33. $\dfrac{2m^3n}{15}$
35. $-27a^4$
37. $\dfrac{-27b}{5a^2}$
39. $6a^3b^4$
41. $\dfrac{2ab^2}{3}$
43. $27a^3b^3$
45. (a) $a^2 - 3a + 4$ (b) $x^3 - x^2 - 2x - 4$ (c) $\dfrac{3x}{y-3}$
47. $y^2 - y + 3$

Exercise Set 7.2

1. 3
3. 1
5. -1
7. 1
9. $3y$
11. y
13. $2x^2$
15. 1
17. $x + 1$
19. $\dfrac{1}{x^3}$
21. $\dfrac{2}{x^2}$
23. $\dfrac{1}{4x^2}$
25. $\dfrac{y}{x^4}$
27. $\dfrac{5y}{x^2}$
29. $\dfrac{3}{x^3}$
31. $\dfrac{3y}{x^3}$
33. $\dfrac{4}{x^2y^2}$
35. $\dfrac{y^3}{8x^3}$
37. $\dfrac{x^2}{3y}$
39. $\dfrac{9x^4}{4y^2}$
41. $\dfrac{1}{3x^4y}$
43. $\dfrac{y^9}{27x^9}$
45. $\dfrac{4}{9}$
47. $\dfrac{a+1}{a^2}$
49. $-xy$
51. $\dfrac{1}{a^2} - \dfrac{2}{ab} + \dfrac{1}{b^2}$
53. $3xy^{-1}$
55. $3xy^{-3}$
57. $\dfrac{5x^{-2}y}{3}$
59. $\dfrac{x^{-4}y^{-2}}{4}$

61. The exponent tells you how many factors of the base there are.

63. $x > -1$
    $y > -1$

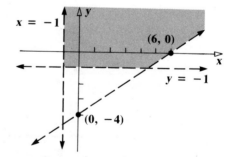

65. (a) $\dfrac{14}{5}$ (b) $\dfrac{x^2 - x - 1}{x^2 + 2x}$

Exercise Set 7.3

1. 3.25×10^2
3. 5.7×10^6
5. 9.25×10^{-3}
7. -2.1×10^4
9. -6.66×10^{-2}
11. 2.68×10^7
13. -6.06×10^{-9}
15. 2710
17. 9.26
19. 41.1
21. $-461{,}000{,}000$
23. 0.00642
25. -0.00442
27. $138{,}000{,}000$
29. 0.0000000316
31. 600,000
33. 0.000003
35. 0.0002
37. $-600{,}000{,}000$
39. 1.5×10^9
41. 2.84×10^{-8}
43. 4×10^5
45. 9.4×10^{-5}
47. 9.43
49. (a) 8 (b) -8 (c) 81 (d) 81 (e) 125 (f) -125
51. (a) 6 (b) $\dfrac{5m+7}{(m+1)(m+2)}$ (c) $\dfrac{x^2+2x+3}{(x-1)(x+1)}$

Exercise Set 7.4

1. 9
3. -2
5. 8
7. 3
9. $\dfrac{1}{2}$
11. -4
13. undefined
15. 32
17. -32
19. 8
21. 125
23. -8
25. undefined
27. $\dfrac{1}{2}$
29. $\dfrac{1}{216}$
31. $4a$
33. n
35. $x^{5/2}$
37. $y^{1/2}$
39. $\dfrac{1}{2}a^{2/3}$
41. x^4
43. $9a^3$
45. $x^2 y^3$
47. $\dfrac{1}{x}$
49. $\dfrac{1}{b^{3/2}}$
51. $\dfrac{x}{9y}$
53. $\dfrac{1}{25x^2 y^{1/2}}$
55. $\dfrac{2a}{3}$
57. $(-36)^{1/2}$ means take the square root of -36, and $-36^{1/2}$ means take the square root of 36 and make it negative.
59. Because there is no number that raised to the fourth power is negative.
61. $4y^2 + 3y - 2$
63. (a) $6r^2 + 25r + 14$ (b) $9a^2 - 25$ (c) $u^2 + 14u + 49$ (d) $9v^2 - 6v + 1$
65. (a) $\dfrac{x^2}{-3}$ (b) $\dfrac{2x}{3}$

Exercise Set 7.5

1. 7
3. -7
5. undefined
7. 2
9. -2
11. -2
13. $y^{1/2}$
15. $y^{1/3}$
17. $(xy)^{1/2}$
19. $(3u)^{1/3}$
21. $a^{3/2}$
23. $a^{2/3}$
25. $(x^2 + y^2)^{1/2}$
27. $\sqrt[4]{x^5}$
29. $\sqrt{c^5}$
31. $2\sqrt{x}$
33. $\sqrt{2x}$
35. $6\sqrt{a}\sqrt[3]{b}$
37. $\sqrt{3}\sqrt{x}$
41. $\dfrac{\sqrt{5}}{\sqrt{m}}$
45. $\sqrt[3]{x^2}$
49. 11
53. (a) distributive (b) commutative (c) associative
55. (a) $A = 40$ (b) $x = 7$
39. $\sqrt[3]{9}\sqrt[3]{y}$
43. $\dfrac{3\sqrt{x}}{\sqrt[3]{t}}$
47. $\sqrt[4]{m^3}$
51. $2x$

Exercise Set 7.6

1. $2\sqrt{5}$
3. $5\sqrt{3}$
5. $3\sqrt{5}$
7. $2xy^2$
9. $y^2\sqrt{2xy}$
11. $\dfrac{\sqrt{2}}{3}$
13. $\dfrac{\sqrt{10}}{2}$
15. $\dfrac{\sqrt{6mn}}{2n}$
17. $\dfrac{3\sqrt{2ab}}{2b}$
19. $2\sqrt{x}$
21. $\dfrac{5\sqrt{6x}}{6x}$
23. $\dfrac{5\sqrt{7}}{7}$
25. $\dfrac{5\sqrt{3}}{9}$
27. $y\sqrt[3]{9xy}$
29. $2xy^2\sqrt[3]{3x^2}$
31. $\dfrac{2\sqrt[3]{x^2}}{x}$
33. $\sqrt[3]{3x^2}$
35. $\dfrac{\sqrt[3]{10}}{2}$
37. $\dfrac{\sqrt[3]{20x}}{2x}$
39. $\dfrac{\sqrt[3]{6ab}}{2b}$
41. $\sqrt{x}$
43. $\sqrt[3]{x^2}$
45. $\dfrac{\sqrt{3}}{3}$ Use the table to find $\sqrt{3}$ and then divide by 3.

47. (a) $4x - y = 8$
$y = 4x - 8$

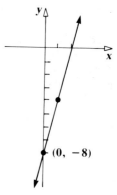

(b) $y = -\dfrac{2}{3}x + 1$ (c) $x = 3$

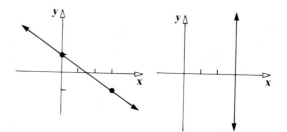

Exercise Set 7.7

1. $6\sqrt{2}$
3. $\sqrt{x}$
5. $\sqrt{5}$
7. $11\sqrt{3}$
9. $\sqrt{5} + 4\sqrt{2}$
11. $\sqrt{7}$
13. $2\sqrt{2}$
15. $\dfrac{9}{2}\sqrt{2}$
17. $\dfrac{19}{3}\sqrt{6}$
19. $13\sqrt{x}$
21. $\sqrt{6} + 4\sqrt{10}$
23. $\dfrac{7\sqrt{3}}{3}$
25. $\dfrac{5}{2}\sqrt{2x} + 8\sqrt{x}$
27. $4\sqrt[3]{3}$
29. $3\sqrt[3]{3}$
31. $4x^2\sqrt{6x}$
33. $9m\sqrt[3]{2t}$
35. $4y\sqrt[3]{4xy}$
37. Yes, there may be other factors that make them not like terms.
39. (a) $3m^2 - 12m$ (b) $8y^2 + 10y - 3$
(c) $p^2 - 36$ (d) $4x^2 - 9$
41. (a) -3.7×10^4 (b) 6.7×10^{-8}

Exercise Set 7.8

1. $15 - 6\sqrt{3}$
3. $2\sqrt{5} + 5$
5. $3 + \sqrt{2}$
7. 26
9. $6 - 7\sqrt{5}$
11. $10 - 4\sqrt{6}$
13. $25 + 11\sqrt{5}$

15. $3\sqrt{2} - 2\sqrt{3} + \sqrt{6} - 2$
17. $2x - \sqrt{x} - 10$
19. $15m + 26\sqrt{m} + 8$
21. 4
23. 19
25. $x - 16$
27. $4a - b$
29. $3\sqrt{5}$
31. $2\sqrt{5} - 1$
33. $\dfrac{5\sqrt{2}}{2}$
35. $8 - \dfrac{2\sqrt{7}}{7}$
37. $3 + \dfrac{8\sqrt{x}}{x}$
39. $\sqrt{3} - 1$
41. $6\sqrt{5} - 12$
43. $2 + \sqrt{3}$
45. $\dfrac{5\sqrt{3} + 3}{22}$
47. $\dfrac{11 - 3\sqrt{15}}{-7}$
49. $\sqrt{42} + 2\sqrt{6}$
51. $\dfrac{x\sqrt{x} + 2x}{x - 4}$
53. $\dfrac{n - 3\sqrt{n}}{n - 9}$
55. *Case 1:* If dividing by a rational number, divide each term by that rational number.

Case 2: If dividing by a radical expression of one term, rationalize the denominator by multiplying the numerator and denominator by the appropriate factor.

Case 3: If dividing by a radical expression of two terms, multiply the numerator and denominator by the conjugate of the denominator.

57. (a) $x^2 - 2x - 8 = 0$
    (b) $4y^2 - y - 8 = 0$
59. $3x - 4y > 6$

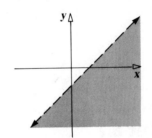

Exercise Set 7.9

1. $15x^3$
2. $\dfrac{-6}{x}$
3. $2x^{2/3}$
4. $\dfrac{1}{u^{2/3}}$
5. $\dfrac{3x^{2/3}}{2}$
6. $\dfrac{x^4}{3y^3}$
7. $y^{1/6}$
8. $-\dfrac{b^2}{2a^{1/3}}$
9. -7.89×10^7
10. 1.06×10^{-4}
11. 0.0000438
12. $-6{,}220{,}000$
13. 2
14. 32
15. -5
16. not a real number
17. 1
18. $\dfrac{1}{9}$
19. $2\sqrt[4]{x^3}$
20. $\sqrt[4]{(2x)^3}$
21. $\sqrt{x^2 + y^2}$
22. $\sqrt[3]{3x - 5}$
23. $2y\sqrt[2]{10x}$
24. $5x^2y\sqrt{2y}$
25. $2b^2\sqrt[3]{4a^2b^2}$
26. $\dfrac{5\sqrt{2x}}{2x}$
27. $2\sqrt{2x}$
28. $\dfrac{\sqrt{xy}}{y}$
29. $\dfrac{\sqrt{10x}}{4x}$
30. $\dfrac{\sqrt[3]{4x}}{2x}$
31. $6\sqrt{3} - \sqrt{5}$
32. $10\sqrt{5}$
33. $19\sqrt{2}$
34. $\sqrt{2x}$
35. $5\sqrt{2} - 6$
36. $x - \sqrt{x} - 12$
37. $2x + 2\sqrt{6xy} + 3y$
38. 2
39. $\dfrac{8\sqrt{7} + 14}{7}$
40. $4\sqrt{6} + 8$
41. $2\sqrt{10} - 4$
42. $-2 + \sqrt{5}$
43. $\dfrac{x - 2\sqrt{xy} + y}{x - y}$

Chapter 7 Review Test

1. (a) $\dfrac{x}{3y^2}$ [7.1] (b) $\dfrac{1}{x^{7/20}}$ [7.4]
   (c) $\dfrac{3y^5}{x^3}$ [7.2]
2. (a) 2.71×10^4 [7.3] (b) 5.1×10^{-4}
3. (a) 0.0000256 [7.3] (b) $75{,}500{,}000$
4. (a) 2 (b) 2 (c) -2
   (d) -2 (e) -2 [7.5]
   (f) undefined (g) 16 (h) 8
5. (a) $4\sqrt[3]{x^2}$ [7.5] (b) $\sqrt[3]{(4x)^2}$
   (c) $\sqrt{x^2+9}$

6. (a) $3ab^2\sqrt{3a}$ (b) $\dfrac{7\sqrt{3x}}{3x}$
   (c) $\dfrac{a\sqrt{ab}}{b}$ (d) $3a\sqrt[3]{b}$ [7.6]
   (e) $\dfrac{\sqrt[3]{10x}}{2x}$
7. (a) $4\sqrt{5} - 9\sqrt{3}$ [7.7]
   (b) $3\sqrt{6} - 6\sqrt{3}$ [7.8]
   (c) $2x - 4$ [7.8]
   (d) $\dfrac{5\sqrt{2} - 5\sqrt{5}}{-3}$ [7.8]
   (e) $\sqrt{2x}$ [7.7]

Exercise Set 8.1

1. $y = -4,\ y = 3$
3. $u = 5,\ u = 0$
5. $x = 4,\ x = 3$
7. $m = -5,\ m = 2$
9. $x = -3,\ x = -1$
11. $c = -5,\ c = 5$
13. $y = \dfrac{1}{2},\ y = 1$
15. $a = \dfrac{5}{2},\ a = -3$
17. $m = -\dfrac{3}{4},\ m = \dfrac{3}{2}$
19. $y = 0,\ y = 3$
21. $w = -\dfrac{1}{4},\ w = 2$
23. $x = -5,\ x = 1$
25. $a = -\dfrac{7}{2},\ a = 1$
27. $v = -3,\ v = -2$
29. $z = \dfrac{3}{2},\ z = -8$
31. no, counter example is if $a = 4,\ b = \dfrac{1}{4}$. The factor theorem only works if the product is zero.
33. $x = 3: L = 4,\ w = 1\ (x = -2$ is meaningless$)$
35. $a = 4: L = 2,\ w = 7\ (a = -\dfrac{3}{2}$ is meaningless$)$
37. (a) 2 (b) 10 (c) 7
39. $r = \dfrac{3t + 12}{2}$

Exercise Set 8.2

1. $x = \pm 5$
3. $y = \pm 6$
5. $u = \pm \dfrac{5}{2}$
7. $x = \pm\sqrt{7}$
9. no solution
11. $x = \pm 3$
13. $x = \pm\sqrt{5}$
15. no solution
17. $x = 5,\ -1$
19. $y = -2,\ -4$
21. $a = 3,\ -7$
23. $u = 11,\ 1$
25. $x = 2,\ -3$
27. $x = -\dfrac{1}{3}$, a double root
29. no solution
31. $a = -2 + \sqrt{5},\ -2 - \sqrt{5}$
33. $y = 2 + 2\sqrt{2},\ 2 - 2\sqrt{2}$
35. $x = 6$
37. $x = \dfrac{-3 + \sqrt{11}}{2}$
39. (a) 36 (b) 36 (c) $\dfrac{1}{9}$

Exercise Set 8.4

(d) $\dfrac{1}{9}$ (e) $\dfrac{9}{16}$ (f) $\dfrac{9}{16}$

41.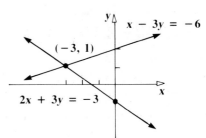

Exercise Set 8.3

1. $x^2 + 2x + \underline{1} = (x + 1)^2$
3. $m^2 - 6m + \underline{9} = (m - 3)^2$
5. $u^2 - 3u + \underline{\dfrac{9}{4}} = \left(u - \dfrac{3}{2}\right)^2$
7. $y^2 + 7y + \underline{\dfrac{49}{4}} = \left(y + \dfrac{7}{2}\right)^2$
9. $x^2 + \dfrac{3}{2}x + \underline{\dfrac{9}{16}} = \left(x + \dfrac{3}{4}\right)^2$
11. $x = 2, -4$
13. $m = 3, 1$
15. $x = 4, -3$
17. $a = -3 \pm \sqrt{5}$
19. $m = 1, \dfrac{-3}{2}$
21. $y = \dfrac{1 \pm \sqrt{13}}{4}$
23. There are no quadratic equations that cannot be solved by completing the square.

25. $x = 0, y = -6$
    $y = 0, x = -10$

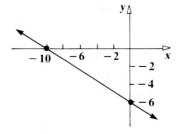

27. (a) $(3b + 1)(2b - 7)$
    (b) $5(x + 2)(x^2 - 2x + 4)$
    (c) $(a^2 + 5)(a + 1)$

Exercise Set 8.4

1. $x^2 - 3x = 0$
   $a = 1, b = -3, c = 0$
3. $x^2 - 9 = 0$
   $a = 1, b = 0, c = -9$
5. $2x^2 - 3x + 8 = 0$
   $a = 2, b = -3, c = 8$
7. $x^2 + 6x - 1 = 0$
   $a = 1, b = 6, c = -1$
9. $x = 4, -4$
11. no real solution
13. $x = 2, 2$
15. $x = 3, -5$
17. $x = 1, \dfrac{1}{2}$
19. $x = 5, -4$
21. $x = \dfrac{-1}{2}, -4$
23. $x = 3 \pm \sqrt{5}$
25. $x = \dfrac{1 \pm \sqrt{2}}{2}$
27. $x = 3, 0$
29. no real solution
31. no real solution
33. Step 1 Clear fractions
    Step 2 Get all terms on one side
    Step 3 Determine a, b, c
    Step 4 Substitute into quadratic formula
    Step 5 Solve
35. (a) $3\sqrt{6}$ (b) $6\sqrt{5}$ (c) $\dfrac{\sqrt{6}}{3}$
    (d) $\dfrac{5\sqrt{2}}{4}$ (e) $5ab\sqrt{a}$ (f) $2a\sqrt[3]{2a^2}$
37. $\dfrac{2x^2 - x}{3 + 2x^2}$

Exercise Set 8.5

1. 2 real rational roots
3. 2 real irrational roots
5. no real roots
7. no real roots
9. double root (one real)
11. $K = 4$
13. $K = 9$
15. $K = 12, -12$
17. $K \leq \dfrac{9}{8}$
19. $K \leq \dfrac{4}{3}$
21. $u = 14, -6$
23. $a = 5, a = -2$
25. $y = \dfrac{3}{2}, y = 4$
27. $t = \dfrac{3 \pm \sqrt{7}}{2}$
29. $-2 \pm \sqrt{7}$
31. no real solution
33. Jane's age $= 56$, John's age $= 32$
35. (a) $m^2 - 3m + 1$ (b) $a^2 - 3a - 4$ (c) $\dfrac{3(c - 2)}{5}$

Exercise Set 8.6

1. 4, 6
3. 3, 4
5. 6 ft
7. 6, 8, 10
9. 5 in., 12 in.
11. Jackie's age $= 12$, Dale's age $= 15$
13. Brett's age $= 7$, Jan's age $= 10$
15. Jason's time $= 6$ hrs, John's time $= 8$ hrs
17. new secretary $= 12$ hrs, experienced secretary $= 6$ hrs
19. John's time $= 2$, Jan's time $= 4$
21. $x = 24$
23. 20
25. length 16, width 12
27. (a) x (b) $2a$ (c) $4a$ (d) $x + 4\sqrt{x} + 4$ (e) $36y$ (f) $6y$ (g) $36 + 12\sqrt{y} + y$ (h) $4x + 12\sqrt{x} + 9$
29. (a) $(3x + 2)(2x - 7)$ (b) $(y - 5)(y^2 + 5y + 25)$ (c) $(a + 2)(2b - 5)$ (d) $(m - 5)(m - 3)$

Exercise Set 8.7

1. 5
3. 7
5. 13
7. no solution, (1 is extraneous)
9. no solution (by inspection)
11. 21
13. 13
15. 3
17. 7, (2 is extraneous)
19. 3, $\left(\dfrac{3}{4}\text{ is extraneous}\right)$
21. $-\dfrac{1}{2}$
23. 1, (-3 is extraneous)
25. 1, (6 is extraneous)
27. 3
29. 6, (-2 is extraneous)
31. -1, (3 is extraneous)
33. If the radical is not isolated, when squaring it will create a "middle" term with another radical in it.
35. (a) 8 (b) 3 (c) 0 (d) -1 (e) 0 (f) 3
37. (a) $2a\sqrt{3a}$ (b) $\dfrac{5\sqrt{6}}{6}$ (c) $\dfrac{\sqrt{5} + 1}{2}$ (d) $10 + \sqrt{5}$

Exercise Set 8.8

1.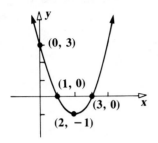

3.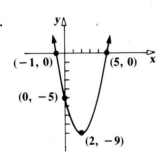

5.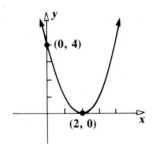

7.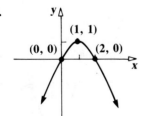

9.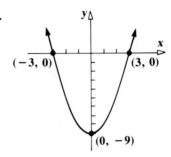

11.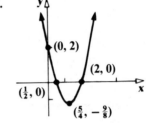

13.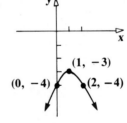

15.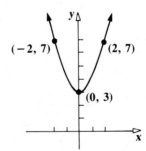

564   Selected Answers

17.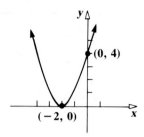

19.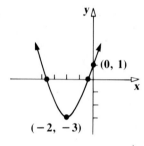

21. The graph of no. 17 is shifted down by 3 units in no. 19.
23. (a) $x = 4$ (b) $x = 4, x = 2$
    (c) $x = 5, x = 1$
    (d) $x = 13$ (e) $x = 9$
25. slope, $m = \dfrac{2}{3}$

Exercise Set 8.9

1. $\dfrac{1}{4}, 2$
2. $0, -3$
3. $-10, 10$
4. ± 9
5. $-\dfrac{1}{2}, -\dfrac{9}{2}$
6. $3 \pm \sqrt{7}$
7. no real solution
8. $-\dfrac{1}{2}, -5$
9. $4 \pm \sqrt{3}$
10. 2 real roots: $-3, -5$
11. 2 real roots: $\dfrac{1}{2}, -1$
12. 2 real roots: $-2 \pm \sqrt{5}$
13. 2 real roots: $3, 0$
14. no real roots: $x = \dfrac{-2 \pm \sqrt{-8}}{2}$
15. $k = 4$
16. $K = \pm 8$
17. $-\dfrac{2}{3}, 3$
18. $-4 \pm \sqrt{5}$
19. $1 \pm \sqrt{7}$
20. 13
21. 3, 1
22. 8, (0 is extraneous)
23. length = 7 cm, width = 5 cm
24. Ralph: 2 hrs, Janie: 4 hrs

25.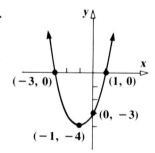

26.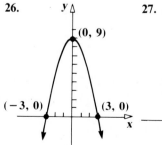

27.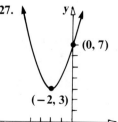

Chapter 8 Review Test

1. (a) $-\frac{2}{3}, \frac{1}{2}$ [8.1]   (b) $0, -6$ [8.1]
2. (a) $\pm\sqrt{7}$ [8.2]   (b) $\frac{7}{2}, -\frac{1}{2}$ [8.2]
3. (a) $-1, -3$ [8.3]   (b) $\frac{2}{3}, -3$ [8.3]
4. (a) $-2 \pm \sqrt{7}$ [8.4]   (b) $-5, -\frac{1}{2}$ [8.4]
5. (a) no real roots [8.5]
   (b) no real roots [8.5]
   (c) one real root [8.5]
   (d) two real roots [8.5]
6. (a) $\frac{-3 \pm \sqrt{7}}{2}$ [8.5]
   (b) $-\frac{9}{2}, -\frac{5}{2}$ [8.5]   (c) $\frac{5}{3}, 4$ [8.5]
7. 5, (0 is extraneous) [8.7]
8. 6, 8, 10 [8.6]
9. [8.8]

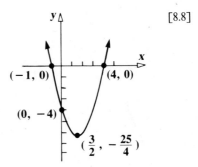

Exercise Set 9.1

1. Function: domain = $\{A, B, C, D\}$
   range = $\{1, 2, 3, 4\}$
3. Function: domain = $\{4, 5, 6\}$
   range = $\{$Bill, Jerry, Edward$\}$
5. domain = $\{1, 2, 3, 4, 5\}$
   range = $\{-3, -2, -1, 0, 1\}$

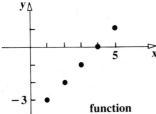
function

7. domain = $\{-4, 0, 4\}$
   range = $\{1\}$

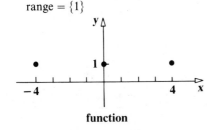
function

9. domain = $\{-2, 2\}$
   range = $\{2, -2\}$

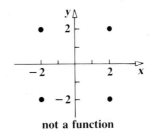
not a function

11.

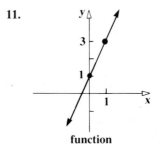
function

13.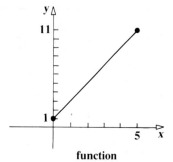
function

15.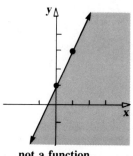
not a function

17.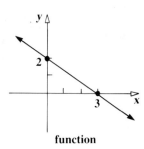
function

19.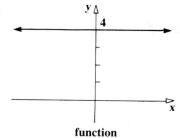
function

21.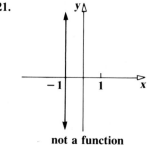
not a function

23. function
25. function
27. not a function
29. function
31. function
33. (a) 3 (b) 15 (c) 9 (d) 23
35.
x	y
−2	−2
0	4
2	10
4	16

37. (a) $y = \dfrac{1}{2}$ or $y = -4$ (b) $t = 1 \pm \sqrt{5}$

Exercise Set 9.2

1. (a) 2 (b) 12 (c) $a = 3$
3. (a) −2 (b) 8 (c) −17
   (d) 23 (e) $-17 + 5h$
   (f) $5x + 5h - 2$
5. (a) −3 (b) 1 (c) 1
   (d) 13 (e) 13 (f) 22
7. (a) 5 (b) 14 (c) 19
   (d) 3 (e) 11 (f) −8
   (g) −5 (h) 3 (i) 3
   (j) −4
9. (a) 4 (b) 9 (c) 16

11. In $f(m + 3)$, the function f operates on the entire quantity $m + 3$, while in $f(m) + 3$, f operates only on m.
$f(m + 3) = 2(m + 3) + 3 = 2m + 9$
$f(m) + 3 = (2m + 3) + 3 = 2m + 6$

13. (a) $2\sqrt{3}$ (b) $5x\sqrt{2x}$
    (c) $\dfrac{\sqrt{6}}{2}$ (d) $\dfrac{\sqrt{14x}}{4}$

15. (a) $x - 3$ (b) $\dfrac{10 - 3\sqrt{2}}{2}$ (c) $3\sqrt{5} - 6$

Exercise Set 9.3

1. Linear

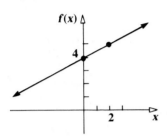

3. Constant

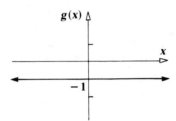

5. Identity

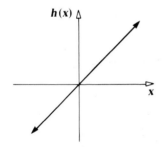

7. Linear

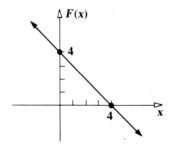

9. Linear

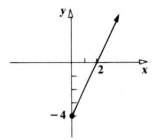

11. Linear

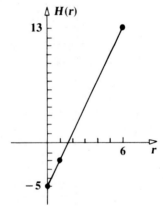

13. Constant

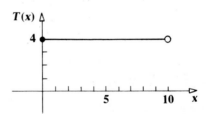

15. Linear

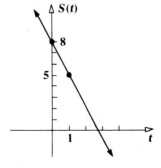

17. Linear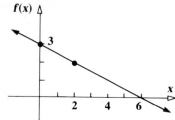

19. Linear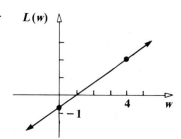

21. Identity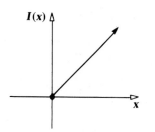

23. Identity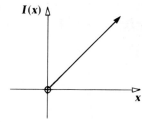

25. (a) 6
(b) $b = -\dfrac{5}{2}$ or $b = 3$
(c) $x = 12$
(d) $y = 2 \pm \sqrt{3}$

27.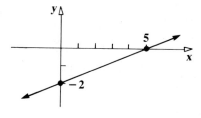

Exercise Set 9.4

1. domain = $\{-2, -1, 0, 1, 2\}$
range = $\{5, 2, 1\}$
function

[9.1]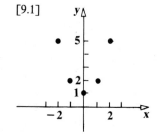

2. domain = $\{5, 2, 1\}$
range = $\{-2, -1, 0, 1, 2\}$
not a function

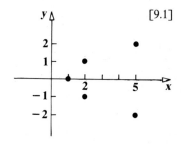 [9.1]

Exercise Set 9.4

3. domain = $\{-3, -2, -1, 0, 1, 2\}$
   range = $\{3, 2, 1, 0\}$
   function

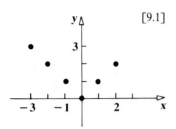
[9.1]

21. identity

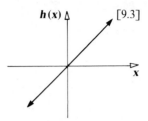
[9.3]

22. linear

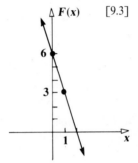
[9.3]

4. function [9.1]
5. function [9.1]
6. not a function [9.1]
7. not a function [9.1]
8. 28 [9.2]
9. 10 [9.2]
10. 12 [9.2]
11. -8 [9.2]
12. 11 [9.2]
13. 9 [9.2]
14. -20 [9.2]
15. -104 [9.2]
16. 7 [9.2]
17. 19 [9.2]
18. 19 [9.2]

19. linear

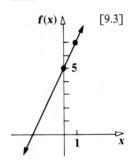
[9.3]

23. linear

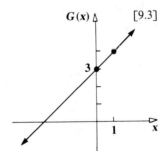
[9.3]

20. constant

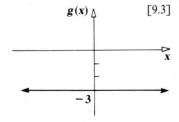
[9.3]

24. linear

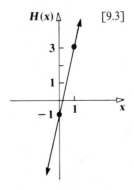
[9.3]

Chapter 9 Review Test

1. (a) domain = $\{-3, -1, 0, 2, 4\}$
   (b) range = $\{-7, -3, -1, 3, 7\}$
   (c) $F(2) = 3$ (d) $x = 4$
   (e)

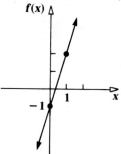

   (f) function
2. (a) function (b) function
   (c) not a function (d) not a function
3. (a) 11 (b) -7 (c) 12
   (d) 18 (e) 8 (f) 15
4. (a)

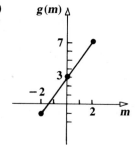

   (b)

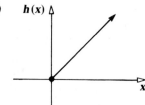

   (c)

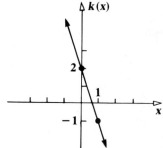

   (d)

INDEX

Abscissa, 143
Absolute value, 17
Addition of
 polynomials, 233
 radical expressions, 422
 rational expressions, 327
 signed numbers, 34
Applications, 117, 207, 356
Associative property, 23
Axes, coordinate, 143
Axis, horizontal, 143
Axis, vertical, 143
Axis of symmetry, 479

Base of exponential, 51, 380
Binary operations, 15
Binomial, 231
Boundary line, 169
Braces, 28
Brackets, 28

Cartesian coordinate system, 143
Cartesian plane, 145
Closed, 6
Closure, 22
Combining like terms, 76
Commutative property, 22
Completing the square, 450
Conjugates, 428
Constant, 50
Constant term, 75, 230
Coordinate, 8
 axes, 143
 first, 142
 of a point, 143
 second, 142
Cube, 52

Degree
 of a polynomial, 231
 of a term, 231
Difference, 6
 of two squares, 280
Discriminant, 461
Disjoint sets, 7
Distance, directed, 13
Distributive property, 25
Dividend, 46
Division of
 polynomials, 256
 radical expressions, 422
 rational expressions, 313, 314
 signed numbers, 46
Divisor, 46
Domain of relation, 493

Element of a set, 1
Elimination method, 194
Empty set, 2
Equality of sets, 2
Equation
 algebraic, 88
 conditional, 89
 containing radicals, 474
 equivalent, 91
 fractional, 341
 linear, in one variable, 92
 linear, in two variables, 147

Equation (*continued*)
  literal, 104
  quadratic, 440
    solving by:
      completing the square, 450–454
      factoring, 440–444
      square root, 445–447
      using the quadratic formula, 455–460
  second degree, 440
  solving, 89
Expanded form, 52
Exponent, 51, 380
  integer, 388
  natural number, 388
  negative, 390
  rational, 406
  zero, 389
Exponential form, 51, 52, 380
Expression, algebraic, 52
Expression, evaluation of, 53
Extraneous solution, 343, 474

Factor, 41, 241, 319
Factor, common, 305
Factoring, 74, 75
  the difference of two cubes, 285–286
  the difference of two squares, 280–282
  by grouping, 286–288
  by removing common factor, 266–268
  summary, 289–291
  the sum of two cubes, 285–286
  trinomials, 270, 275
FOIL, 249
Formula, 104
Fraction
  clearing, 97
  complex, 335
  fundamental principle of, 304
  reducing, 306
  simple, 335
Function, 495
  composition, 502
  constant, 507
  identity, 508
  linear, 505
Function machine, 500
Function notation, 499
Fundamental theorem of arithmetic, 320

Graph of
  equation, 147
  inequality, 109
  ordered pair, 143
  parabola, 480
  point, 8
Grouping symbols, 27, 28

Half-plane, 169, 214
Higher terms, 308, 309
Horizontal line, 151
Hypotenuse, 470

Identity, 89, 346
  additive, 23
  multiplicative, 23
Index of radical, 411
Inequality, 108
  double, 113
  equivalent, 110
  properties of, 109
  sense of, 111
  solving, 110
  in two variables, 167
Infinite set, 5
Intelligence quotient, 349
Intercepts, 155
Inverse
  additive, 24
  multiplicative, 24

Least common denominator, 82, 325, 329
Least common multiple, 81, 322
Like radicals, 422
Like terms, 76, 233
Linear equation in one variable, 92
Linear equation in two variables, 147
Literal part, 75, 230
Lowest terms, 304

Member of equation, 88
Member of set, 1
Monomial, 231
Multiple, common, 81
Multiples, 81, 319
Multiplication of
  polynomials, 241
  radical expressions, 425–428
  rational expressions, 311, 312
  signed numbers, 41

Index

Negative, 15
nth power, 380
nth root, 406
Null set, 2
Number
  composite, 320
  counting, 5
  irrational, 7
  natural, 5
  prime, 320
  rational, 6
  real, 7
  signed, 33
  whole, 5
Number line, 8
Numerical coefficient, 75, 230

Ordered pair, 142
Ordered set, 10, 109
Order of operations agreement, 27
Ordinate, 143
Origin
  of Cartesian plane, 143
  of number line, 8
Overbar, 7

Parabola, 479
Parentheses, 28
  removing 73, 74
Perfect square trinomial, 282
Plot of
  ordered pair, 143
  point, 8
Point-slope form, 174
Polynomial
  addition, 233–238
  classification, 230
  completely factored form, 266
  definitions, 229
  degree, 231
  division, 256–264
  factoring, 266–292
  multiplication, 241–253
  prime, 266
  subtraction, 233–238
Power of a product, 381
Power of a quotient, 382
Power to a power, 383
Powers of 10, 400
Prime factored form, 320

Prime factorization, 320
Product, 6, 41, 241
Product of sum and difference, 252
Product rule for exponents, 381
Proper subset, 3
Proportion, 348, 349
Pythagorean theorem, 470

Quadrant, 143
Quadratic formula, 456
Quotient, 6, 46
Quotient rule for exponents, 384, 391

Radicals, properties of, 413, 414
Radical expression, 413
  addition, 422
  division, 427
  multiplication, 425
  subtraction, 422
Radical form, 411
Radical sign, 411
Radicand, 411
Range of relation, 493
Ratio, 348
Rational expression, 303
  addition, 327
  division, 313
  multiplication, 311
  reducing, 304
  subtraction, 327
Rationalize denominator, 417, 427, 428
Real numbers, properties of, 22
Reciprocal, 17
Rectangular coordinate system, 145
Relation, 493
Rise, 159
Root
  cube, 406
  double, 443
  principal, 407
  square, 406
Root of
  equation, 89
  number, 406
  quadratic equation, 461
Rules of exponents, summary of, 396
Run, 159

Scale, 143
Scientific notation, 399

Sense of inequality, 111
Set, 1
Set builder notation, 2
Sides of equation, 88
Signs, properties of, 305
Similar terms, 76
Similar triangles, 354
Simplest radical form, 416
Slope, 159, 160
Slope-intercept form, 163
Solution, extraneous, 343, 474
Solution of
  equation, 89
  equation in two variables, 142
  system of equations, 189
Square, 52
Square of a binomial, 251
Subscripted variable, 106
Subset, 3
Substitution method, 200
Subtraction of
  polynomials, 233
  radical expressions, 422
  rational expressions, 327
  signed numbers, 37
Sum, 6
Sum and difference of two cubes, 285

System of equations, 188
  equivalent, 193
System of inequalities, 215

Term, 33, 75
Terminating decimal, 7
Transitive property, 109
Trichotomy, 109
Trinomial, 231

Unary operations, 15

Variable, 51
Vertex of parabola, 479
Vertical line, 150
Vertical line test, 496

x-axis, 145
x-intercept, 155
x-y plane, 145, 149

y-axis, 145
y-intercept, 155

Zero exponent, 389
Zero factor theorem, 441